GENETICS IN AQUACULTURE II

GENETICS IN AQUACULTURE II

Proceedings of the Second International Symposium on Genetics in Aquaculture, sponsored by the University of California, Davis, the California Sea Grant College Program, and the International Association for Genetics in Aquaculture

Held at the University of California, Davis, U.S.A., 23—28 June 1985

Edited by

G.A.E. GALL
Department of Animal Science, University of California, Davis, CA 95616 (U.S.A.)

and

C.A. BUSACK
Department of Biology, University of Mississippi, University, MS 38677 (U.S.A.)

This volume is reprinted from the journal *Aquaculture*, vol. 57.

ELSEVIER   Amsterdam — Oxford — New York — Tokyo 1986

ELSEVIER SCIENCE PUBLISHERS B.V.
Sara Burgerhartstraat 25
P.O. Box 211, 1000 AE Amsterdam, The Netherlands

Distributors for the United States and Canada:

ELSEVIER SCIENCE PUBLISHING COMPANY INC.
52, Vanderbilt Avenue
New York, NY 10017, U.S.A.

Library of Congress Cataloging-in-Publication Data

International Symposium on Genetics in Aquaculture
    (2nd : 1985 : University of California, Davis)
    Genetics in aquaculture II.

    "Reprinted from the journal Aquaculture, vol.57."
    Bibliography: p.
    Includes index.
    1. Fish-culture--Congresses. 2. Shellfish
culture--Congresses. 3. Fishes--Genetics--
Congresses. 4. Shellfish--Genetics--Congresses.
I. Gall, G. A. E. II. Busack, C. A. (Craig A.)
III. University of California, Davis. IV. University
of California (System). Sea Grant College Program.
V. International Association for Genetics in
Aquaculture. VI. Title. VII. Title: Genetics in
aquaculture 2. VIII. Title: Genetics in aquaculture
two.
SH151.I58 1985   639.3   86-24402
ISBN 0-444-42732-5 (U.S.)

ISBN 0-444-42732-5

© Elsevier Science Publishers B.V., 1986

All rights reserved. No part of this publication may be reproduced, stored in a retrieval system or transmitted in any form or by any means, electronic, mechanical, photocopying, recording or otherwise, without the prior written permission of the publisher, Elsevier Science Publishers B.V./Science & Technology Division, P.O. Box 330, 1000 AH Amsterdam, The Netherlands.

Special regulations for readers in the USA — This publication has been registered with the Copyright Clearance Center Inc. (CCC), Salem, Massachusetts. Information can be obtained from the CCC about conditions under which photocopies of parts of this publication may be made in the USA. All other copyright questions, including photocopying outside of the USA, should be referred to the copyright owner, Elsevier Science Publishers B.V., unless otherwise specified.

Printed in The Netherlands

CONTENTS

Prologue and Acknowledgments
G.A.E. Gall (Davis, CA, U.S.A.) and C.A. Busack (University, MS, U.S.A.) ix

Reviews of Selected Topics
Monitoring genetic change
    H. Fredeen (Lacombe, Alta., Canada) 1
Effective population size and selection in variable aquaculture stocks
    R.W. Doyle and A.J. Talbot (Halifax, N.S., Canada) 27
Growth and reproduction in fish and shellfish
    B. Gjerde (Ås, Norway) 37
Ploidy manipulation and performance
    G.H. Thorgaard (Pullman, WA, U.S.A.) 57
Developing a commercial breeding program
    F.T. Shultz (Sonoma, CA, U.S.A.) 65

Domestication and Broodstock Development
Breeding plan for sea ranching
    T. Gjedrem (Ås, Norway) 77
Application of a genetic fitness model to extensive aquaculture
    J.E. Lannan (Newport, OR, U.S.A.) and A.R.D.
    Kapuscinski (St. Paul, MN, U.S.A.) 81
Duration of feeding and indirect selection for growth of tilapia
    C.T. Villegas (Tigbauan, Philippines) and R.W. Doyle (Halifax, N.S., Canada) 89
Replicate variance and the choice of selection procedures for tilapia (*Oreochromis niloticus*) stock improvement in Thailand
    S. Uraiwan (Bangkok, Thailand) and R.W. Doyle (Halifax, N.S., Canada) 93
Genetic aspects of hatchery rearing of the scallop, *Pecten maximus* (L.)
    A.R. Beaumont (Menai Bridge, Great Britain) 99
Controlled mating of the European oyster, *Ostrea edulis*
    G.F. Newkirk (Halifax, N.S., Canada) 111
Calluses, cells, and protoplasts in studies towards genetic improvement of seaweeds
    M. Polne-Fuller and A. Gibor (Santa Barbara, CA, U.S.A.) 117

Genetic and Environmental Effects
Genetic and environmental components of variation for growth of juvenile Atlantic salmon (*Salmo salar*)
    J.K. Bailey and E.J. Loudenslager (St. Andrews, N.B., Canada) 125
The genetics of production characters in the blue mussel *Mytilus edulis*. I. A preliminary analysis
    A.L. Mallet, K.R. Freeman and L.M. Dickie
    (Dartmouth, N.S., Canada) 133
The effect of strain crossing on the production performance in rainbow trout
    G. Hörstgen-Schwark, H. Fricke and H.-J. Langholz (Göttingen, Federal Republic of Germany) 141
Genotype-environment interactions for growth of rainbow trout, *Salmo gairdneri*
    R.N. Iwamoto, J.M. Myers and W.K. Hershberger (Seattle, WA, U.S.A.) 153

Selection Response

Response of channel catfish to multi-factor and divergent selection of economic traits
  K. Bondari (Tifton, GA, U.S.A.) 163
Genetic selection for shell traits in the Japanese pearl oyster, *Pinctada fucata martensii*
  K.T. Wada (Nansei, Japan) 171
Mass selection for growth rate in the Nile tilapia (*Oreochromis niloticus*)
  G. Hulata, G.W. Wohlfarth and A. Halevy (Hof Hacarmel, Israel).......................... 177

Stock Comparison

Factors affecting growth in rainbow trout (*Salmo gairdneri*) stocks
  L. Siitonen (Jokioinen, Finland) 185
Comparison of strains, crossbreeds and hybrids of channel catfish for vulnerability to angling
  R.A. Dunham, R.O. Smitherman, R.K. Goodman and P. Kemp (Auburn, AL, U.S.A.) 193
Interspecific hybridization of the species of *Macrocystis* in California
  R.J. Lewis (Goleta, CA, U.S.A.), M. Neushul (Santa Barbara, CA, U.S.A.) and B.W.W. Harger (Goleta, CA, U.S.A.) 203
Genetic differentiation among seasonally distinct spawning populations of chum salmon, *Oncorhynchus keta*
  R.F. Tallman (Nanaimo, B.C., Canada) 211
Variability of embryo development rate, fry growth, and disease susceptibility in hatchery stocks of chum salmon
  W.W. Smoker (Corvallis, OR, U.S.A.) 219

Use of Electrophoretic Markers

Genetic studies connected with artificial propagation of cod (*Gadus morhua* L.)
  K.E. Jørstad (Bergen, Norway) 227
Bottleneck effects and the depression of genetic variability in hatchery stocks of *Penaeus japonicus* (Crustacea, Decapoda)
  V. Sbordoni, E. De Matthaeis, M. Cobolli Sbordoni, G. La Rosa and M. Mattoccia (Rome, Italy)... 239
The enzyme gene variation of ten Finnish rainbow trout strains and the relation between growth rate and mean heterozygosity
  M.-L. Koljonen (Helsinki, Finland) 253
Genetic heterozygosity and growth rate in Louisiana oysters (*Crassostrea virginica*)
  D.W. Foltz (Baton Rouge, LA, U.S.A.) and M. Chatry (Grand Isle, LA, U.S.A.).......... 261

Application of Ploidy Manipulation

Triploidy induction by thermal shocks in the Japanese oyster, *Crassostrea gigas*
  E. Quillet and P.J. Panelay (Brest, France)... 271
Tetraploid induction in *Oreochromis* spp.
  J.M. Myers (Seattle, WA, U.S.A.) 281
Androgenetic rainbow trout produced from inbred and outbred sperm sources show similar survival
  P.D. Scheerer, G.H. Thorgaard (Pullman, WA, U.S.A.), F.W. Allendorf and K.L. Knudsen (Missoula, MT, U.S.A.) 289
Commercial methods for the control of sexual maturation in rainbow trout (*Salmo gairdneri* R.)
  V.J. Bye and R.F. Lincoln (Lowestoft, Great Britain) 299

Broodstock development for monosex production of grass carp
   W.L. Shelton (Norman, OK, U.S.A.) 311
Color, growth and maturation in ploidy-manipulated fancy carp
   N. Taniguchi, A. Kijima, T. Tamura, K. Takegami and I. Yamasaki (Nankoku, Japan) 321
Comparative growth and development of diploid and triploid coho salmon, *Oncorhynchus kisutch*
   O.W. Johnson, W.W. Dickhoff and F.M. Utter (Seattle, WA, U.S.A.) 329
Increased resistance of triploid rainbow trout × coho salmon hybrids to infectious hematopoietic necrosis virus
   J.E. Parsons, R.A. Busch (Buhl, ID, U.S.A.), G.H. Thorgaard and P.D. Scheerer (Pullman, WA, U.S.A.) 337
Female brown trout × Atlantic salmon hybrids produce gynogens and triploids when backcrossed to male Atlantic salmon
   K.R. Johnson and J.E. Wright (University Park, PA, U.S.A.) 345

Abstracts of Poster Papers 359

Author Index 381

Prologue and Acknowledgments

Aquacultural genetics had its birth in the early 20th century when private individuals and public agencies around the world began raising salmonids and tropical fish of special breeding. It coalesced into an identified field of science at the First International Symposium on Genetics in Aquaculture held at University College, Galway, Ireland in 1982 (*Aquaculture,* Volume 33). The first symposium examined the status of aquacultural applications of various subdisciplines of genetics (inbreeding, quantitative inheritance, etc.). The outcome was an up-to-date review of current knowledge of genetics applied to aquaculture organisms.

The Second International Symposium on Genetics in Aquaculture was held in 1985 at the University of California, Davis, U.S.A. It built on the foundation laid down at the first symposium by addressing specific problems and developments in aquaculture genetics. The proceedings are divided into eight sections. The first presents reviews of specific topics, six contain short papers on current research results and the final section consists of abstracts of papers presented as posters.

The review of selected topics provides historical insight into modern animal breeding, an examination of the potential of biotechnology in aquaculture, and an overview of elements critical to a commercial breeding program. The first section of short papers addresses genetic aspects of domestication and the development of broodstocks using salmon, tilapia, scallops, and oysters as examples. This is followed by papers dealing with genetic and environmental factors affecting performance, a section on results of selection experiments, and papers demonstrating levels of genetic variability observed among different stocks.

A section dealing with the use of electrophoretic markers outlines results of experiments using markers as an adjunct to standard quantitative tools and results of research evaluating the importance of heterozygosity to performance. The potential of biotechnology in aquaculture is exemplified in the section on Applications of Ploidy Manipulation. The wide range of topics covered, from methods of induction of triploidy to increasing disease resistance, is an indication of future developments in this very active area of aquaculture genetics.

The final section contains an alphabetic presentation of abstracts of poster session papers. It is unfortunate that the reader of this special issue does not have the benefit of the full poster presentation. The posters were of excellent quality and for many, were a highlight of the symposium.

The content of the proceedings of the Second Genetics Symposium demonstrates the tremendous breadth of research activity as well as the complexity of issues in aquaculture genetics. The depth of understanding and progress achieved in the very short time since the first symposium reflects not only the quality and quantity of effort in aquaculture genetics, but also expresses the dedication and enthusiasm of individuals and organizations world-wide to this extremely important area of aquacultural science. We look forward to the Third Symposium, scheduled for Norway in 1988, with great anticipation.

We wish to thank all symposium participants for their contributions and the program hosts at the University of California, Davis, for providing a stimulating atmosphere. Special thanks are extended to sponsors of the symposium: the College of Agricultural and Environmental Sciences, the Aquaculture and Fisheries Program, and the Department of Animal Science at the University of California, Davis, and the California Sea Grant College Program. The service and consultation provided by Program Committee members Fred T. Shultz, Dennis Hedgecock, Fred S. Conte, and Donald E. Campton is also gratefully acknowledged.

Graham A.E. Gall
Craig A. Busack
May 1986

Monitoring Genetic Change

HOWARD FREDEEN

Vanag Consulting Services, Box 1810, Lacombe, Alta. T0C 1S0 (Canada)

ABSTRACT

Fredeen, H., 1986. Monitoring genetic change. *Aquaculture*, 57: 1-26.

Control procedures which permit evaluation of treatment effects in terms of a known standard (i.e., a zero treatment) are an integral component of sound experimental design. The control technique appropriate for genetic experiments will depend on the specific objectives to be served by the research (e.g., measuring absolute vs. relative differences), by the biological attributes of the species, and by the relative costs of alternative techniques. Rarely, if ever, will it be possible to exclude non-genetic variation through rigid control of the environment. At the same time, it is only prudent to impose on all experiments, whether or not a genetic control per se is used, the most rigorous environmental management feasible with the species in order to minimize the variance attributable to non-genetic sources. Conceptually, random-mating genetic control populations will ensure genetic/phenotypic stability. For real-life finite populations, this concept must be modified to accommodate genetic drift and inbreeding effects. These will be a function of population size and structure. Genotype—environment interactions must also be considered; such interactions are likely to increase in importance as the selected lines diverge genetically from the control. These points are illustrated by three examples derived from selection studies with beef cattle, pigs and poultry. Each of these experiments employed systems of complete pedigree identification for all parents and progeny, thereby permitting pedigree control of the mating systems for each treatment (control and selected) and statistical estimation of the inbreeding effects contributing to time trends in performance. For these purposes, the importance of a comprehensive system for the unique and permanent identification of every individual produced in the experimental populations cannot be overemphasized.

INTRODUCTION

Genetic improvement, most simply stated, is the process of replacing a given population of genotypes with another that subtends superior phenotypic performance. Selection, by imposing differential reproductive opportunities for individuals of the population, provides directional impetus to genetic change.

Lush (1945) likened selection to the architectural process for creating a building of original design from materials already available. It is an apt analogy. Animal breeders define their design in terms of a desired phenotype and their building materials are the alleles, distributed among the genomes of individuals in the populations, which qualitatively or quantitatively condition expression of their phenotypes. Selection does not create new alleles. It merely expedites the propagation of favored allelic combinations and promotes the progressive elimination of those with less favorable effects.

It would be erroneous to infer that selection within or among populations is the only process subtending directional genetic change. (Note that the word "directional" excludes the random, and presumably rare, events of mutations and chromosomal rearrangement. It also excludes the process of random drift which will be described in more detail later.) The other non-random process which must be considered is natural selection. Feral populations represent the culmination of untold generations of natural selection under environmental conditions which, with rare exceptions, were unlikely to have remained static over time. Thus, adaption to environmental change has been an evolutionary genetic process, propagation rates for individuals in successive generations of the population tending to favor those whose genomes provided the optimum homeostatic balance.

Human intervention in this evolutionary process had a two-fold effect. It replaced the natural environment with intensive husbandry practices and it invoked artificial selection to enhance the rate of genetic change for traits of specific (economic) interest. These consequences of domestication obviously did not create new alleles. They merely altered the rules of the game, loading the dice in favor of allelic combinations which were not optimum under feral conditions. Neither did they totally supplant natural selection which, even under a presumably stable environment, will assert itself in terms of differential reproductive rates according to the "fitness" of individuals in the population.

To set the stage for discussion of the basis for, and design of, techniques for monitoring genetic change it is appropriate to review briefly the evolution of quantitative genetics theory and its application with livestock species.

THE EVOLUTION OF QUANTITATIVE GENETICS

Classical Mendelian genetics originated with the work of the Austrian monk, Gregor Mendel, in the mid eighteen-hundreds. His work gathered dust until the turn of the present century when others, working as had Mendel with qualitative traits, "rediscovered" the principles of inheritance which he had enunciated. Several biologists of this era sought to describe the genetics of quantitative (continuous) variation but their statistical tools were blunt and progress was delayed until more effective biometrical techniques had evolved (Fisher, 1918; Wright, 1921). To J.L. Lush of Iowa goes the credit for adapting the techniques of Fisher and Wright to the study and advancement of population genetics. His lifetime contributions to this field, documented in a symposium held in his name (Shrode, 1973), are aptly summarized in the words of Shrode: "In effect, the field of Animal Breeding is a program of intellectual 'line breeding' to Lush". In the context of the present symposium, A.B. Chapman's contribution to this 1973 publication, "Selection theory and experimental results", provides valuable background information.

Recent advances in molecular genetics and genetic engineering have immensely broadened our understanding of the ways in which genes operate and opened new vistas for accelerating the rate of genetic change. It seems unlikely, however, that these advances will supplant the selection techniques which have derived from the teachings of Lush and his colleagues.

QUANTITATIVE GENETICS AND SELECTION

Continuous variation characterizes most of the metric traits of interest in genetic improvement programs. Such traits are termed "quantitative" as distinct from the qualitative traits which exhibited discrete or discontinuous variation. They are under the influence of many genetic loci distributed throughout the genome with the potential for each locus to be represented by several alleles.

Metric traits may be influenced by "major" genes, major in the sense that their presence causes a substantial discontinuity in the population distribution. Such genes are amenable to study by classical Mendelian techniques and provide grist to the mill of genetic engineering as illustrated by the transfer to mice of a growth-promoting "gene" from the rat. However, such genes exert their influence in the genetic "environment" provided by the "minor" genes segregating at other loci (also termed polygenic variation) and it is the combined effects of these minor genes that are the focus of attention in the genetic manipulation of populations. The individual effects of "minor" genes are too small to permit identification of their segregation patterns and the specific biometrical techniques evolved for their study gave rise to the specialized field known variously as biometrical, quantitative or population genetics.

Central to the concept of population genetics is the knowledge that all metric traits are subject to non-genetic as well as genetic variation. This non-genetic variation, superimposed on the genetic background, "smooths" any discontinuities that might result from gene segregation and produces the normal distribution patterns which characterize all metric traits. The biometrical techniques employed in population genetics were devised to quantify the proportion of the observed phenotypic variation which might reasonably be attributed to genetic causes. This proportion is referred to as heritability (Lush, 1940).

Since heritability (h^2) is calculated as a proportion of the total phenotypic variance, it is self-evident that its magnitude for a specific trait in a specific population will be a function of the non-genetic variation present. Thus, an estimate derived from a population under strict environmental control is expected to be larger than one drawn from a population subject to wide environmental fluctuations (provided of course that genetic and non-genetic effects are not confounded).

The concept of heritability is pivotal to selection programs. When multiplied by the selection differential applied to the population (the mean phenotypic difference between the selected individuals and the contemporary groups from which they came) it produces an estimate of the genotypic gain expected in the progeny of the selected parents.

Another important concept of population genetics is that of genetic correlation. That an individual allele may have multiple (pleiotropic) effects is well documented in the case of certain "major" genes and there is no basis for believing that minor genes lack this capability. Since net merit is usually a function of two or more metric traits, and since genetic correlations may range from positive to negative in respect of merit, estimates of genetic correlations have direct relevance to the prediction of selection responses. Techniques for their estimation and their contribution to observed phenotypic correlations were first provided by Hazel (1943) in a paper which described the application of discriminant functions (selection indices) to the genetic improvement of net merit in animal populations. The parameters he employed for his selection index (heritabilities, genetic and phenotypic correlations, and economic weights for the traits comprising net merit) have endured as the basis for modern selection practices.

Early enthusiasm for the biometrical approach to population genetics was dampened by the evidence that livestock populations, presumed to have been under long and continuous selection, showed no evidence of cumulative genetic change (Craft, 1951). The industry was not prepared to consider the possibility that selection may have been negligible. Indeed some scientists, convinced that selection had been reasonably intense and consistent, argued that the absence of progress was evidence of a substantial negative genetic correlation between the metric traits and important attributes of fitness (reproduction). Other scientists, arguing that "field data" were inadequate to resolve hypotheses concerning selection intensities and genetic correlations, proposed that selection experiments be conducted to test the validity of the biometrical approach. They visualized that the realized genetic parameters derived from adequately designed experiments would bear directly on this issue.

Two problems had to be faced in designing appropriate experiments: one was the finite dimensions of the resources available at each research institution; the other was the need for controlling and/or monitoring the influence of non-genetic variation on the performance of experimental populations. The relevance of suitable controls was never in doubt; the problem lay in providing such controls without unduly compromising the finite, usually very limited, resources. From this beginning there has evolved a substantial body of theory on the design of genetic controls appropriate to the specific objectives of individual selection experiments.

PURPOSES SERVED BY SELECTION EXPERIMENTS

There is no singular definition of genetic merit. In feral populations the "optimum" genotypes will be those which confer superiority in terms of propagation and the definition of "optimum" will be specific to the environmental circumstances which dictate mating successes and survival. Propagation rates are also important to domesticated populations but superimposed on this attribute will be characteristics deemed of importance for economic or other reasons. Since "economic" merit is a composite of many different productivity traits, and since perceptions of the relative importance of these component traits will differ among breeders and will be subject to change over time, the definition of total genetic merit for any domestic species will be both variable and dynamic.

Theoretical population genetics has provided the biometrical framework for defining selection approaches and mating programs appropriate for achieving desired genetic change of quantitative traits and for estimating the rate of genetic change which might be subtended by application of specific components of desired phenotypes. To prove that this mathematical approach has biological relevance requires empirical studies based on real-life populations.

Selection experiments were initially perceived in this context. But their importance to the science of quantitative genetics has derived in far greater measure from their failures than from their successes. By providing insight into the limitation of theory, the failures have encouraged biometricians to seek ways to improve both the specifics and the generalities of the theoretical framework and to sharpen the tools for statistical analyses of data.

Artificial selection takes a variety of forms. The diversity of options includes the following.

Direct vs. indirect (i.e., based directly on the trait of interest vs. use of a physiological indicator of that trait). For example, direct selection for muscular hypertrophy in pigs involves visual appraisal of the phenotype; indirect selection can use the criterion of reaction to the anaesthetic halothane.

Single vs. multiple trait. The latter is usually accommodated through use of an index. Many index options are available, each one possessing unique attributes. Indices are also employed for combining data records obtained from different sources (e.g., own record plus records for sibs, parents or collateral relatives).

Part vs. complete record. If the phenotype of interest is expressed at an age beyond sexual maturity it is desirable to use a part record to accelerate the rate of generation turnover (and hence rate of genetic change).

Own performance record vs. progeny record vs. records of collateral relatives vs. some combination of these. Sex-limited traits (e.g., lactation, egg production) require a different form of selection in each sex. Traits which can be measured only after slaughter (e.g., organoleptic properties of meat) require selection based on collateral relatives and/or progeny.

Permutations among these options provide an immense array of selection alternatives. Thus one of the purposes of selection experiments is to identify the selection alternative(s) most relevant to a particular phenotypic objective within a specific population.

Other purposes of selection experiments include:

(1) validation of selection theory and the statistical estimates of genetic parameters (heritabilities, genetic correlations, estimates of heterotic effects, etc.);

(2) examination of genotype×environment interactions (e.g., comparison in different environments of a line selected under one environment or comparison in one environment of lines selected under different environments);

(3) comparing techniques for performance evaluation in terms of their relative effectiveness for defining the genotype(s) of central interest (e.g., what performance evaluation methodology is most appropriate for defining genotypic differences in the efficiency of lean tissue feed conversion?);

(4) providing contrasting genotypes (e.g., using divergent selection to produce lines of contrasting performance) for empirical study of physiological-/biochemical/hormonal indicators of phenotypic differences.

None of these purposes can be effectively served unless the experimental design has incorporated some form of standard against which comparisons may be made. In the absence of such standards, any interpretations subtended by results are wide open to the challenge "compared with what?" Before considering the types of controls which may be invoked, it is desirable to examine the relevance of random genetic drift to both selected and control populations.

INBREEDING, GENETIC DRIFT AND EFFECTIVE POPULATION SIZE

In each generation of any population, the individuals which reproduce represent less than the total of the contemporaries with which they were born. Thus, any alleles unique to those which fail to reproduce will not be represented in subsequent generations. This, of course, is the essence of the selection process. Selection, whether natural or artificial, results in the propagation of favored genotypes with a corresponding reduction in the frequency of less desired alleles.

Random chance, as distinct from selection (either natural or artificial), contributes to changes in gene frequencies and its contribution will be inversely related to population size. Since all populations, both feral and domestic, comprise a multitude of sub-populations whose boundaries are defined by geographic location and/or environmental conditions, genotypic differentiation due to random chance is the rule rather than the exception. However, it may have no discernible impact on the total population since the events (e.g., accidents, luck of the draw) which dictate the mating opportunities and/or mating successes for each sub-population will be randomly distributed among them. This randomness ensures that, while each sub-population will have its own unique progression of change in gene frequencies, any cumulative effect on the total population will be of a stochastic nature.

This observation does not apply to the finite populations employed for genetic research. For such populations, changes in gene frequencies which result from random chance (termed random genetic drift) may be of a magnitude sufficient to impede interpretation of the genetic consequences subtended by the experimental treatments applied. Another factor which must be considered in interpreting results from experiments involving finite populations is inbreeding. Although inbreeding and genetic drift are not independent processes, they will be discussed separately.

Inbreeding

In the propagation of finite populations the genetic covariances (relationships) among the parents increase with each generation of breeding. While the rate of increase can, within limits, be modified by the mating system employed, the central fact is that the proportion of genes shared in common among members of the population increases, and the proportion of heterozygous loci decreases, as the genetic covariances increase. Wright (1921) quantified the rate of increase in homozygous loci as F (F defined as the inbreeding coefficient) and an approximation of this quantity is given by the formula

$$dF = \frac{1}{8m} + \frac{1}{8f}$$

where m and f represent the number of breeding males and females, respectively, contributing to a given generation in a bisexual, cross-fertilizing population. Two points to be noted from this formula are

(1) the sex contributing the fewest breeding individuals in any generation (males in most species) will have the greatest influence on dF;

(2) dF is cumulative over time. Thus any increases associated with a "bottleneck" generation (i.e., a diminished number of either sex) will have a dominant effect.

It should be noted that F (the cumulative inbreeding of a population) is not an absolute statistic but is always relative to the base population. As will be noted later, dF has connotations in respect of genetic drift (e.g., minimizing dF will minimize the drift variance) but the specific point of concern here is the phenomenon known as inbreeding depression. Enforced homozygosity (the asymptotic result of long-continued inbreeding) is generally detrimental to animal populations. Random chance determines which loci will become homozygous in each generation of inbreeding and the specific alleles which become fixed may or may not be beneficial to the organism. Inbreeding depression is a manifestation of the extent to which unfavorable alleles have become homozygous and/or favorable allelic combinations have been disrupted.

Inbreeding depression is most strongly manifested for traits contributing to total reproductive performance. Indeed, many studies with mammals have indicated that the inbreeding-related changes in post-weaning performance traits (other than reproduction) are mediated not by the inbreeding of the individual but by a depression of the maternal environment provided by its inbred dam. Whether the consequences of inbreeding on the trait under consideration are direct or indirect, the fact remains that sound genetic comparisons among individuals of finite populations require adjustments for population differences in inbreeding.

Genetic drift

The concept of genetic drift can be illustrated by considering two random mating populations derived from a common base population. Replacement breeders for one population are designated in each generation by a strictly random process (i.e., without regard for equivalence of family representation). For the second population, the random selection process is modified to ensure equivalence of family representation. It is intuitively obvious that, while perpetuation of the original gene frequencies cannot be guaranteed for either of these derived populations, the opportunities for random changes will be least in the second population. These random changes, defined as genetic drift, will be cumulative over time. They will occur in all finite populations, selected or otherwise, but they will have specific pertinence to populations which are being maintained as genetic controls.

All estimation procedures relating to genetic drift deal with relative, not absolute, values and they are couched in terms of the drift variance, σ_d^2. In each generation of an idealized random mating population, the value of this variance for a quantitative trait with additive genetic variance of σ_G^2 will be given by the expression

$$\sigma_d^2 = \sigma_G^2/N_e$$

where N_e is the effective population size. Over g generations, the drift variance will approximate

$$\sigma_d^2 = g\sigma_G^2/N_e$$

Thus, drift variance for this trait per unit of time (t) will be minimized by increasing N_e and/or decreasing g (i.e., increasing the generation interval).

In bisexual populations, N_e is estimated by the formula

$$\frac{1}{N_e} = \frac{1}{4m_e} + \frac{1}{4f_e}$$

where m_e and f_e are the effective numbers of male and female parents, respectively. Note that $1/N_e = 2F$, where F is the approximation of Wright's inbreeding formula described earlier. Thus, the formula for drift variance may also be expressed as

$$\sigma_d^2 = 2F\sigma_G^2$$

Values for N_e will be determined by a variety of biological constraints (e.g., variance of family size, sex ratios, family structure, variance among families in fertility and/or survival). In the highly simplistic situation where the ratio of males to females is one to one, and each mating rears exactly one male and one female to breeding age, $N_e = N(m+f) = 2N$. In real life, these simplistic assumptions will not be met and N_e values will always be less than $2N$. However, as indicated by the relationship between N_e and F given above, it is evident that the selection-mating strategies appropriate for reducing the rate of inbreeding will also be appropriate for increasing N_e.

An introduction to the extensive literature dealing with inbreeding, genetic drift and effective population size is provided by the comprehensive reviews authored by Dickerson (1969), Wright (1969), Kimura and Crow (1963) and Hill (1971, 1972a,b).

MONITORING TECHNIQUES

The monitoring technique appropriate to any given genetic experiment will depend on:

(1) the specific objectives of that experiment. Speaking in broad terms, the objectives will fall into two categories — those which seek to establish the relative magnitude of genetic differences among treatments (or populations), and those which seek to quantify the absolute magnitude of the genetic differences;

(2) the feasibility of comprehensive control of non-genetic influences throughout the duration of the experiment;

(3) the relative cost of alternative forms of controls.

All experimenters in the field of population genetics must acknowledge the reality of finite resources. They must estimate the total breeding population which can be accommodated by available resources and then, bearing in mind the effective population size (N_e) required within each sub-population defined by their experimental design (treatments, replicates), determine the number of sub-populations which can be accommodated. This planning process must consider the resources required for maintenance of the breeding populations as well as those required for the performance evaluation of their progeny. In this context (i.e., the performance evaluation process) the importance of comprehensive evaluation cannot be over-emphasized. All too frequently an experiment fails to achieve its potential not because of inadequate design but because performance data relevant to the full interpretation of correlated genetic responses were not recorded.

Evaluating relative genetic differences

Many genetic studies are concerned with relative rather than absolute genetic differences. Two or more populations are compared to determine their relative ranking in terms of specific performance criteria. In livestock breeding such comparisons may involve breeds, lines within breeds, breed or line crosses (mating systems) or lines developed through the application of different performance evaluation/selection criteria. Specialized control populations are not required, since any one population of those compared has the potential to serve as the standard, and attention will focus on limiting the opportunity for variation associated with non-genetic sources. This implies two levels of control, one being the emphasis on reasonable environmental constancy through the life of the experiment, and the other, the principle of strict contemporary comparisons. Concern for these levels of control is much broader than physical control of the environment (climatic standardization, lighting, nutrition, parasites and health); it must include standardization of the maternal environment (e.g., dam age and parity) and sex of the progeny evaluated. In some instances at least, standardization might be achieved by appropriate statistical techniques.

It is pertinent at this point to consider the concept of a control based strictly on constancy of the environment. Such control can be visualized in the case of experimental populations of species maintained under strict laboratory conditions (e.g., *Drosophila*, *Tribolium*). However, experience has failed to demonstrate that this concept is viable since its "successes" have been limited to traits where it is needed least, namely, those known to be particularly insensitive to changes in the environment (e.g., bristle number in *Drosophila*; Clayton et al., 1957). While this observation demonstrates that the concept of an environmental control per se is unlikely to have general validity, it does not detract in any way from the principle that reasonable control of the environment must be exercised in all genetic experiments.

The data from experiments designed to measure relative genetic differences may be used for statistical estimation of genetic parameters but, speaking generally, they have limited potential for use in the validation of population genetic theory. One exception is the comparison of populations with documented contemporary histories of high—low selection. This particular case will be dealt with in the following sections.

Evaluating absolute genetic differences

A pre-requisite for the measurement of absolute genetic differences is the provision of some means for separating the genetic and environmental components of the observed population differences. In essence, this must involve contemporaneous comparisons of the experimental population(s) with a stable (genetically unchanging) control population. If this control population is adequately structured to minimize inbreeding and genetic drift, its time trends in performance may be interpreted as a direct measure of the environmental component and its deviation from the performance of the selected population(s) will be a direct measure of the genetic component.

Genetic controls are, in essence, the replication of the same genetic material in successive generations. This can take a variety of forms, each with its own advantages and deficiencies.

(1) Genotype and/or gamete storage. Genotype storage is exemplified by plant species where, through storage of seed or through vegetative reproduction, it is possible to reproduce the foundation population at any desired point throughout a long-term genetic experiment. This method is not generally available to those who work with species other than plants but progress is being made in techniques for low-temperature storage of fertilized ova. It is self-evident that any correlation between the quantitative traits of genetic interest and the viability of the stored genotypes would compromise the validity of this form of control. Gamete storage, specifically the deep-freezing of semen, has been available to the cattle industry for many years and is now considered a viable technique for pigs. Application of this genetic control technique is exemplified in the cattle industry (dairy and beef) by the use of reference sires. Progeny produced by their semen are compared with contemporary progeny from a current crop of young sires to provide estimates of one-half of the additive genetic change that has accrued in the interval since the reference sires were born.

Crucial to the genetic interpretation of any comparisons involving this form of genetic control is the question of the criteria employed in designating the parents whose gametes were stored. This question has particular relevance to the use of reference sires by the livestock industry since such sires, if they are to be acceptable to the potential users, are unlikely to have been chosen at random. In using this control technique (gamete storage) in research, the possibility that progeny produced from stored gametes may be less inbred than those from parents of an advanced generation of selection cannot be ignored. This caution also applies to genotype storage.

Highly inbred lines, a specialized form of genotype storage, have limited relevance to large animal breeding. They are costly to develop and maintain and those which survive intense inbreeding are likely to possess attributes that are not characteristic of the populations with which they (or their crosses) are to be compared.

(2) Repeated mating techniques. These techniques represent a form of genotype storage over a short period (two or three generations) of time. Contemporary progeny are produced from new replacement (selected) breeders and from second (or later) parities of their parents and their relative performance for traits devoid of maternal age effects provides a measure of genetic response subtended by the selection applied. Since few traits are devoid of direct or indirect maternal age effects, this simplistic approach is seldom relevant and highly complex experimental designs have been evolved (e.g., Goodwin et al., 1960; Hickman and Freeman, 1969) to permit a clear separation of the genetic and maternal effects. For some species gamete storage has specific relevance to utilization of repeated mating techniques.

Repeated mating techniques, specifically those limited to cycles of two or three generations, ensure a high degree of relationship between the control and the experimental populations. This minimizes the likelihood of confounding genetic and inbreeding effects and also minimizes any contributions to line differences associated with genotype—environment interactions.

The principle of repeated matings is commonly employed by animal geneticists to estimate, from field records, the genetic component of performance trends occurring in livestock populations. Such populations are characterized by overlapping generations and some populations, or sub-sets thereof, may utilize reference sires. The estimation processes employed are characterized by statistical complexity and the inferences they subtend generally relate to relative rather than absolute genetic trends.

(3) Segregating control populations. These represent a dynamic form of genotype storage with the original (foundation) genotypes reproduced anew each generation through the normal reproductive process. Genetic segregation occurs in each generation but genetic change, whether due to drift, inbreeding, natural selection or artificial selection, is minimized by specific constraints that are designed to maximize effective population size (N_e) and to maintain zero selection in the designation of replacement breeding stock. The latter (zero selection) requires that replacements be designated at random from the progeny available (hence the common designation of such populations as random bred controls). In practice, replacements will be designated at random within family (full-sibs for females and half-sibs for males) and the individuals chosen will be those deviating the least from their family mean. The topic of effective population size (N_e) was dealt with briefly in a previous section. As was noted there, N_e will be optimized by the same procedures which serve to minimize the rate of inbreeding.

(4) High-low (divergent) selection. Divergent selection was visualized as a means of eliminating environmental bias from estimates of genetic response without expending resources for the maintenance of a control population per se. With two populations, selected simultaneously in opposite directions, line deviations at any point represent the total genetic response to the cumulative selection estimated as the sum of the intensities applied separately to each line.

Hill (1972c,d) has shown that divergent selection is the most efficient design for estimation of genetic parameters. It should be noted, however, that for traits of economic importance (e.g., growth rate or carcass merit), selection downwards in the context of economic merit is of limited relevance to the industry. The question of whether the trait or traits of interest respond equally to both upwards and downwards selection must also be considered. While the answer to this question must be sought through an experiment involving divergent selection, it will not be answered unless the experimental design includes an adequate genetic control population. Another unanswered question in respect of divergent selection experiments relates to line differences in their environmental responses. This may be of little importance in the short term, but in instances where substantial genetic differentiation has been achieved, whether due to the intensity of selection or long duration of the experiment, the possibility of a substantial genotype—environment interaction cannot be overlooked.

Choice of an appropriate genetic control

As was indicated at the outset of this section, the objectives of a particular experiment will determine the complexity of the procedures appropriate for monitoring genetic change. If use of a random-bred control population is indicated, the choice made among the several alternatives will be dictated in part by biological considerations (e.g., amenability to genotype or gamete storage), in part by the relative efficiency of alternative designs, and in part by their relative costs. The relative efficiency of alternative designs (e.g., divergent selection vs. unidirectional selection with a control vs. unidirectional selection without a control) for the estimation of realized heritabilities has been evaluated in statistical terms by Hill (1971, 1972c,d). Smith (1976) has compared various alternative designs of control populations in terms of their relative costs. In the context of costs, it is pertinent to note that a single well-designed genetic control population can serve for the simultaneous monitoring of several different (but contemporary) selected populations.

SPECIFIC CASE STUDIES INVOLVING CONTROL POPULATIONS

The literature citations provided by references given in this paper are evidence that virtually all of the information relating to the design and use of control populations has derived from studies involving laboratory species. Further evidence of the wealth of information generated by such studies may be found in the many review articles on population genetics (e.g., Robertson, 1956; Roberts, 1965). This emphasis has been no accident. Laboratory species offer the potential for rapid generation turnover (e.g., 12 per year for *Tribolium*), facilities of modest cost, high reproductive rates, and a reasonably complete control of the environment, all of which bear directly on the likelihood that meaningful experimental results can be obtained within a reasonable time frame.

These attributes do not apply to most large animal species. Beef cattle, for example, have a long generation interval (realistically 3 years) and low reproductive rates (0.8 progeny reared per mating under optimum conditions), and research with this species involves high maintenance costs, extensive facilities and little latitude for control of the environment. Also characteristic of large animal species is the fact that many traits of economic interest are measurable only on one sex (e.g., the female traits of egg or milk production) or must be measured after slaughter, with slaughter occurring near the age of sexual maturity (carcass traits) which negates their direct evaluation for mass selection. Because of these contrasts with laboratory species, experimental designs for large animals must incorporate techniques to minimize generation interval (e.g., part-record selection for traits such as milk production) and must utilize separate testing procedures for performance vs. product evaluation (carcass merit). Considerable reliance must also be placed on complex statistical procedures to compensate for imperfect control of the environment, to adjust for variation in family size and sex distribution, and to adjust for variation associated with maternal age, sequential culling and/or overlapping generations.

The experimental designs employed with large animals, and some of the statistical procedures invoked for data anlyses, have been dealt with and/or referenced in several of the literature citations already quoted (e.g., Dickerson, 1969). Rather than try to summarize these reviews, I will close by citing examples from the experiments conducted with my colleagues during my 37 years in animal breeding research at Lacombe.

Beef cattle. The selection objective was rate of growth to 1 year of age. A herd of 500 breeding cows derived from a common foundation population was divided into two replicates (250 cows). Within each replicate, a random third of the cows was designated as a control. Sires for the control line were chosen at random from all yearlings in generation zero; those for the selected line were identified on the basis of the selection criterion (yearling weight) and, of course, included some already randomly identified for use in the control. In subsequent generations the lines were maintained as separate populations with sires in the selected line replaced annually (with yearling bulls) and females replaced at a rate approximating 30% per year. The control was maintained in a static condition, the same matings being repeated each year except for changes (replacements) as necessitated by attrition. Such replacements were required to be sons or daughters of the control animal being replaced with yearling weights within one standard deviation of that for their contemporary control line average.

The total population size (500 cows) was determined on the basis of sampling theory (i.e., the anticipated error of estimate for the realized heritability). It was several times larger than could be accommodated by any of the locations available and the compromise solution was to divide the two replicates between five locations. This was reduced to two locations by the midyear of this 10-year experiment. This compromise, plus the generation differences in age structure of the control vs. the selected line, required that specialized statistical procedures be applied in genetic analyses. These procedures and the results obtained were described by Newman et al. (1973).

The relevance of the control population in this experiment is illustrated in Fig. 1A which shows, for the two locations common to all years, the annual fluctuations in actual performance of progeny of the select and control lines. The substantial and erratic environmental component of the performance time trend for the select line was inferred only from the behavior of the control line. Thus, an estimate of the progressive genetic change subtended by selection was deduced from the time trend in the relative performance of the two lines (Fig. 1B).

Pigs. This selection experiment was designed to evaluate the relative efficiency of single trait vs. index selection for the improvement of lean tissue growth rate in pigs. Four lines were developed from a common foundation:
(1) selected for maximum post-weaning growth rate (20–90 kg);
(2) selected for minimum backfat at 90 kg (live animal measurement);
(3) selected for a phenotypic index combining these traits;
(4) a random-bred control.

The selected lines were replicated and advanced by one generation annually. The control was maintained as a single line with repeat matings and minimum replacement of breeding stock (long generation interval). Limitations of facilities imposed the following restraints.

(1) Separate performance testing of the two sexes, males reared under confinement and females under extensive management. This ensured more precise evaluation of the sex amenable to the most intense selection (males) and provided females with a rearing environment (exercise) deemed desirable for sound structural development.

(2) Annual propagation of the control could not be accommodated in the period contemporary with performance testing (selected lines). As a compromise, its progeny were produced contemporary with second litters (repeat mat-

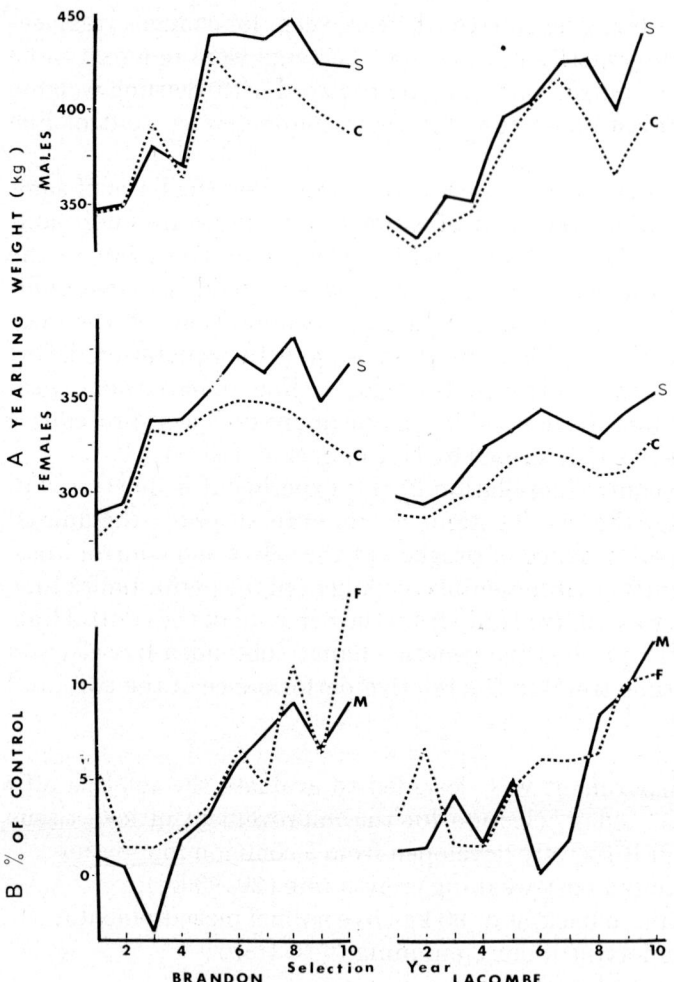

Fig. 1. Time trends in yearling weight of control (C) and selected (S) line beef cattle at two locations. A, Absolute weights; B, selected line as a percentage of the control. (M, males; F, females).

ings) taken from the select lines with all progeny placed on post-weaning test for feedlot and carcass evaluation (slaughter test).

(3) Each replicate of the selected lines was of limited size (breeding population of six males and 18 females) with the result that inbreeding progressed steadily, reaching approximately 20% by the final generation. In this period (10 years), the larger control population with its lower rate of generation turnover attained a final inbreeding coefficient of 5%.

These restraints imposed parallel restraints on data analyses. Selection intensities had direct relevance only to the performance traits measured on the contemporaries evaluated (first litters) and the genetic trends they subtended

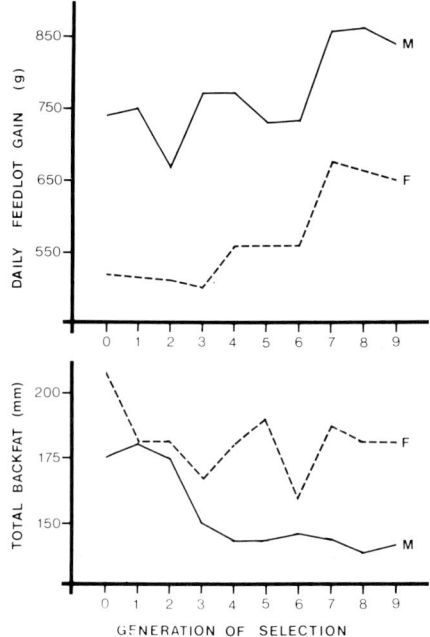

Fig. 2. Time trends in average daily gain and backfat for males and females of the index line.

had to be estimated in a relative sense since no contemporary controls were evaluated. Comparisons made with the contemporary control required recalculation of selection intensities (i.e., restricted to those parents producing second litters), adjustments for the consequences of inbreeding (Mikami et al., 1977) and consideration of differences in test management and the traits actually measured for first and second litters. Indeed, it would not be unreasonable to view the traits measured for second litters as correlated responses to the selection applied to first litters.

Sex differences in growth and fat as evaluated by the performance test (first litters) are illustrated for the index line in Fig. 2. Females grew less rapidly and were fatter than their contemporary males and, while growth rate trended upwards for both sexes, females exhibited no consistent time trend in weight-adjusted backfat after the first generation. In other words, it appeared that genetic change occurred for both sexes in growth rate but only for males in backfat.

These differences were not independent of the differences in the test management given the two sexes. Male tests, carried on under confinement rearing, were terminated at a constant weight (90 ± 2.5 kg) and required minimum statistical adjustment to standardize the fat measurements. Female tests, with rearing on pasture, were terminated at a date which was constant for all lines in any year (regardless of birth date) but varied erratically among years in accordance with availability of labor. This introduced annual differences in age (Fig. 3) and weight, both tending to increase with generation. Note that these differences did not affect line comparisons made within generations (Fig. 4). Since fat was adjusted to the average weight of contemporaries within each line-replicate sub-group, the averages for weight-adjusted fat tended to increase for females while declining for males (Fig. 2).

In the absence of a contemporary control, genetic trends for first-litter traits (performance evaluation) were estimated relative to the phenotypic changes observed in the index line. Inbreeding was ignored since the three selected lines were essentially equivalent in rate of inbreeding. As indicated in Fig. 5, the time trends for the growth line relative to the index were essentially zero for growth, progressively positive (fatter) for fat, with substantial agreement between sexes for both traits. In contrast, the minimum fat line declined progressively in growth rate and, on average for the two sexes, remained equivalent to the index line for backfat (Fig. 6). Subsequent statistical analyses established that the apparent sex difference in the time trend for fat was actually an artifact introduced by inadequacy of the statistical procedures employed for adjusting fat to a constant terminal weight with accuracy inversely related to the variance of terminal weight (Fredeen and Mikami, 1986a).

Time trends for growth rate and fat tended to be positive in respect of merit for both the control and index lines (Fig. 7) but the presence of large and erratic annual differences, while generally parallel for these two lines, demonstrated the presence of substantial non-genetic variation. Inbreeding effects also contributed to these time trends, reducing growth rate and increasing fatness in the index line as generations progressed. Adjusting the data for differences in inbreeding and expressing line performance as a percentage of the contemporary control permitted a more rational picture of the genetic differences among lines (Fig. 8). Fig. 8, when compared with Fig. 6 (performance relative to the index line), clearly demonstrates the broader inferences provided by absolute over relative comparisons among lines (i.e., with or without a formal control population). The results provided by these second-litter comparisons have been described in detail by Fredeen and Mikami (1985b, c).

Poultry. This experiment was designed to examine the genetic consequences of part-record selection for egg production (to 275 days of age) on full-record (500 day) egg production. The foundation population was derived from the large (40 males and 200 females per generation) random mating population known as the Ottawa control strain (Gowe et al., 1959). Eggs from this control population were shipped to Lacombe for incubation and rearing on range. At 147 days of age, a random sample of 40 males and 1000 females was housed. After 128 days of production (275 days of age), 200 females representing those with the highest part-record egg production and 40 males, one from the full-sib family with the highest average female-sib record for egg production within

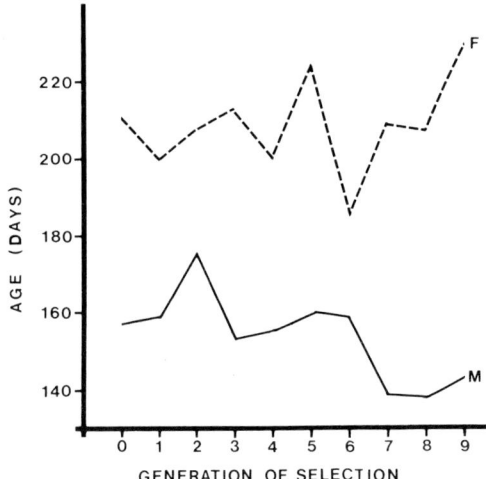

Fig. 3. Time trends in age at test termination of males and females of the index line.

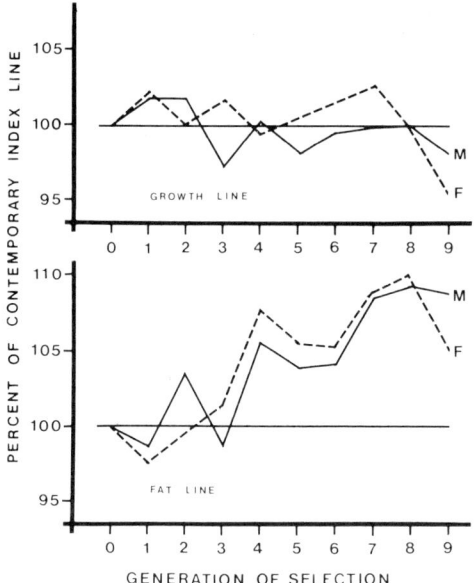

Fig. 4. Time trends in age at test termination of males and females of the growth and minimum fat lines expressed as percentages of the contemporary index line.

each half-sib family, were designated as breeders to reproduce the first generation. All matings were by artificial insemination. Eggs for hatching were stored a maximum of 3 weeks then incubated along with a sample of 400 eggs obtained from the foundation control strain. Selected and control-line chicks were intermingled for brooding and range rearing. At 147 days of age, a random sample

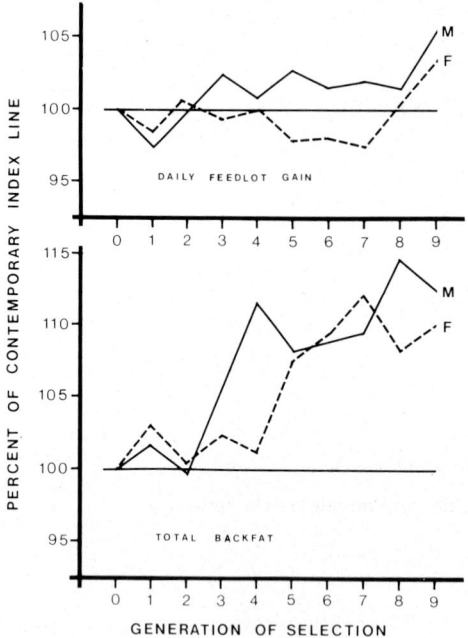

Fig. 5. Time trends of growth rate and fat of the growth line relative to the contemporary index line.

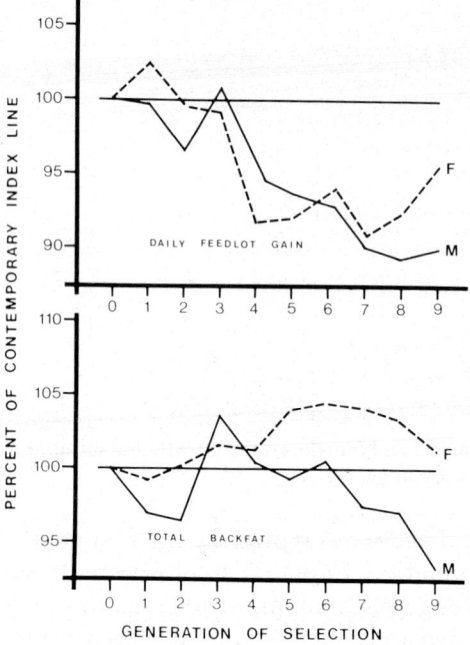

Fig. 6. Time trends of growth rate and fat of the minimum fat line relative to the contemporary index line.

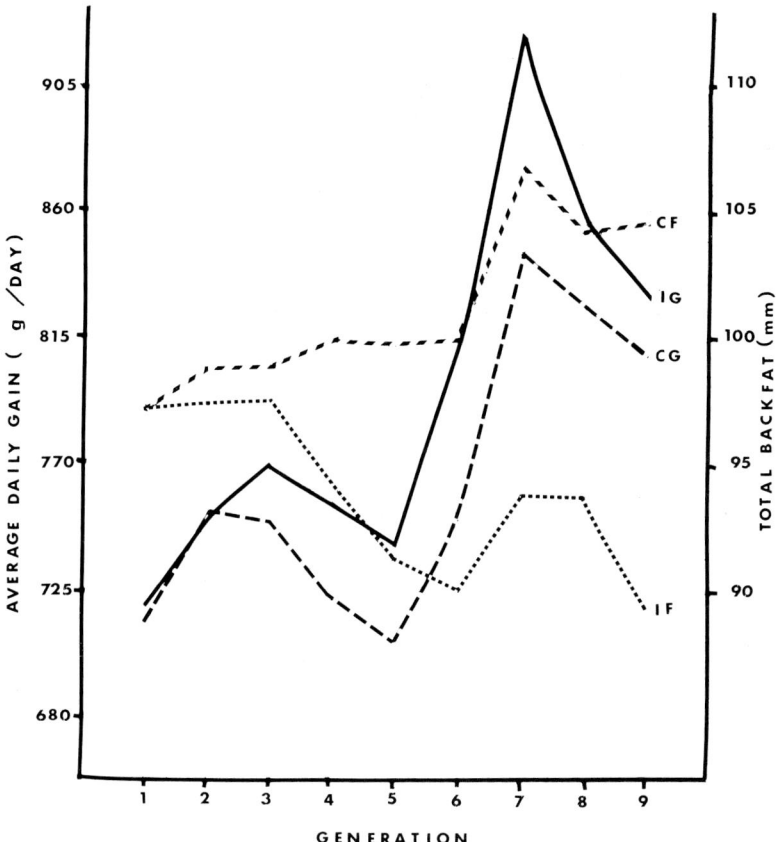

Fig. 7. Time trends in daily gain (G) and total backfat (F) of index (I) and control (C) lines (season 2, sexes combined).

of each line (40 males and 800 females of the select line, 200 females of the control) was housed for production estimation. As was the case through rearing, the lines were intermingled in the laying house. Selection and propagation procedures in the second and subsequent generations were as described for generation 1 with a new sample of control eggs obtained annually from the foundation population.

The primary traits measured through 17 generations of selection were age at sexual maturity for females (first recorded egg), production to 275 days of age (part record) and production through the subsequent 225 days to measure the complete record. Secondary traits included age and cause of mortality, morbidity, fertility, embryonic loss, body weight, egg weight, and several criteria of egg quality. The present discussion will focus on one of the potential inadequacies of the control line with specific attention to time trends for the three primary traits during the first 14 generations of selection.

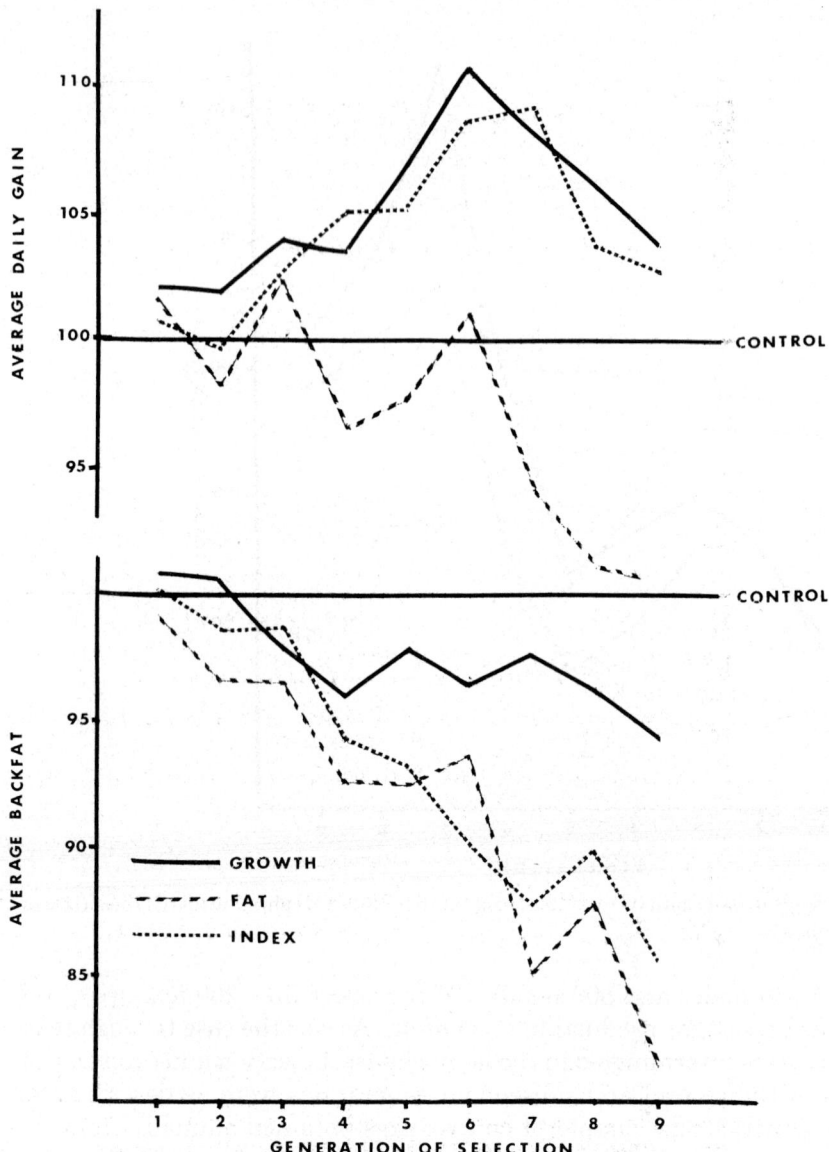

Fig. 8. Time trends in performance of the selected lines adjusted for inbreeding and expressed as percentages of the contemporary index.

As was the case for the other selection studies described, the time trends for the control line demonstrated that the environmental control applied to this species — obviously much more rigorous than could be practiced with large animals — was incapable of eliminating some major non-genetic influences (e.g., disease). Each trait recorded exhibited the substantial annual variation

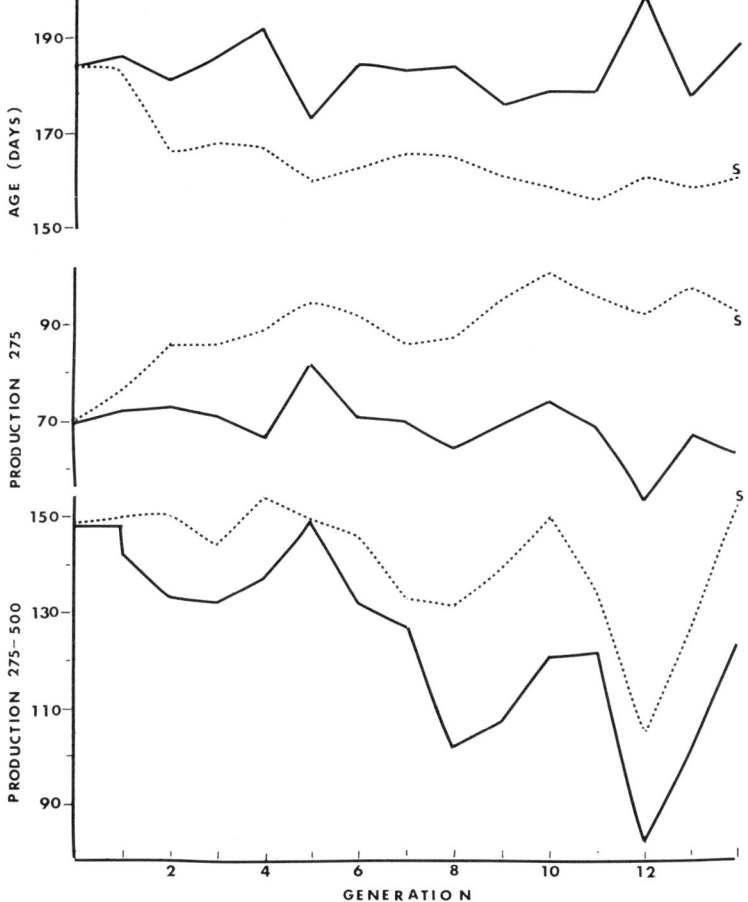

Fig. 9. Performance trends for the control and selected poultry lines.

observed for the three primary traits (Fig. 9). As a result, no genetic interpretations would have been possible in the absence of the control population.

While certain deductions might be drawn from visual inspection of Fig. 9, interpretation is greatly facilitated by portraying select-line performance as a percentage of the contemporary control (Fig. 10). Relative to the control, the select line showed a favorable and reasonably continuous response in egg production to 275 days (the object of direct selection), a decrease in age at first recorded egg to generation 4 with little change thereafter, and a modest response in production from 275 to 500 days, with this response essentially confined to a single generation (generation 8). The "plateau" in age at sexual maturity simply reflected the inadequacy of the range trap nests to attract early layers, an inadequacy which was first recognized at generation 4. Thus, in all subsequent generations this trait was clearly overestimated for the selected line and production to 275 days was obviously underestimated.

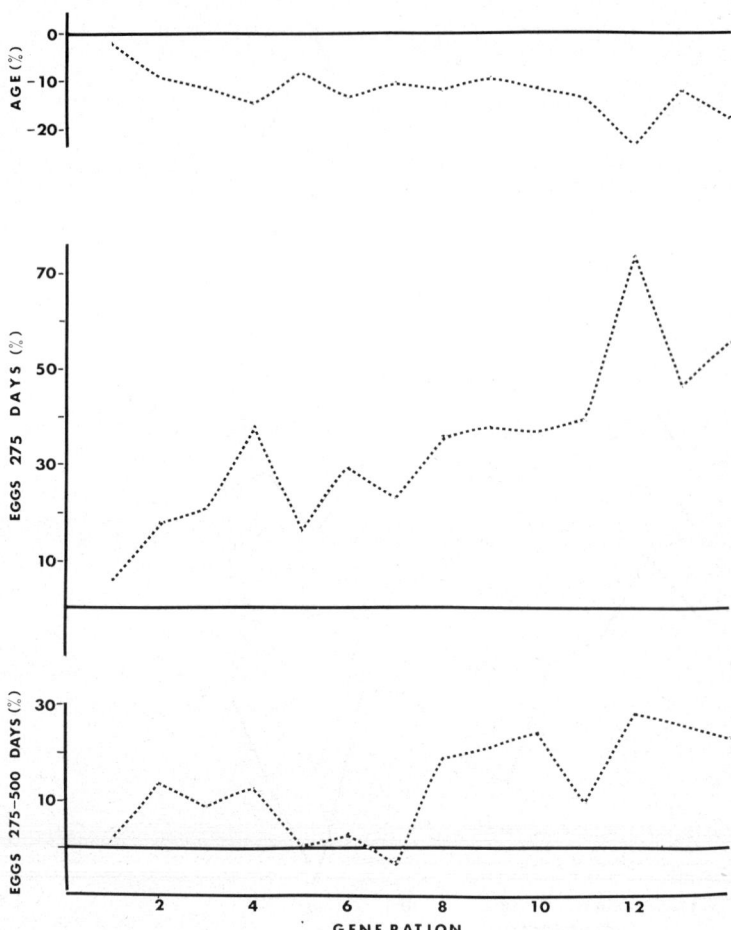

Fig. 10. Performance of the selected line expressed as a percentage of the contemporary control population.

The annual irregularities featuring these trends were undoubtedly associated with management factors. For example, a severe outbreak of coccidiosis occurred in generation 5, selection intensities were enhanced commencing with generation 8 by a 25% increase in the number of eggs incubated, and floor-housing of intermingled strains was replaced by individual caging in generation 12. Differences in strain reactions to these or other less obvious changes in the environment could not be assessed although a distinct strain difference in morbidity was observed following the disease outbreak in generation 5.

Of particular concern, however, were the strain differences observed in behavior, the select line being more aggressive at housing time, and less aggressive during the post-selection period, than the control. That this may have influenced the relative egg production of the two strains was indicated by comparative tests conducted after the installation of individual cage facilities but retrospective interpretation of the results obtained was limited by the substantial line differences in inbreeding when the comparisons were made (generation 12).

REFERENCES

Chapman, A.B., 1973. Selection theory and experimental results. In: R. Shrode (Editor), Proceedings, Animal Breeding and Genetics Symposium in Honor of Dr. Jay L. Lush. Am. Soc. Anim. Sci., Champaign, IL, pp. 42–53.
Clayton, G.A., Morris, J.A. and Robertson, A., 1957. An experimental check on quantitative genetical theory. 1. Short-term responses to selection. J. Genet., 55: 131–151.
Craft, W.A., 1951. Fifty years of progress in swine breeding. J. Anim. Sci., 17: 960–980.
Dickerson, G.E., 1969. Techniques for research in quantitative animal genetics. In: Techniques and Procedures in Animal Science Research, revised edition. Am. Soc. Anim. Sci., New York, NY, pp. 36–79.
Fisher, R.A., 1918. The correlation between relatives on the supposition of Mendelian inheritance. Trans. R. Soc. Edinburgh, 52: 399–433.
Fredeen, H.T. and Mikami, H., 1986a. Mass selection in a pig population. Experimental design and responses to direct selection for rapid growth and minimum fat. J. Anim. Sci., 62: 1492–1508.
Fredeen, H.T. and Mikami, H., 1986b. Mass selection in a pig population. Realized heritabilities. J. Anim. Sci., 62: 1509–1522.
Fredeen, H.T. and Mikami, H., 1986c. Mass selection in a pig population. Correlated responses in carcass traits. J. Anim. Sci., 62: 1546–1554.
Goodwin, K., Dickerson, G.E. and Lamoreux, W.F., 1960. An experimental design for separating genetic and environmental changes in animal populations under selection. In: O. Kempthorne (Editor), Biometrical Genetics. Union Int. Sci. Biol. Ser. B, pp. 117–138.
Gowe, R.S., Robertson, A. and Latter, B.D., 1959. Environment and poultry breeding problems. 5. The design of poultry control strains. Poult. Sci., 38: 462–471.
Hazel, L.N., 1943. The genetic basis for constructing genetic indexes. Genetics, 28: 476–490.
Hickman, C.G. and Freeman, A.E., 1969. New approach to experimental design for selection studies in dairy cattle and other species. J. Dairy Sci., 52: 1044–1054.
Hill, W.G., 1971. Design and efficiency of selection experiments for estimating genetic parameters. Biometrics, 27: 293–311.
Hill, W.G., 1972a. Estimation of genetic change: genetic theory and design of control populations. Anim. Breed. Abstr., 40: 1–15.
Hill, W.G., 1972b. Estimation of genetic change: experimental evaluation of control populations. Anim. Breed., Abstr., 40: 193–213.
Hill, W.G., 1972c. Estimation of realized heritabilities from selection experiments. 1. Divergent selection. Biometrics, 28: 747–766.
Hill, W.G., 1972d. Estimation of realized heritabilities from selection experiments. 2. Selection in one direction. Biometrics, 28: 767–780.
Kimura, M. and Crow, J.F., 1963. The measurement of effective population number. Evolution, Lawrence, Kans., 17: 279–288.
Lush, J.L., 1940. Intra-sire correlations or regressions of offspring on dam as a method of estimating heritability of characteristics. Proc. 23rd Annu. Meet. Am. Soc. Anim. Prod., pp. 293–308.
Lush, J.L., 1945. Animal Breeding Plans, 3rd edition. Iowa State College Press, Ames, IA, viii + 443 pp.
Mikami, H., Fredeen, H.T. and Sather, A.P., 1977. Mass selection in a pig population. 2. The effects of inbreeding within the selected populations. Can. J. Anim. Sci., 57: 627–634.

Newman, J.A., Rahnefeld, G. and Fredeen, H.T., 1973. Selection intensity and response to selection for yearling weight in beef cattle. Can. J. Anim. Sci., 53: 1–12.
Roberts, R.C., 1965. Some contributions of the laboratory mouse to animal breeding research. Anim. Breed., Abstr., 33: 339–353; 515–526.
Robertson, F.W., 1956. The use of *Drosophila* in the experimental study of animal breeding problems. Anim. Breed., Abstr., 24: 218–224.
Shrode, R.R. (Editor), 1973. Proceedings, Animal Breeding and Genetics Symposium in Honor of Dr. Jay L. Lush. Am. Soc. Anim. Sci., Champaign, IL.
Smith, C., 1976. Alternative forms of genetic controls. Anim. Prod., 23: 403–412.
Wright, S., 1921. Systems of mating. I, II, III, IV, V. Genetics, 6: 111–178.
Wright, S., 1969. Evolution and the Genetics of Populations. Vol. 2. The Theory of Gene Frequencies. University of Chicago Press, Chicago, IL, vii+511 pp.

Effective Population Size and Selection in Variable Aquaculture Stocks

ROGER W. DOYLE and ANDRE J. TALBOT

Department of Biology, Dalhousie University, Halifax, NS. B3H 4J1 (Canada)

ABSTRACT

Doyle, R.W. and Talbot A.J., 1986. Effective population size and selection in variable aquaculture stocks. *Aquaculture*, 57: 27–35.

Effective population size and inbreeding rates are derived in terms of the variability of broodstock replacement; the parameters are more useful for extensive aquaculture than the usual formulation. It is shown that response to mass selection may improve if variability in birth date and age is reduced by weight-specific rather than age-specific selection for growth rate. Current studies on domestication selection in Asia suggest that traditional broodstock management may be causing stocks to deteriorate.

INTRODUCTION

The orderly disposition of genetic tools for stock improvement in aquaculture is greatly hindered by variability at all levels of the analysis. Individual fish vary too much, replicates and controls vary too much, and the results of similar experiments vary too much for the easy application of standard quantitative genetic techniques. The most important problems stem from fluctuations in the size of the broodstock, and variation in spawning time and condition among the breeders. Broodstock fluctuation makes it difficult to calculate effective population sizes, and rates of inbreeding and genetic drift of local stocks. Spawning variability greatly increases unwanted phenotypic variance in selection programs. Some unconventional solutions to this problems are considered here in the context of extensive aquaculture.

EFFECTIVE POPULATION SIZES IN EXTENSIVE AQUACULTURE

It is generally not easy to measure effective population size to confirm or disprove a hatchery operator's belief that inbreeding is the cause of reduced growth or increased incidence of disease in his fish. Inbreeding is blamed for problems that may, in fact, be caused by domestication selection in undesirable directions or introgression from inferior stocks. The danger for extensive aquaculture is that a solution may be sought where there is no problem, and genetic progress (e.g. adaptation to the local aquaculture environment) may be thrown away by introducing wild stock into the population. The inbreeding problem must be disposed of first, both in reality and in the mind of the aquaculturist, before other genetic stock improvement measures are undertaken.

Inbreeding is usually expressed in terms of the effective size, N_e, of a random-mating population. One of the practical difficulties of the usual parameterization is that N_e is not well defined in the farms and fisheries stations involved in routine broodstock management for extensive aquaculture. When fish breeders have special ponds for broodstock it is often only on a temporary basis, or the number of fish varies widely from season to season. Usually there is no clear idea of how many fish are currently in the broodstock pond and most farmers are reluctant to let a fisheries scientist do any seining to find out. The "population" in such cases is not a real entity and it is expedient to treat its effective size as a derived parameter rather than one that can be directly observed.

Hill (1979) expressed N_e as a function of the number of new members of each sex entering the breeding population each year and the variance of reproductive success. Hill's formulation is closely related to "variance effective number" of Crow and Kimura (1970) with the additional consideration of overlapping generations and differential reproductive variance between the sexes. We attempt below to apply these ideas to extensive aquaculture situations where stock sizes fluctuate and information is uncertain.

Mature fish added to an aquaculture broodstock in any given year can be treated as consisting of two groups: those that are eventually used as spawners and those that are not. A proportion (P) of the spawners in the broodstock will replenish the breeding population while the rest ($1-P$) will not (i.e., their offspring will all be sold). The net effect is that each spawned fish used to replenish broodstock must replace both itself and a proportionate share of the other fish in the broodstock in the next generation.

Estimates of the following parameters are required.

Data from the farmer (on a per annum basis):

(a) N_f, N_m = the number of new breeders of each sex added to the broodstock pond;

(b) N_{sf}, N_{sm} = the number of females and males spawned, respectively;

(c) P = the proportion of spawners that have some probability ($P>0$) of contributing to the broodstock;

(d) D = average duration of stay in the broodstock.

Derived parameters:

(e) $N_t = (N_f + N_m)$ = the total number of new breeders added;

(f) $L = (D+1)/2$ = approximately the effective generation length;

(g) N_{rf}, N_{rm} = (b) (c) = the number of new females and males (per year) that eventually contribute to replacing the broodstock [note that (g) = (a) if (b) (c) > (a)];

(h) $N_{pf}, N_{om} = $ (a) – (g) = the number of new males and females that never replace broodstock;

(i) Var(f), Var(m) = variance of replacement success for each sex calculated separately.

The sum-of-squares (SS) for each sex consists of the weighted SS of the infertile $(1-P)$ and fertile (P) fish around the population mean plus the SS of the members of each group around their group means. For example,

$$SS(\text{females}) = N_{of}\left(\frac{N_t}{N_f}\right)^2 + N_{rf}\left(\frac{N_t}{N_{rf}} - \frac{N_t}{N_f}\right)^2 + N_t \qquad (1)$$

where the Poisson variance of fertile females $= N_t/N_{rf}$. Var(f) = SS(females)/N_f and Var(m) = SS(males)/N_m. The derived effective population size N_e can be defined as follows, simplified from Hill (1979, Eqn. 9)

$$1/N_e = \frac{1}{16N_fL}\left[2 + \text{Var}(f)\right] + \frac{1}{16N_mL}\left[2 + \text{Var}(m)\right] \qquad (2)$$

We have made two assumptions in deriving the expressions above. If the number of fish of a particular sex that contribute to the next generation exceeds the number of fish of that sex added in a year, we assumed that there is no variation in fertility, only in family size (i.e., $P=1.0$). We also assumed replacement success follows a Poisson distribution among the fish that contribute to the next generation. Broodstock population size is unobservable and fluctuating, yet the calculations gives an estimate of the asymptotic N_e with a stable age structure (Hill, 1979). Hill suggested that small disturbances in age structure or replacement rate will not substantially affect the results. In real aquaculture the estimate of N_e will vary from year to year, but it is not clear how successive estimates should be combined. In the case of non-overlapping generations (no age structure) the long-term effective size is the harmonic mean of the generational N_e values, which is closer to the smallest value than to the arithmetic average.

Once N_e is calculated, estimated or guessed at, the annual rate of accumulation of inbreeding, ΔF, follows as usual from

$\Delta F = 0.5(1/N_e)$ (Falconer, 1981)

An example of the calculation is shown in Table 1 for data obtained from the KBB and Kabini Indian carp hatcheries near Mangalore, India (Eknath and Doyle, 1985a,b). In both farms the offspring of the first "spawning sets" (one female and two males spawned together) of the breeding season are sold to farmers and do not contribute to the broodstock. A portion of the progeny from later sets is retained. The value of P is the proportion of sets that were spawned

TABLE 1

Calculation of effective population size (N_e) in two Indian fish farms for three species of Indian major carp, *Catla catla*, *Labeo rohita*, *Cirrhinus mrigala*. The subscript s refers to male (M) or female (F) sex as indicated (e.g., row designated N_s contains N_m and N_f values). For explanation of the variables, see text

	Farm							
	KBB						Kabini	
	Catla		Rohu		Mrigal		Rohu	
	M	F	M	F	M	F	M	F
N_s	14	6	14	6	14	6	10	5
N_{1s}	48	24	188	94	170	85	70	35
P	0.21	0.21	0.23	0.23	0.59	0.59	0.10	0.10
D	3.5	3.5	3.5	3.5	3.5	3.5	3.5	3.5
N_t	20	20	20	20	20	20	15	15
L	2.25	2.25	2.25	2.25	2.25	2.25	2.25	2.25
N_{rs}	10	5	14	6	14	6	7	3.5
N_{0s}	4	1	0	0	0	0	3	1.5
Var (s)	2.22	5.45	1.43	3.33	1.43	3.33	2.46	6.86
N_e	23.3		31.8		31.8		16.2	
ΔF	0.021		0.016		0.016		0.031	

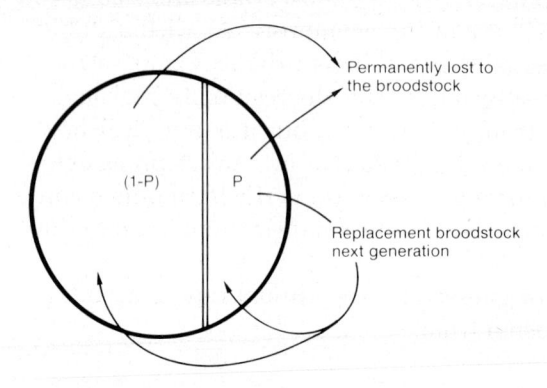

Fig. 1. The meaning of the parameter P. The breeding population is enclosed in a circle that might represent the banks of a real pond or some more abstract closure in space and time. The breeders are separated into two groups by a "fence" that might be an actual fence, a division into two seasons or some other partitioning determined by hatchery procedures. A proportion $1-P$ of the breeders have no chance of contributing to the next generation because they spawn on the wrong side of the fence or the wrong season, are sold for grow-out, etc.; their self-replacement rate is therefore zero. A proportion P have some chance of replacing the broodstock; their actual success rate follows a Poisson distribution with zero values allowed.

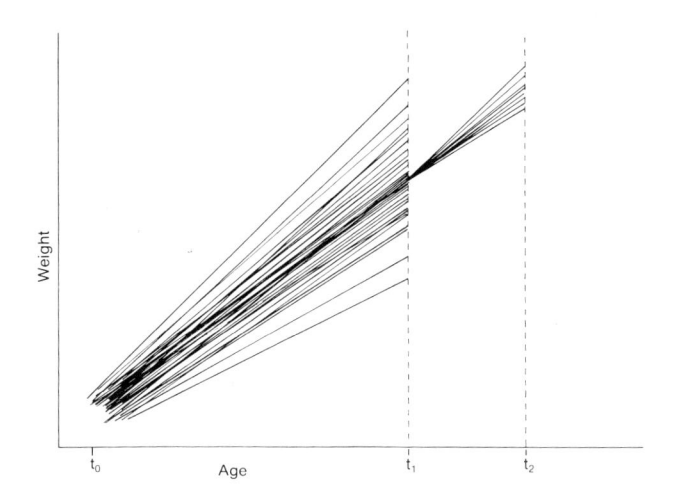

Fig. 2. Weight specific selection. Hypothetical growth trajectories of fish that are nominally born at t_0 but which in fact show considerable variation in birth weight and date. Variance is culled at t_1, removing all of the size-at-age variance, and the fish from the middle of the distribution continue to grow until t_2 when selection for size takes place.

after the date at which fish began to be retained for broodstock (Fig. 1). The estimated inbreeding rates of 1–3% per year are not excessive. However, if farms of this size have an effective N_e of 30 fish or less, it is probable that smaller operations are rapidly inbreeding their stocks. Stocks with a small effective size can also be expected to diverge rapidly through random drift, but a discussion of the relationship between drift and N_e is beyond the scope of this paper.

WEIGHT-SPECIFIC ARTIFICIAL SELECTION

A possible solution to the problem of excessive variation in selection programs is to change the operational definition of the selected variable. Instead of selecting for weight-at-age, which is the usual practice, selection can be performed on weight-specific growth rate. This consists operationally of starting with fish having the same standard weight after an initial grow-out period, and selecting them on the basis of their subsequent rate of growth (Fig. 2). By beginning the selection process at a known time with fish of identical weight, one is in effect giving them a "rebirth" that is free from the phenotypic variation associated with the original spawning and hatching.

Linear growth

If weight increases linearly with time the weight of an individual fish at time t_1, W_1, is a function of three parameters

$$W_1 = W_0 + gt \tag{3}$$

where W_0 is initial weight, g is the growth rate and t is the age of the fish. Although all the fish are measured at the same time t_1, W_0 and t are both variable because of variation in hatching date t_0. This variation will be magnified by competition, thus reducing the realized heritability. The variable of interest for selection purposes is g, which itself has both genetic and environmental components. Taking variances we get approximately

$$\text{Var}(W_1) = \text{Var}(W_0) + g^2 \text{Var}(t) + t^2 \text{Var}(g) - \text{Var}(g)\text{Var}(t) \tag{4}$$

Except for the third term, the variances and covariances in the equation arise from historical reasons only, that is, if Var (W_1) is artificially reduced to zero by selecting a subset of animals that are all the same weight (the "rebirth"), these variances and covariances have no way of reappearing later on in the life of the population, e.g., at t_2. Var (g), however is caused by persistent biological differences in growth rate among the animals and should soon re-establish weight differenced in the selected subset. These weight differences will, of course, be influenced by competition, but the spurious effects of the variable age and starting weight will theoretically have been removed forever. After culling out the large and small animals at t_1

$$\text{Var}(W_2) = 0 + 0 + (t_2 - t_1)^2 \text{Var}(g) - 0 \tag{5}$$

It can be seen that at t_2 the variance of weight is directly proportional to the variance in growth rate only, rendering selection much more efficient (i.e., increasing the realized heritability of weight at age t_2.

Exponential growth

The argument applies equally well to weight during exponential growth when

$$W_1 = W_0 e^{gt} \tag{6}$$

and

$$\text{Var}(\ln W_2) = 0 + 0 + (t_2 - t_1)^2 \text{Var}(g) - 0 \tag{7}$$

where selection would be performed on $\ln W_2$ at t_2 and g is the exponential growth rate.

Some of the genetic variance of growth rate may be lost during the culling of the tails of the weight distribution at t_1. The proportion lost will depend on the non-genetic variance at t_0, the length of the interval $t_1 - t_0$ and the value of W_1 chosen for the "rebirth". In Fig. 2 the starting weight for the second grow-out period is located close to the middle of the distribution of W_1. The midpoint will give the greatest number of individuals of similar weights, thus permitting higher selection intensities, and can be expected to sample the progenies of the greatest number of parents. The genetic variance of growth rate in the population may be somewhat reduced by weight-specific selection, but the realized heritability may be considerably higher.

The temptation to choose a "rebirth weight" near the top of the W_1 distribution should be avoided because the upper and lower tails may be dominated by a few broods with unusually large maternal effects — the very phenomenon one is trying to avoid. Selecting these progenies would amount to family selection with a small but unknown number of families and might decrease Var(g) rather drastically, besides adding to inbreeding.

Weight-specific selection is presently being tested on *Oreochromis niloticus* in Thailand by the National Inland Fisheries Institute. A large population of mature fish is spawned over the course of 2 weeks (t_0), whereupon the fry are collected and placed in a rearing pond for 6 weeks of growth. At that time (t_1) the fingerlings are inspected and the upper and lower tails of the weight distribution are discarded. In practice this means discarding about 50% of the fingerlings: ideally as many as possible should be thrown out so that the remainder are as uniform as possible. The remaining young fish, now nominally the same weight, are grown for a further 6-week period. At this time (t_2) growth rate differences have produced weight differences, Var(lnW_2) unconfounded with differences in grow-out time or initial weight, and the largest 15% are selected to be the parents of the next generation. After three generations of selection the estimated heritability of weight-specific exponential growth rate is 0.19 (Mrs. Parnsri Jarimopas, personal communication, 1985).

DOMESTICATION SELECTION FOR GROWTH RATE

Domestication selection is indirect selection that is an incidental byproduct of grow-out and broodstock management practices in genetically closed populations. It can be analyzed as a type of natural selection that operates in domestic environments that are created and controlled by the farmer. Domestication selection is unavoidable in practice, and it has been shown that selection intensity on traits of commercial interest can be very large in the environments of extensive aquaculture, rivalling the intensities that might be expected from artificial selection (Doyle, 1983).

Eknath and Doyle (1985a, b) have attempted to measure selection intensities in the extensive cultivation of Indian carp in the two hatcheries mentioned earlier. This work is based on analysis of physical measurements of mature fish at several stages of the ripening and spawning cycle, and various physical characteristics of scales collected from the broodstock. The three variables of greatest interest for a selection study, pre-maturation growth rate, age at maturation and present age of the spawners, could not be observed directly because hatcheries do not keep records on individually identified fish. Values of the unobservable variables could, however, be estimated indirectly from the scales and other data by using a statistical technique related to common factor analysis.

The conclusion of the study was that present broodstock management practices exert strong, negative selection on early growth rate. This was true in both farms for all three species studied: *Labeo rohita*, *Catla catla* and *Cirrhinus mrigala*. As in the *Macrobrachium* population previously analyzed in Thailand (Doyle et al., 1983), an easy solution to this negative domestication selection is to select broodstock when they are young and only the fastest-growing individuals have matured.

The genetics unit at the National Inland Fisheries Institute, Thailand, has undertaken a study of domestication selection on tilapia and common carp in one of the northern fisheries stations (Chiangmai) and a group of nearby farms. The farms and station are constantly producing and selling fish of all ages to the market, to other farmers and to each other. With the cooperation of the farmers, fish were measured before and after any "episode" in which they were seined out to be transferred, spawned or sold. All of the episodes were size-selective. Each selection episode was considered positive if the larger fish had at least some chance of contributing to the next generation, and negative if the larger fish were sold to the market or otherwise permanently excluded from the broodstock. The preliminary analysis (Atchara Wongsaengchan, NIFI internal report, 1985) shows that almost all of the more than 30 episodes observed were negative, except when fish were chosen for immediate spawning.

Furthermore, Miss Wongsaengchan mapped the transfers of the stock and found that the farms and station were so interconnected that any negative genetic response to selection that occurs in one place will accumulate and spread throughout the population. Fortunately, the populations of the species observed are large enough to preclude significant inbreeding.

The very limited quantitative analyses of domestication selection presently available all point in the same direction: selection of inferior stock, at least in regard to growth rate.

EVOLUTION OF TAMENESS

There is widespread concern among aquaculture geneticists that selection for rapid growth in populations of competing fish will favor the most aggressive animals, not those that have any innate growth superiority (Kinghorn, 1983). The assumption is that successful competitors that obtain food will grow larger and become even more successful, thus amplifying growth-rate variance in a positive feedback loop. The intuitive expection is that competitive behaviour is favored by selection. The end result will merely be more aggressive, not faster growing fish.

Doyle and Talbot (1986) recently used game-theory techniques to establish the direction of selection on interactive behavior of fish artificially selected for rapid growth. The common-sense view that selection for large size exerts correlated selection on aggressiveness does not adequately describe the usual aquaculture situation. The main finding was that in natural environments success in competition may be required to obtain a full share of resources, but in managed environments this is less true. Game-theoretical analysis suggests that there is no reason to fear that selection for high-yielding strains will inevitably result in increased aggressiveness.

ACKNOWLEDGEMENTS

This work was supported by grants from the Natural Sciences and Engineering Research Council and the International Development Research Centre, Canada.

REFERENCES

Crow, J.F. and Kimura, M., 1970. An Introduction to Population Genetics Theory. Harper and Row, New York, NY 591 pp.

Doyle, R.W., 1983. An approach to the quantitative analysis of domestication selection in aquaculture. Aquaculture, 33: 167–185.

Doyle, R.W. and Talbot, A.J., 1986. Artificial selection on growth and correlated selection on competitive behaviour in fish. Can. J. Fish. Aquat. Sci., 43: 1059–1064.

Doyle, R.W., Singholka, S. and New, M.B., 1983. Indirect selection for genetic change: a quantitative analysis illustrated with *Macrobrachium rosenbergii*. Aquaculture, 30: 237–247.

Eknath, A.E. and Doyle, R.W., 1985a. Maximum likelihood estimation of 'unobservable' growth and development rates using scale data: application to carp aquaculture in India. Aquaculture, 49: 55–71.

Eknath, A.E. and Doyle, R.W., 1985b. Indirect selection for growth an life-history traits in Indian major carp aquaculture. 1. Effects of broodstock management. Aquaculture, 49: 73–84.

Falconer, D.S., 1981. Introduction to Quantitative Genetics, 2nd edition. Longman, New York, NY, 340 pp.

Hill, W.G., 1979. A note on effective population size with overlapping generations. Genetics, 92: 317–322.

Kinghorn, B.P., 1983. A review of quantitative genetics in fish breeding. Aquaculture, 31: 283–304.

Growth and Reproduction in Fish and Shellfish

BJARNE GJERDE

The Agricultural Research Council of Norway, Institute of Aquaculture Research, N-1432 Ås-NLH (Norway)

ABSTRACT

Gjerde, B., 1986. Growth and reproduction in fish and shellfish. *Aquaculture*, 57: 37–55.

This paper reviews research findings relevant to the additive genetic improvement of growth rate and reproductive traits of commercially farmed fish and shellfish. Based on estimates of heritabilities and realized responses to selection, the possibilities for improving growth rate are very promising in many species. In salmonids the potential to improve age at maturity is also very promising, while little information is available concerning this trait in other species. The advantage of applying family selection to improve age at maturity is stressed.

In Atlantic salmon and rainbow trout both the phenotypic and genetic correlations between egg volume and egg number are large and positive, and body weight has a strong positive influence on both egg traits. There seems to be no strong unfavorable relationship between growth rate and age at maturity and growth rate and egg size in these species. More research is needed, though, to better understand both the phenotypic and genetic relationships between growth, survival and fecundity traits and age at maturity in fish and shellfish.

Fecundity traits should only be considered in a breeding program if the genetic correlation between egg size or egg quality with early survival or growth rate is negative. It is, however, important to keep records of these traits to study whether or not they are changing over a period of time.

INTRODUCTION

Reliable estimates of phenotypic and genetic parameters are needed in order to predict response to selection, plan breeding programs, and estimate breeding values. Two important parameters are the values of the additive and the non-additive genetic variance in relation to the total variance of the trait. The magnitude of these two parameters largely determines which breeding method (purebreeding and/or crossbreeding) and selection method (individual and/or family selection) should be applied. Furthermore, reliable estimates of the phenotypic and genetic correlations between the traits are a prerequisite for planning breeding programs.

From a practical point of view, it should be stressed that in planning a breeding program including only one trait it is sufficient to know whether the heritability is low, medium or high, as long as the estimate is reliable. However, in order to optimize a breeding scheme including several traits, more precise estimates both of heritabilities and genetic and phenotypic correlations between the traits included in the program are required. In the following, estimates of additive genetic variation for growth and reproduction traits for commercially farmed fish and shellfish are summarized. The relationship between growth and reproduction is also reviewed.

GROWTH

Growth rate is considered a trait of great economic importance for all species used in aquaculture. Rapid growth speeds up the turnover of production, and frequently larger animals attain a higher price per unit of weight compared to smaller ones. Growth is also easy to estimate through measurement of body weight or length. In addition, although selection for rapid growth rate can have adverse effects in some types of meat production where a correlated response in mature weight causes an increase in the cost of maintaining broodstock, this factor is negligible in fish and shellfish production due to their high reproductive rate. Large (1976) has estimated the percentage of total food which is eaten by female broodstock to be 72% in sheep raised for meat production, 52% in beef cattle, 33% in pigs, and 10% in poultry. The figures for fish and shellfish probably range from less than 1% to about 5% (Kinghorn, 1983). Thus the objective of high growth rate is uncomplicated in fish.

Gunnes and Gjedrem (1978) reported significant differences in growth rate between strains of Atlantic salmon and particularly between full-sib families within a strain (Fig. 1). Significant differences in growth rate between strains were also reported by Nævdal et al. (1978). Genetic variation in growth rate was also reported between strains of rainbow trout (Gall and Gross, 1978; Klupp, 1979; Ayles and Baker, 1983; Linder et al., 1983), brook trout (Cooper, 1961),

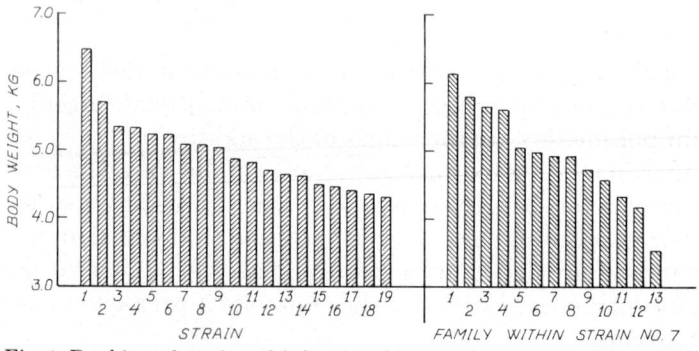

Fig. 1. Ranking of strains of Atlantic salmon and full-sib families within a strain for body weight after 2 years in cages in the sea (year-class 1973).

carp (Moav et al., 1975; Suzuki and Yamaguchi, 1980) channel catfish (Smitherman et al., 1983), oyster (Haskin, 1978, as reported by Newkirk, 1980; Newkirk, 1978), blue mussel (Newkirk et al., 1980), and freshwater prawn (Sarver et al., 1979). In founding a population for the genetic improvement of growth rate it is therefore important to test all available strains under commercial farming conditions and select the best ones for future selection programs. However, due to the very large variation between full-sib families within a strain as shown in Fig. 1, it might be equally important to select the very best families irrespective of strain. The large variation in body weight between families within a strain also clearly illustrates the importance of obtaining a broad genetic base from each strain to evaluate its potential for growth rate.

Estimates of heritabilities and phenotypic variation in quantitative traits in species of economic importance in aquaculture were reviewed by Gjedrem (1983). According to his Table II, reliable estimates of additive genetic variance for growth rate are available only for rainbow trout and Atlantic salmon and to some extent for channel catfish. In other species, such as Pacific salmon, carp, tilapia, oyster, blue mussel, prawn and lobster, some having great economic importance, most heritability estimates have been derived from full-sib analyses or from half-sib analyses with very few degrees of freedom left for sire effects. Due to common environmental, maternal and nonadditive effects, such estimates are generally biased. In addition, most estimates of heritabilities in these species are based on data recorded very early in the life cycle of the animal. Such estimates are in general not valid for their use in making selection decisions at an older age. Larger-scale experiments are therefore needed in order to evaluate the potential for selective breeding for additive gene effects in these species.

It should be stressed that estimates of genetic parameters and selection in general should be based on growth data recorded close to marketing time. In this context, marketing must be properly understood. For example, in farming of salmonids growth rate in the freshwater period is of great economic importance for a smolt producer. However, for a farmer who raises the fish from smolt until they reach maturity, the growth rate until the fish are slaughtered is of greatest importance. It is therefore important to improve the growth rate in both the freshwater and seawater periods. The genetic correlation between growth rate in these two periods is not known in salmonids. The magnitude of this correlation largely determines how much selection pressure should be put on the growth rate in these two periods of the salmonid's life. The author is not aware of any genetic correlation between the growth records in the different periods of life in other fish species, either.

In general, body weight shows a very high phenotypic variation in all species of fish and shellfish (Gjedrem, 1983), and due to the species' high fecundity, it is possible to practice a very high selection intensity. Therefore, even at low heritabilities the potential for genetic improvement of growth rate through selection for additive gene effects is promising. In Atlantic salmon and rainbow trout, the heritability for body weight of juveniles is rather low (around 0.10) while medium heritabilities (0.2–0.4) have been found for the body weight of adults. In channel catfish, medium heritabilities have been found for body weight of both young and older fish. Thus, in these species, the potential for genetic improvement of growth rate is very promising.

Predicted responses to selection usually prove to be overestimates, probably because most sources of bias cause overestimation of heritability values. Therefore realized heritabilities are arguably more relevant than information from sib analyses in the prediction of genetic improvement. However, it should be borne in mind that realized responses to selection are subject to sampling errors, especially where few animals have been used in breeding, and where only data from few generations are available.

Some estimates of realized responses to selection for growth rate found in the literature are listed in Table 1. Most estimates became available after only one to three generations of selection. The study of Donaldson and Olson (1955),

TABLE 1

Realized response to selection for growth rate in fish and shellfish; I – individual slection, F = family selection, W = within-family selection

Species/ testing environment	Body wt. of control	Selection method	No. of generation	Total gain (%)	Gain/ year (%)	Authors
Rainbow trout						
Earthen ponds	No. contr.	I	3	+	+	Lewis (1944)
Lakes and streams	No. contr.	I	23 years	+	+	Donaldson and Olson (1955)
Tanks	4.3 g	F	3	30	5	Kincaid et al. (1977)
Cages in the sea	4 kg	F+W	2	26	4.3	B. Gjerde (unpubl. data, 1985)
Atlantic salmon						
Tanks	6.5 g	F+W	1	30	7.7	Gjedrem (1979)
Cages in the sea	4.5 kg	F+W	1	14	3.6	B. Gjerde (unpubl. data, 1985)
Common carp						
Earthen ponds	~600 g	W	5	0, –		Moav and Wohlfarth (1976)
Cages in ponds	~600 g	W	3	+		Moav and Wohlfarth (1976)
Earthen ponds	~600 g	F	2	+		Moav and Wohlfarth (1976)
Channel catfish						
Tanks	179 g	F+W	1	+21, −21	+7, −7	Bondari (1983)
Cages	698 g (♀)	F+W	1	+ 8, −23	+2.7, −7.7	Bondari (1983)
	890 g (♂)	F+W	1	+29, −25	+9.7, −8.3	Bondari (1983)
Earthen ponds	413 g	I	1	+16	+5.3	Dunham and Smitherman (1983)
Oysters						
Nets	26.6 g	I	1	23	11.5	Newkirk and Haley (1982)
Nets	24 g (P)	I	2	6.8	1.7	Newkirk and Haley (1983)
Nets	22 g (C)	I	2	16.8	4.2	Newkirk and Haley (1983)
Mosquitofish						
Laboratory	242 mg (♀)	W	4	12		Busack (1983)
	84 mg (♂)	W	4	19		Busack (1983)
Laboratory	264 mg (♀)	I	5	6.8		D.E. Campton (unpublished data, 1986)

in which individual selection had been carried out over a period of 23 years, is an exception. However, in this experiment, and also in that reported by Lewis (1944), the reported selection response is confounded by changes in management and facilities as well as the fact that control populations were not maintained.

Of the 16 selection experiments for improving growth rate reported here, 15 yielded positive responses to selection. The selection experiment involving common carp was not consistent. Moav and Wohlfarth (1976) reported a lack of response to within-spawn selection for high growth rate in ponds over five generations, but they indicated a relatively strong response to selection for slow growth rate the first three generations. They suggested that selection for fast growth rate had reached a plateau. However, during three generations where replicated groups were held in cages in ponds, a significant response to selection for high growth rate was found. Family selection over two generations also showed significant increase in growth rate. Therefore, it can be concluded as reported by Kinghorn (1983) that the report of no response to selection for high growth rate in common carp is not conclusive and that there seems to be a need for an objective assessment of these results in this very important species.

The realized responses to selection for growth rate listed in Table 1 are generally very high compared to what has been reported in species of farm animals. An annual rate of improvement of the order of 1% for most traits appears relatively frequently in those species (Cunningham, 1983).

REPRODUCTION

Complete control of the reproductive cycle is a prerequisite for the development of efficient breeding and selection schemes for any species. In fact, one of the major obstacles to aquaculture development is that artificial reproduction and controlled mating has not been achieved in many species, due to the difficulty or impossibility of hatching and feeding larvae or fry. This has been the case, for instance, in the farming of eel, yellowtail and halibut.

Most studies of reproductive traits and their relation to growth rate or body weight are derived from data recorded on natural populations or from individuals reared under different environmental conditions. Thus, the genetic and environmental components of variance are often confounded, and body weight is often confounded with the age of the individual under consideration. For this reason, available literature about genetic variation in reproductive characters and their relation to growth rate in particular is very limited.

Fecundity

Fish and shellfish usually have a very high reproductive rate: salmonids have more than 1000 eggs/kg body weight; channel catfish, 5000–10 000 eggs/kg;

common carp, more than 100 000 eggs/kg; tilapia, 1000 eggs/kg; halibut, 15 000 eggs/kg; and oysters, million eggs/spawn . This makes the expense of maintaining broodstock quite low (see Growth). Therefore, it is not important to consider egg number in a selection program for these species. One exception might be a production system where the eggs are the end product, such as for caviar. However, egg size and egg quality could be important as selection criteria if they are correlated with early survival or growth rate. Moreover, traits under selection may be genetically correlated with other traits not considered. It is therefore important to keep records of traits not necessarily of economic importance today, and to study whether or not they change over a period of time.

Significant difference in egg size have been reported between strains of both Atlantic salmon (Pope et al., 1961; Aulstad and Gjedrem, 1973; Glebe et al., 1979) and rainbow trout (Gall and Gross, 1978). These differences in egg size are believed to be mainly of genetic origin. However, the Atlantic salmon data were all recorded from wild fish caught in rivers. This means that no reliable conclusion can be drawn from these data because salmon from different rivers have followed various migration routes and have fed on different types of food. However, Gjerde and Refstie (1984) reported significant differences in egg diameter between strains of Atlantic salmon reared under the same farming conditions in cages in the sea.

Heritability estimates for egg size, egg volume and egg number in rainbow trout and Atlantic salmon are given in Table 2. The results of Gall (1975) and Gall and Gross (1978) are all based on data recorded on full sibs. Therefore, these data might be biased due to maternal or nonadditive genetic effects. In addition, the results of Halseth (1984) and Haus (1984) might be biased because the data were recorded on females that had been subjected to between and within full-sib family selection for body weight. Due to these possible biases, these heritabilities must be looked upon as gross heritabilities, although they all are based on rather extensive data. The data give strong evidence of a significant additive genetic variance for egg size, egg number and egg volume. One exception is egg size in rainbow trout (Haus, 1984), where the sire components of variance were found to be not significantly different from zero.

Kanis et al. (1976) reported highly significant differences in mortality of eyed eggs and alevins between strains of Atlantic salmon. However, they found evidence of generally low heritabilities for these traits both in Atlantic salmon, sea trout and rainbow trout, based on the sire component of variance. Gjerde and Refstie (1984) reported significant differences in the average (additive) effect of strains of Atlantic salmon for survival of fry and fingerlings but not for survival of eyed eggs and alevins. Ayles (1974) reported significant additive genetic variance for survival of splake hybrid alevins, but this did not apply to survival of uneyed eggs and eyed eggs.

Gall (1974) reported a positive correlation in rainbow trout between egg size and egg quality measured as hatching percentage and also between egg size and growth up to 75 days of age. Similar results were reported by Pitman (1979). In both these experiments, confounding between the effect of age of female and egg size makes it difficult to separate the two effects. Reagan and Conley (1977), working with channel catfish, and Chevassus and Blanc (1979), working with rainbow trout, reported a positive correlation between egg size and growth up to 1 and 4 months of age, respectively. After that time the correlation was not significantly different from zero. Similarly, Springate and Bromage (1985) found that in rainbow trout the significant correlation between egg and fry size at hatching was lost 4 weeks after the time of first feeding. They also reported no significant correlation between initial egg size and survival rates to eying, hatching and swim-up and as 3-month fed fry.

Given its low additive genetic variance and low economic value, the best possibilities for improving survival lie in improving the environment in the early stages of life. However, it is a good idea to record the traits important to survival to avoid a negative correlated response when selecting for other traits. More research is needed to study the relationship between egg size or egg quality with survival and growth rate.

TABLE 2

Heritabilities with standard errors for egg size, egg volume and egg number in rainbow trout and Atlantic salmon estimated from the sire (s), dam (d) or full-sib family (f) component of variance

Species/trait	h_s^2	h_d^2 or h_f^2	Authors
Rainbow trout			
Egg size		0.20 ± 0.05	Gall (1975)[a]
Egg volume		0.20 ± 0.05	Gall (1975)
Egg number		0.19 ± 0.06	Gall (1975)
Egg size		0.32	Gall and Gross (1978)[b]
Egg volume		0.52	Gall and Gross (1978)
Egg number		0.44	Gall and Gross (1978)
Egg size	-0.02 ± 0.16	0.65 ± 0.20	Haus (1984)[c]
Egg volume	0.44 ± 0.20	0.34 ± 0.15	Haus (1984)
Egg number	0.33 ± 0.20	0.50 ± 0.17	Haus (1984)
Atlantic salmon			
Egg size	0.44 ± 0.18	0.42 ± 0.16	Halseth (1984)[d]
Egg volume	0.13 ± 0.15	0.40 ± 0.17	Halseth (1984)
Egg number	0.30 ± 0.16	0.37 ± 0.16	Halseth (1984)

[a]Data from two year-classes. Total of 148 full-sib families and 1604 observations.
[b]Data from four year-classes and three strains. Total of 140 full-sib families and 861 observations.
[c]Data from six year-classes. Total of 51 sires, 124 dams and 717 observations.
[d]Data from six year-classes and 24 strains. Total of 91 sires, 218 dams and 854 observations.

Age at sexual maturity

When salmonids become sexually mature, their growth rate decreases, flesh quality deteriorates (Aksnes et al., 1986) and mortality generally increases. The fish must be marketed before external sex characters have developed. The size of fish at slaughtering is partly dependent on their age at maturation. Thus, when producing a large fish, age at maturation becomes an important economic trait.

The potential for genetic improvement of age at maturity in salmonids has been well documented. Ricker (1972) reviewed the literature and concluded that genetic factors play a major part in the determination of age at maturation of adult Pacific salmon. In chinook salmon, Ricker (1980b) estimated the heritability to be 0.30 based on data from L.R. Donaldson and associates. Grilse parents of Atlantic salmon were found to produce higher proportions of grilse than two-sea-winter parents (Elson, 1973; Piggins, 1974; Ritter and Newbould, 1977; Gjerde, 1984). Ellis and Noble (1961) reported similar results in chinook salmon.

Significant responses to individual selection for age at maturity have been reported in chinook salmon (Donaldson and Menasveta, 1961), in rainbow trout (Donaldson and Olson, 1955) and in Atlantic salmon (Gjerde, 1984). Lewis (1944) developed a stock of rainbow trout that spawned in October rather than during the normal spring spawning season, and Kato (1978) succeeded in getting one stock to spawn twice a year. He also reported significant response for early spawning within the same season. D.E. Campton (unpublished data, 1986) reported considerable response to selection for late maturity in the mosquitofish.

Significant differences in age at maturity have been found between strains of Atlantic salmon (Gunnes, 1978; Nævdal et al., 1978; Gjerde and Refstie, 1984). Nævdal (1983) reported significant differences in age at maturity between full-sib groups of Atlantic salmon and rainbow trout. Significant differences in the percentage of immature fish have been found between sire groups of Atlantic salmon and rainbow trout (Figs. 2 and 3). Significant differences in frequency of precocious males were reported between strains of Atlantic salmon by Schiefer (1971), Nævdal et al. (1975), and Saunders and Sreedharan (1978). Thorpe et al. (1983) and Gjerde (1984) found that precocious male parents produced higher proportions of precocious male offspring than did two- and three-sea-winter parents. Glebe et al. (1980) and Gjerde (1984) found strong evidence that age at maturation of parr is independently inherited from age at maturation in the sea.

In species with very short generation intervals, age at maturity can be measured in days as a continuous trait. Thus complete records of age at maturity can be obtained. This was done in the study on mosquitofish by Busack and

Gall (1983), who estimated the heritability for age at maturity as 0.41, based on the full-sib component of variance.

In species with a generation interval where age at maturity has to be measured in years, complete records are difficult to obtain. Therefore, age at maturity most frequently has to be treated as an all-or-none trait. This causes difficulties in the statistical analysis of such a trait, because of the interdependence of the variance and the frequency.

Heritability estimates for age at maturity for different year-classes of Atlantic salmon and rainbow trout are given in Table 3. On the average continuous scale heritability for early maturity is 0.15 and 0.05, and 0.37 and 0.20 for late maturity for Atlantic salmon and rainbow trout, respectively. It can be seen that the heritability on an all-or-none (h_p^2) and the heritability on a continuous scale (h_x^2) are quite different, particularly when the percentage of immature fish is close to 0 or 100%. As shown by Dempster and Lerner (1950), the non-

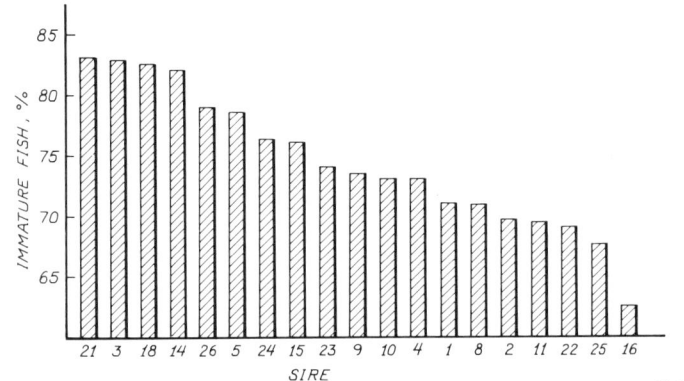

Fig. 2. Ranking of sires of Atlantic salmon for percent immature fish after 1 year in cages in the sea (year-class 1982, farm no. 02).

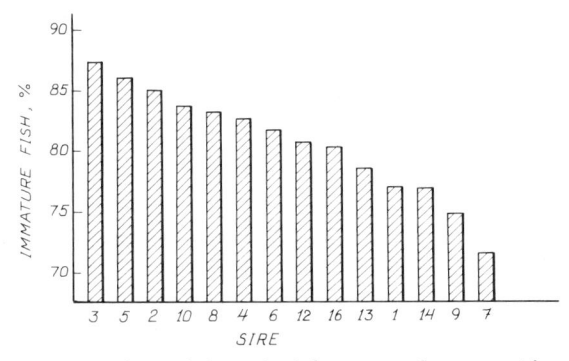

Fig. 3. Ranking of sires of rainbow trout for percent immature fish during the first year in cages in the sea (year-class 1981, farm no. 02).

additive genetic portion may constitute most of the total genetic variance where the heritability on the continuous scale is very high or the incidence is close to 0 or 1. This clearly illustrates the importance of taking the incidence level into consideration when comparing heritabilities for all-or-none traits. It can be concluded that there is considerable additive genetic variation in this trait in salmonids. Thus, the potential for improving age at maturity in salmonids through selection is very promising.

Large variation in percentage of immature Atlantic salmon and rainbow trout

TABLE 3

Estimates of heritability from the sire component of variance for age at sexual maturity in Atlantic salmon and rainbow trout
h_p^2 = heritability on the all-or-none scale; h_x^2 = heritability on the underlying continuous scale when transformed according to Robertson and Lerner (1949)

Species	Year-class	No. of sires	Immature (%)	h_p^2	h_x^2	Authors
Early maturity[a]						
Atlantic salmon	1980	17	72.1	0.14	0.25	
	1981	14	89.7	0.04	0.12	
	1982	19	74.0	0.04	0.07	B. Gjerde (unpublished data, 1985)
Rainbow trout	1981	15	81.5	0.03	0.06	
	1982	34	94.1	0.01	0.04	
Late maturity[b]						
Atlantic salmon	1973	18	89.4	0.14	0.40	
	1974	30	99.0	0.06	0.84	Gjerde and Gjedrem (1984)
	1975	35	47.1	0.39	0.61	
	1980	17	27.7	0.11	0.20	
	1981	14	23.4	0.02	0.04	B. Gjerde (unpublished data, 1985)
	1982	19	38.7	0.09	0.15	
Rainbow trout	1973	20	2.9	−0.05	—	
	1974	18	22.9	0.16	0.31	Gjerde and Gjedrem (1984)
	1975	18	14.1	0.28	0.68	
	1978	17	81.4	0.07	0.15	
	1979	14	62.7	0.14	0.23	
	1980	19	58.9	0.03	0.05	B. Gjerde (unpublished data, 1985)
	1981	14	14.2	0.03	0.07	
	1982	34	39.8	0.08	0.13	

[a]Immature fish after 1 year (Atlantic salmon) and during the first year (rainbow trout) in cages in the sea.
[b]Immature fish after 2 years (Atlantic salmon) and during the second year (rainbow trout) in cages in the sea. Fish which matured in the previous year are not included.

has been found in a comparison between fish farms rearing fish from the same full-sib families and between year-classes of fish reared on the same farm (Table 4). Extensive variation was found also between fish reared in different cages on the same farm. It is reasonable to conclude that a series of environmental sources of variations (water temperature, salinity, latitude, amount of food and type of food given, densities, etc.) must be responsible for this large variation between farms within the same year-class of fish, while differences between year-classes within a farm might also reflect a genetic component. R. Alderson (unpublished data, 1986) also reported considerable variation in the percent mature rainbow trout between farms rearing fish from the same families, and Saunders et al. (1983) reported a higher grilse percentage when sea-ranching a certain stock of Atlantic salmon than when cage-rearing it. It must be concluded that environmental factors also have a great influence on age at sexual maturity in salmonid species.

This large variation in percentage of immature fish between farms is of great importance in a breeding program where age at maturity is an important trait. This is illustrated in Fig. 4. The combination of incidence level with the proportion of selected individuals has a pronounced effect on the response when applying individual selection for an all-or-none trait. In particular, selection for a frequent character (or against one of low incidence) will be impaired by low effective heritability coupled with low selection intensity. Only when the proportion selected is identical with the incidence, will the predicted response be the same as when the selection was based on a continuous visible scale. Therefore, family selection for age at maturity in salmonids has several advan-

TABLE 4

Percent immature[a] fish in different year-classes of Atlantic salmon and rainbow trout and at different fish farms located along the Norwegian coast

Species/fish farm	Year-class			
	1979	1980	1981	1982
Atlantic salmon				
02	—	72.1	89.7	74.0
05	—	—	—	82.5
06	—	—	—	97.5
15	88.9	91.4	88.5	96.6
19	—	—	82.9	74.2
Rainbow trout				
02	96.9	96.0	81.5	94.1
15	87.2	—	85.5	87.8
17	95.0	—	—	—
18	90.5	85.8	83.7	86.0

[a]After 1 year (Atlantic salmon) and during the first year (rainbow trout) in cages in the sea.

tages as compared to individual selection: (1) selection on the underlying scale can be practiced and a high selection intensity independent of frequency can be obtained; (2) direct selection can be practiced as opposed to indirect selection; (3) the generation interval need not be prolonged as it would be when individual selection for late maturity is practiced.

GROWTH AND FECUNDITY

Estimates of phenotypic and genetic correlation between egg size, egg volume, egg number and female body weight in Atlantic salmon and rainbow trout are given in Table 5. In the four experiments referred to, egg number was calculated as the product of egg volume and egg size. Haus (1984) found that the sire component of variance for egg size was not significantly different from zero. Therefore, no genetic correlations between egg size and the other traits are listed for this experiment. The estimates made by Gall (1975) and Gall and Gross (1978) were all derived from full-sib components of variance and covariance. These include additive as well as nonadditive genetic and common environmental effects. The estimates made by Haus (1984) and Halseth (1984) may be biased because the data were from females that had been subjected to between and within full-sib family selection for body weight. In addition, the genetic correlation estimates all have very large standard errors. Therefore, the genetic correlations among these traits must be looked upon as gross correlations.

The observed phenotypic and genetic correlations between egg volume and egg number are large and positive. In contrast, the correlations between egg size and egg number are small (except the genetic correlation in Atlantic salmon). As stated by Gall and Gross (1978), it can be concluded that egg

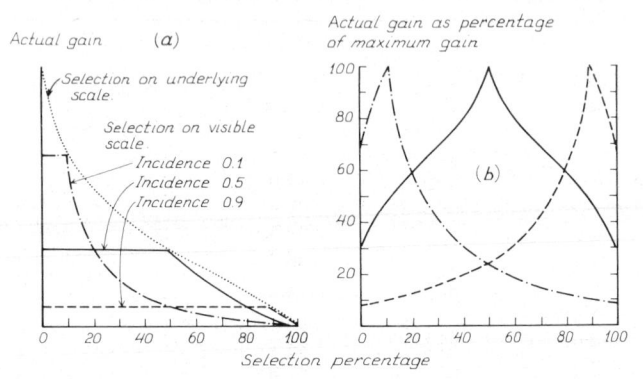

Fig. 4. Effect on the underlying scale achieved by individual selection for a single all-or-none trait, based on the procedure of Dempster and Lerner (1950): (a) actual gains, (b) gains expressed as a percentage of what could be achieved from direct selection on the underlying scale. For further explanation, see (a) (Danell and Rønningen, 1981).

volume is the principal determinant of egg number, and that body weight has a strong positive influence on egg volume. In rainbow trout, however, Haus (1984) found a small negative genetic correlation between egg volume and body weight.

The phenotypic correlations between egg size and body weight were all found to be small but positive. The genetic correlations in rainbow trout were also found to be positive or zero, while in Atlantic salmon a small negative genetic correlation was found. However, this negative correlation between egg size and body weight was not significantly different from zero.

In general it can be concluded that a favorable or at least no strong unfavorable correlation does exist between egg size, egg number, egg volume and female body weight in rainbow trout and Atlantic salmon. In a selection progam where selection for increased growth rate is practiced, there seems to be no need to select for increased egg volume and egg number. However, the genetic correlation between egg size or quality and growth rate should be more thoroughly investigated. Even a small negative genetic correlation between egg size/qual-

TABLE 5

Genetic (above diagonal) and phenotypic (below diagonal) corrlations between egg size, egg volume, egg number and female body weight in rainbow trout and Atlantic salmon

The first, second and third line for each trait combination refers to estimates in rainbow trout reported by Gall (1975), Gall and Gross (1978) (simple averages for three strains) and Haus (1984), respectively, while the fourth line refers to estimates reported by Halseth (1984) for Atlantic salmon

	Egg size	Egg volume	Egg number	Body wt.
Egg size		0.55	−0.09	0.46
		0.43	0.18	0.40
		—	—	—
		−0.56	−0.71	−0.27
Egg volume	0.45		0.77	0.56
	0.37		0.93	0.80
	0.16		1.00	−0.19
	0.25		0.94	0.98
Egg number	−0.15	0.78		0.39
	−0.10	0.86		0.73
	−0.27	0.89		0.00
	−0.09	0.90		0.70
Body weight	0.19	0.43	0.35	
	0.24	0.63	0.53	
	0.18	0.36	0.27	
	0.24	0.66	0.56	

ity and growth rate might cause a negative correlated response in egg size /quality and thereby affect early survival and growth rate. It might therefore be necessary to take egg size and egg quality into account in a selection scheme for these two species.

GROWTH AND AGE AT MATURATION

Phenotypic correlation

In Table 6, the average body weight for maturing males and females and immature fish are given for two year-classes of Atlantic salmon and rainbow trout. For both species, body weight was recorded about 4 months before the fish were expected to spawn. These results agree with several studies of salmonids (Kato, 1975, 1978; Nævdal et al., 1979; Thorpe et al., 1983; Gjerde, 1984; Gjerde and Refstie, 1984) which have shown that maturing fish are heavier than immature fish. Gjerde and Gjedrem (1984) studied the frequency of mature Atlantic salmon and rainbow trout after 2 and 1.5 years in cages in the sea, respectively. They reported a significant negative phenotypic correlation between body weight and age at maturity in Atlantic salmon and a low negative correlation in rainbow trout. According to these results, there seems to be a negative phenotypic correlation between growth rate and age at maturation in salmonids.

Genetic correlation

Ricker (1980a, b) reported a correlated response in age at maturation in chum and chinook salmon resulting from selective fishing on size. The results indicated a positive genetic correlation between age at maturity and growth

TABLE 6

Mean body weight (kg) of maturing males and females and immature fish for three year-classes of Atlantic salmon and two year-classes of rainbow trout reared for about 2 and 1.5 years in cages in the sea, respectively, at fish farm no. 02

Species/ year class	Maturing males	Maturing females	Immatures
Atlantic salmon			
1980	5.83	4.89	3.14
1981	6.21	5.10	3.94
1982	5.83	4.76	3.04
Rainbow trout			
1981	4.74	4.31	3.71
1982	4.10	3.77	3.18

rate in chum salmon and a negative correlation in chinook salmon. Gjerde and Gjedrem (1984) reported a highly negative genetic correlation (-0.52) between body weight and age at maturation in Atlantic salmon and a lower correlation (-0.11) in rainbow trout. It should, however, be stressed that the body weight of the fish was taken 4–5 months prior to spawning, which means that the development of gonads has already started. If sex hormones stimulate growth in maturing fish, this will impose an automatic correlation between these two traits which will tend to give negative phenotypic as well as genetic correlations.

The correlation coefficients between sire solutions for age at maturation and body weight at slaughter for two year-classes of Atlantic salmon and rainbow trout are given in Table 7. The body weights of maturing males and females were multiplicatively adjusted for the effect of sex. These estimates should be close to the true sample genetic correlations where the number of offspring per family is large and where the nonadditive genetic, maternal and environmental effects common to full sibs are small.

Of the total of 20 correlations between age at sexual maturity and body weight at slaughter given in Table 7, 15 positive estimates were found; 10 out of 12 in Atlantic salmon and five out of eight in rainbow trout. These results indicate that the genetic correlation between age at maturation and growth rate is low in Atlantic salmon and rainbow trout. The large difference in body weight observed between maturing and immature fish might be largely caused by sex hormones accelerating the growth rate. Evidence of a low genetic correlation between growth rate and age at sexual maturity has been found also in the mosquitofish (D.E. Campton, unpublished data, 1986). More research is needed to better understand both the phenotypic and genetic relationship between

TABLE 7

Correlation coefficients between sire solutions for the percentage of immature fish and body weight at slaughter for three year-classes of Atlantic salmon and two year-classes of rainbow trout reared for about 2 and 1.5 years in cages in the sea, respectively, at fish farm no. 02

Traits	Atlantic salmon			Rainbow trout	
	1980	1981	1982	1981	1982
% Imm. (1) and % imm. (2)	0.38	0.77**	0.14	0.15	0.60***
% Imm. (1) and weight of imm.	0.10	0.31	0.57*	0.22	−0.31
% Imm. (1) and weight of mature	0.27	0.37	0.39	0.27	−0.31
% Imm. (2) and weight of imm.	−0.54*	0.22	−0.06	0.05	0.04
% Imm. (2) and weight of mature	0.04	0.05	0.14	0.04	−0.32

*$=P<0.05$, **$=P<0.01$, ***$=P<0.001$.
%Imm. (1) = % Immature fish after 1 year (Atlantic salmon) and during the first year (rainbow trout) in cages in the sea.
% Imm. (2) = % Immature fish after 2 years (Atlantic salmon) and during the second year (rainbow trout) in cages in the sea. Fish which matured in the previous year are not included.

growth rate and age at maturation in fish and shellfish. In such experiments the data have to be based on repeated body weight records (growth curves) of individual tagged fish from many families.

CONCLUDING REMARKS

In general, investment in selection programs gives very high returns, expressed in terms of interest on invested capital. Full investment appraisals have recently been determined by Barlow (1982) and Mitchell et al. (1982). They covered sheep, pigs and beef cattle production and showed net benefit /cost ratios ranging from 5/1 to 50/1.

According to Gjedrem (1983), selection programs dealing with fish and shellfish should be based on a combination of individual and family merit. Individual selection alone is only of value when growth rate is the only trait of economic importance and is highly heritable. If nonadditive genetic variance is of importance for some of the traits included in the breeding program, selection for additive gene effects should be combined with one or another system of crossbreeding to exploit this component of genetic variance. And, as pointed out by Chevassus (1983), the different techniques for production of monosex or sterile populations should be integrated into more general genetic improvement programs.

REFERENCES

Aksnes, A., Gjerde, B. and Roald, S.O., 1986. Biological, chemical and organoleptic changes during maturation of farmed Atlantic salmon, *Salmo salar*. Aquaculture, 53: 7-20.

Aulstad, D. and Gjedrem, T., 1973. The egg size of salmon (*Salmo salar*) in Norwegian rivers. Aquaculture, 2: 337-341.

Ayles, G.B., 1974. Relative importance of additive genetic and maternal sources of variation in early survival of young splake hybrids (*Salvelinus fontinalis* × *S. namaycush*). J. Fish. Res. Board Can., 31: 1499-1502.

Ayles, G.B. and Baker, R.F., 1983. Genetic differences in growth and survival between strains and hybrids of rainbow trout (*Salmo gairdneri*) stocked in aquaculture lakes in the Canadian prairies. Aquaculture, 33: 269-280.

Barlow, R., 1982. Benefit-cost analyses of genetic improvement programs for sheep, beef cattle and pigs in Ireland. Ph.D. Thesis, University of Dublin. Ref. by E.P. Cunningham, 1983.

Bondari, K., 1983. Response to bidirectional selection for body weight in channel catfish. Aquaculture, 33: 73-81.

Busack, C.A., 1983. Four generations of selection for high 56-day weight in the mosquitofish (*Gambusia affinis*). Aquaculture, 33: 83-87.

Busack, C.A. and Gall, G.A.E., 1983. An initial description of the quantitative genetics of growth and reproduction in the mosquitofish, *Gambusia affinis*. Aquaculture, 32: 123-140.

Chevassus, B., 1983. Hybridization in fish. Aquaculture, 33: 245-262.

Chevassus, B. and Blanc, J.M., 1979. Genetic analysis of growth performance in salmonids. Communication to the Third European Ichthyological Congress, Warszawa, 18-25 September 1979, 4 pp.

Cooper, E.L., 1961. Growth of wild and hatchery strains of brook trout. Trans. Am. Fish. Soc., 90: 424-438.

Cunningham, E.P., 1983. Present and future perspectives in animal breeding research. In: Proceedings of the XV International Congress of Genetics, New Delhi, India, 12-21 December 1983. Vol. IV, Applied Genetics, pp. 169-180.

Danell, Ö. and Rønningen, K., 1981. All-or-none traits in index selection. Z. Tierz. Züchtungsbiol., 98: 265-284.

Dempster, E.R. and Lerner, I.M., 1950. Heritability of threshold characters (with appendix by A. Robertson). Genetics, 35: 212-236.

Donaldson, L.R. and Menasveta, D., 1961. Selective breeding of chinook salmon. Trans. Am. Fish. Soc., 90: 160-164.

Donaldson, L.R. and Olson, P.R., 1955. Development of rainbow trout broodstock by selective breeding. Trans. Am. Fish. Soc., 85: 93-101.

Dunham, R.A. and Smitherman, R.O., 1983. Crossbreeding channel catfish for improvement of body weight in earthen ponds. Growth, 47: 97-103.

Ellis, C.H. and Noble, R.E., 1961. Return of marked fall chinook salmon to the Deschutes River, Deschutes River genetic experiment. Wash. State Fish. Dept. Annu. Rept. for 1960, 70: 72-75.

Elson, P.F., 1973. Genetic polymorphism in Northwest Miramichi salmon, in relation to season of river ascent and age at maturation and its implications for management of the stocks. ICNAF Res. Doc. 73/76, 6 pp.

Gall, G.A.E., 1974. Influence of size of eggs and age of female on hatchability and growth in rainbow trout. Calif. Fish Game, 60(1): 26-36.

Gall, G.A.E., 1975. Genetics of reproduction in domesticated rainbow trout. J. Anim. Sci., 40: 19-28.

Gall, G.A.E. and Gross, S.J., 1978. Genetic studies of growth in domesticated rainbow trout. Aquaculture, 13: 225-234.

Gjedrem, T., 1979. Selection for growth rate and domestication in Atlantic salmon. Z. Tierz. Züchtungsbiol., 96: 56-59.

Gjedrem, T., 1983. Genetic variation in quantitative traits and selective breeding in fish and shellfish. Aquaculture, 33: 51-72.

Gjerde, B., 1984. Response to individual selection for age at sexual maturity in Atlantic salmon. Aquaculture, 38: 229-240.

Gjerde, B. and Gjedrem, T., 1984. Estimates of phenotypic and genetic parameters for carcass traits in Atlantic salmon and rainbow trout. Aquaculture, 36: 97-110.

Gjerde, B. and Refstie, T., 1984. Complete diallel cross between five strains of Atlantic salmon. Livest. Prod. Sci., 11: 207-226.

Glebe, B.D., Appy, T.D. and Saunders, R.L., 1979. Variation in Atlantic salmon (*Salmo salar*) reproductive traits and their implications in breeding programs. ICES, C.M., 1979/M23, 11 pp.

Glebe, B.D., Eddy, W. and Saunders, R.L., 1980. The influence of parental age at maturity and rearing practice on precocious maturation of hatchery-reared Atlantic salmon parr. ICES, C.M., 1980/F8, 8 pp.

Gunnes, K., 1978. Genetic variation in production traits of Atlantic salmon. Eur. Assoc. Anim. Prod., Commission of Animal Genetics, Stockholm Meeting, 7 pp.

Gunnes, K. and Gjedrem, T., 1978. Selection experiments with salmon. IV. Growth of Atlantic salmon during two years in the sea. Aquaculture, 15: 19-33.

Halseth, V., 1984. Estimates of phenotypic and genetic parameters for egg size, egg volume and egg number in Atlantic salmon. Dissertation at Department of Animal Genetics and Breeding. Agricultural University of Norway, May 1984, 50 pp. (in Norwegian).

Haus, E., 1984. Estimates of phenotypic and genetic parameters for egg size, egg volume and egg number in rainbow trout. Dissertation at Department of Animal Genetics and Breeding, Agricultural University of Norway, May 1984, 49 pp. (in Norwegian).

Kanis, R., Refstie, T. and Gjedrem, T., 1976. A genetic analysis of egg, alevins and fry mortality in salmon (*Salmo salar*), sea trout (*Salmo trutta*) and rainbow trout (*Salmo gairdneri*). Aquaculture, 8: 259-268.

Kato, T., 1975. The relation between the growth and reproductive characters of rainbow trout, *Salmo gairdneri*. Bull. Fresh. Fish. Res. Lab., 25(2): 83-115.

Kato, T., 1978. Relation of growth to maturity of age and egg characteristics in kokanee salmon (*Oncorhynchus nerka*). Bull. Freshwater Fish. Res. Lab., 28(1): 61-75.

Kincaid, A.L., Bridges, W.R. and Von Limback, B., 1977. Three generations of selection for growth rate in fall-spawning rainbow trout. Trans. Am. Fish. Soc., 106(6): 621-628.

Kinghorn, B.P., 1983. A review of quantitative genetics in fish breeding. Aquaculture, 31: 283-304.

Klupp, R., 1979. Genetic variance for growth in rainbow trout (*Salmo gairdneri*). Aquaculture, 18: 123-134.

Large, R.V., 1976. The influence of reproduction rate on the efficiency of meat production in animal populations. In: D. Lister, D.N. Rhodes, V.R. Fowler and M.F. Fuller (Editors), Meat Animals — Growth and Productivity. Plenum Press, New York, NY and London, pp. 43-55.

Lewis, R.C., 1944. Selective breeding of rainbow trout at Hot Creek Hatchery. Calif. Fish Game, 30: 95-97.

Linder, D., Sumari, D., Nyholm, K. and Sirkkomaa, S., 1983. Genetic and phenotypic variation in production traits in rainbow trout strains and strain crosses in Finland. Aquaculture, 33: 129-134.

Mitchell, G., Smith, C., Makower, M. and Bird, P.J.W.N., 1982. An economic appraisal of pig improvement in Great Britain. Genetic and production aspects. Anim. Prod., 35: 215-224.

Moav, R. and Wohlfarth, G., 1976. Two-way selection for growth rate in the common carp (*Cyprinus carpio* L.). Genetics, 82: 83-101.

Moav, R., Hulata, G. and Wohlfarth, G., 1975. Genetic differences between the Chinese and European races of the common carp. I. Analysis of genotype-environment interactions for growth rate. Heredity, 34(3): 323-340.

Nævdal, G., 1983. Genetic factors in connection with age at maturation. Aquaculture, 33: 97-106.

Nævdal, G., Holm, M., Møller, D. and Østhus, O.D., 1975. Experiments with selective breeding of Atlantic salmon. ICES, C.M., 1975/M22, 10 pp.

Nævdal, G., Holm, M., Lerøy, R. and Møller, D., 1978. Individual growth rate and age at first maturity in Atlantic salmon. Fiskeridir. Skr., Ser. Havunders., 16: 519-529.

Nævdal, G., Holm, M., Lerøy, R. and Møller, D., 1979. Individual growth rate and age at sexual maturity in rainbow trout. Fiskeridir. Skr., Ser. Havunders., 17: 1-10.

Newkirk, G.F., 1978. Interaction of genotype and salinity in larvae of the oyster, *Crassostrea virginica*. Mar. Biol., 48: 227-234.

Newkirk, G.F., 1980. Review of the genetics and the potential for selective breeding of commercially important bivalves. Aquaculture, 19: 209-228.

Newkirk, G.F. and Haley, L.E., 1982. Progress in selection for growth rate in the European oyster, *Ostrea edulis*. Mar. Ecol. Prog. Ser., 10: 77-79.

Newkirk, G.F. and Haley, L.E., 1983. Selection for growth rate in the European oyster, *Ostrea edulis*: response of second generation groups. Aquaculture, 33: 149-155.

Newkirk, G.F., Freeman, K.R. and Dickie, L.M., 1980. Genetic studies of the blue mussel, *Mytilus edulis*, and their implications for commercial culture. Proc. World Maricult. Soc., 11: 596-604.

Piggins, D.J., 1974. The results of selective breeding from known grilse and salmon parents. Annu. Rep. Salmon Res. Trust Ireland, XVII: 35-39.

Pitman, R.W., 1979. Effects of female age and egg size on growth and mortality in rainbow trout. Prog. Fish Cult., 41(4): 202-204.

Pope, J.A., Mills, D.H. and Shearer, W.M., 1961. The fecundity of Atlantic salmon (*Salmo salar* Linn.). Dept. Agric. Fish. Scot. Freshwater Salmon Fish. Res., 26, 12 pp.

Reagan, R.E., Jr. and Conley, C.M., 1977. Effect of egg diameter on growth of channel catfish. Prog. Fish Cult., 39(3): 133–134.

Ricker, W.E., 1972. Hereditary and environmental factors affecting certain salmonid populations. In: The Stock Concept in Pacific Salmon. H.R. MacMillan Lectures in Fisheries, University of British Columbia, Vancouver 8, B.C., pp.27–160.

Ricker, W.E., 1980a. Changes in the age and size of chum salmon (*Oncorhynchus keta*). Can. Tech. Rep. Fish. Aquat. Sci. No. 930, 99 pp.

Ricker, W.E., 1980b. Causes of the decrease in age and size of chinook salmon (*Oncorhynchus tshawytscha*). Can. Tech. Rep. Fish. Aquat. Sci. No. 944, 25 pp.

Ritter, J.A. and Newbould, K., 1977. Relationships of percentage and smolt age to age at first maturity of Atlantic Salmon (*Salmo salar*). ICES, C.M., 1977/M32, 5 pp.

Robertson, A. and Lerner, I.M., 1949. The heritability of all-or-none traits. Viability of poultry. Genetics, 34: 395–411.

Sarver, D., Malecha, S. and Onizuka, D., 1979. Development and characterization of genetic stocks and their hybrids in *Macrobrachium rosenbergii*: physiological responses and larval development rates. Proc. World Maricult. Soc., 10: 880–892.

Saunders, R.L. and Sreedharan, A., 1978. The incidence and genetic implications of sexual maturity in male Atlantic salmon parr. NASRC Rep. No. 2, 1978, 8 pp.

Saunders, R.L., Henderson, E.B., Glebe, B.D. and Loudenslager, E.J., 1983. Evidence of a major environmental component in determination of the grilse: large salmon ratio in Atlantic salmon (*Salmo salar*). Aquaculture, 33: 107–118.

Schiefer, K., 1971. Ecology of Atlantic Salmon With Special Reference to Occurrence and Abundance of Grilse in North Shore Gulf of St. Lawrence Rivers. Ph.D. Thesis, Univ. of Waterloo, Ontario, 129 pp.

Smitherman, R.D., Dunham, R.A. and Tave, D., 1983. Review of catfish breeding research 1969–1981 at Auburn University. Aquaculture, 33: 197–205.

Springate, J.R.C. and Bromage, N.R., 1985. Effects of egg size on early growth and survival in rainbow trout (*Salmo gairdneri* Richardson). Aquaculture, 47: 163–172.

Suzuki, R. and Yamaguchi, M., 1980. Improvement of quality in the common carp by crossbreeding. Bull. Jpn. Soc. Sci. Fish., 46: 1427–1434.

Thorpe, J.E., Morgan, R.I.G., Talbot, C. and Miles, M.S., 1983. Inheritance of developmental rates in Atlantic salmon, *Salmo salar* L. Aquaculture, 33: 119–128.

Ploidy Manipulation and Performance

GARY H. THORGAARD

Department of Zoology and Program in Genetics and Cell Biology, Washington State University, Pullman, WA (U.S.A.)

ABSTRACT

Thorgaard, G.H., 1986. Ploidy manipulation and performance. *Aquaculture*, 57: 57–64.

The chromosome set manipulation techniques of induced polyploidy, gynogenesis and androgenesis are likely to have significant applications in aquaculture. Induced triploidy is valuable for production of sterile individuals and increasing hybrid survival, and may be useful for producing individuals with increased heterozygosity. Gynogenesis (all-maternal inheritance) has considerable potential for rapid generation of inbred lines and can be used to generate all-female populations and in gene-transfer studies. Androgenesis (all-paternal inheritance) might be used for rapid generation of inbred lines, in the recovery of genotypes from cryopreserved sperm and in studying the effects of mitochondrial genotype on performance.

INTRODUCTION

The chromosome set manipulation techniques of induced polyploidy, gynogenesis and androgenesis have been extensively investigated in fish in recent years. Much of the work has been similar to studies done earlier with amphibians (e.g., Fankhauser, 1945). The major goal of these studies has been to develop methods for the control of sex and for rapid inbreeding in fish but, as will be discussed, other applications of these techniques are becoming apparent.

A number of recent articles have reviewed chromosome set manipulation studies in fish (Cherfas, 1981; Chourrout, 1982a; Donaldson and Hunter, 1982; Purdom, 1983; Thorgaard, 1983; Yamazaki, 1983; Thorgaard and Allen, 1986). In this article, the preferred methods that have developed for inducing polyploidy, gynogenesis and androgenesis in fish will be discussed. Probable and possible aquacultural applications of these manipulation techniques will be outlined, and critical problems for future investigation proposed.

INDUCED POLYPLOIDY

Induced polyploidy refers to the production of individuals with extra sets of chromosomes. This can be done by treating fertilized eggs with either temper-

TABLE 1

Large-scale production of triploid rainbow trout by heat shock
A 29°C heat shock lasting 10 min was initiated at 10 min post-fertilization (Chourrout, 1980). Proportion of triploid embryos was analyzed using chromosome preparations as previously described (Thorgaard et al., 1981)

Item	Treatment group (date)	
	Spokane, WA (7 Dec. 1982)	Arlee, MT (8 Dec. 1982)
Number of eggs treated	81 900	134 242
Proportion triploid	33/35	24/25
Survival to initiation of feeding (control survival)	0.79	0.48 (0.84)
Survival to 84 days post-fertilization (control survival	—	0.42 (0.82)
Survival to 150 days post-fertilization	0.62	—

ature shock, hydrostatic pressure or chemical treatments. If the treatments are applied shortly after fertilization, triploids can be produced due to retention of the second polar body of the egg. It the treatments are applied shortly before the first cleavage division, tetraploids can be produced.

Temperature-shock treatments are particularly advantageous for situations in which large volumes of eggs need to be treated. High proportions of triploid rainbow trout have been produced with high survival after heat-shock treatments (Chourrout and Quillet, 1982; also see Table 1). Similar results can be obtained with other species if appropriate treatment regimes are identified.

Hydrostatic pressure has been used for induction of both triploidy and tetraploidy. It has been particularly useful for blocking first cleavage (e.g., Chourrout, 1984; Parsons and Thorgaard, 1985). A limitation of using hydrostatic pressure is that appropriate pressure vessels may not accommodate large volumes of eggs.

Production of triploids by crossing tetraploids with diploids is a new development that holds promise as a simple, economical method (D. Chourrout, personal communication, 1985). It requires that viable, fertile tetraploids can be produced for the species of interest.

The primary interest in induced triploid fish lies in their sterility and in the possibility that this may lead to extended growth and/or survival in mature fish. Data on the performance of sterile triploids are still accumulating; our results with rainbow trout suggest that mature triploids may maintain their growth much better than diploids (Table 2). Sterility is also advantageous in situations where the control of reproduction is desirable. Triploid grass carp are being adopted in aquatic weed control programs (Wattendorf and Anderson, 1986) and triploids may be desirable for species where overpopulation and

TABLE 2

Superior growth in triploid rainbow trout after the age of sexual maturity
Individual fish being reared at the Spokane Trout Hatchery (Washington Department of Game) were identified with jaw tags, classed as triploid or diploid using flow cytometry (Thorgaard et al., 1982) and weighed at ages 2 and 3.5 years. Diploid fish in this group matured at 2 years. Five of six fish in the triploid group were females

Ploidy	Mean weight ± standard error (g)		
	2 years	3.5 years	Growth
Triploid ($n=6$)	638 ± 29	945 ± 94	307 ± 82
Diploid ($n=17$)	640 ± 27	700 ± 29	60 ± 31

associated stunting occur (Thorgaard and Allen, 1986). Another application of induced triploidy lies in the fact that triploid hybrids are typically much more viable than diploid hybrids (Allen and Stanley, 1981; Chevassus et al., 1983; Scheerer and Thorgaard, 1983). This may make it possible to combine desirable characters from two species in a sterile hybrid (e.g., Parsons et al., 1986).

A possible application of induced triploidy may lie in the fact that triploids have higher heterozygosity than diploids (Allendorf and Leary, 1984). This has been shown to be associated with higher developmental stability as measured by fluctuating asymmetry (Leary et al., 1985). Bingham (1980) proposed that increased heterozygosity was a primary advantage of using polyploids in plant breeding programs. Heterozygosity might be maximized in induced triploids by using hybrids between two strains as the female parent and crossing to males of a third strain.

As triploids become more intensively investigated for use in aquacultural programs, several critical questions will need to be answered: (1) How important are treatment effects as compared to triploidy per se in the early survival and growth of induced triploids? Some studies have found inferior early growth in triploid salmonids (Solar et al., 1984) while others have found no difference (Benfey and Sutterlin, 1984). Treatment effects may account for some of the differences between studies. (2) Is the appealing approach of producing triploids by using tetraploid broodstock applicable to many species? (3) How do sterile triploids perform relative to fish sterilized by hormone treatments (Donaldson and Hunter, 1982; Hunter and Donaldson, 1983; Yamazaki, 1983)?

GYNOGENESIS

Gynogenesis is a form of all-maternal inheritance in which eggs are activated to develop using sperm which does not genetically contribute to the embryo. Gynogenetic diploids can then be produced using treatments similar to those in induced polyploidy to induce retention of the second polar body or block first cleavage.

Radiation can be used to inactivate sperm genetically. Because of its convenience and the fact that it does not result in residual paternal inheritance when used properly, ultraviolet light is a preferred method of sperm inactivation (Chourrout, 1982b). Gamma radiation has the advantage of penetrating opaque sperm solutions, thus allowing treatment of large volumes of sperm. At high doses, it may result in little or no residual paternal inheritance (Onozato, 1982a; Thorgaard et al., 1985).

The original application of induced gynogenesis in fish was as a method of sex control; all-female populations are expected to result after gynogenesis in female homogametic species (Stanley, 1976). A disadvantage of using gynogenesis for sex control, however, is that the gynogenetic offspring are inbred. This is not a disadvantage if they are intended to be hormonally sex-reversed and used as "XX males" in the production of all-female populations (Hunter and Donaldson, 1983; Yamazaki, 1983).

Gynogenesis has been adopted as a method for the rapid production of inbred lines in fish (Streisinger et al., 1981; Nagy and Csanyi, 1984). It is particularly valuable when totally homozygous gynogenetic diploids are used as female parents to produce an "inbred line" of genetically identical offspring (Streisinger et al., 1981). Inbred lines produced by gynogenesis are certain to be extremely valuable in fish research, based on the experience with inbred lines in mice (Festing, 1979). In aquaculture, for example, F_1 crosses between inbred lines could be useful as uniform test animals in growth and disease studies. The value of inbred lines in selective breeding programs is an open question.

Partially inbred gynogenetic diploids produced by retention of the second polar body are less valuable in the rapid generation of inbred lines than are homozygous gynogenetic diploids. Studies of heterozygosity in such gynogenetic diploids reveal that some loci remain heterozygous in a high proportion of offspring (Cherfas and Truweller, 1978; Nagy et al., 1979; Thompson, 1983; Thorgaard et al., 1983; Guyomard, 1984). This apparently reflects a distal position of these loci in relation to their centromeres; such information can be useful in making evolutionary comparisons between genes and between species (Thorgaard et al., 1983).

A novel possible application of gynogenesis is as a gene-transfer technique in cases in which sperm inactivation is incomplete and some residual paternal inheritance takes place (Thorgaard et al., 1985). Such residual paternal inheritance may take place in cases in which foreign sperm is used to activate egg development (Table 3). Because the paternal genes appear to be on chromosome fragments in these situations, the residual inheritance is unstable and paternal genes are lost during development. This is likely to limit the practical application of this incomplete gynogenesis as a gene-transfer method, although other applications may emerge.

TABLE 3

Survival and pigmentation of gynogenetic rainbow trout embryos
Albino rainbow trout eggs were fertilized with rainbow trout (RT) or brook trout (*Salvelinus fontinalis*, ST) sperm irradiated with 5.4×10^4 R from a ^{60}Co gamma source, and were heat-shocked 10 min after fertilization to induce retnetion of the second polar body (Thorgaard et al., 1985). Pigmentation in embryos indicates residual paternal inheritance; all embryos showing pigmentation subsequently developed as mosaics (Thorgaard et al., 1985)

Sperm radiation treatment (R)	Proportions of embryos surviving to day		Proportion of survivors showing eye pigmentation
	9	41	Day 41
RT—5.4×10^4 [a]	0.29	0.12	0.025
ST—5.4×10^4 [b]	0.32	0.12	0.006

[a]2767 eggs fertilized 30 Nov. 1983 (Thorgaard et al., 1985).
[b]7000 eggs fertilized 11 Jan. 1985.

Several important questions will be answered as research on gynogenesis as it applies to aquaculture proceeds. (1) Can viable, fertile homozygous gynogenetic diploids be produced in adequate numbers for the generation of inbred lines in most fish species? (2) Are inbred lines a useful approach for genetically improving fish stocks?

ANDROGENESIS

Androgenesis is inheritance in which all genetic material in the offspring comes from the male parent. It can be induced in fish by irradiating eggs with gamma radiation before fertilization (Romashov and Belyaeva, 1964; Purdom, 1969; Arai et al., 1979; Parsons and Thorgaard, 1984). In amphibians, ultraviolet light or pressure treatments have also been used to inactivate the maternal genetic material (Gillespie and Armstrong, 1980; Briedis and Elinson, 1982). Egg genome inactivation is followed by fertilization with normal sperm and suppression of first cleavage as is used for generation of tetraploids and homozygous gynogenetic diploids (Onozato, 1982b in Yamazaki, 1983; Parsons and Thorgaard, 1985). The resulting offspring may be half XX females and half YY males in male heterogametic species.

The probable application of androgenesis is in the rapid generation of inbred lines. The approach is identical in principle to that using homozygous gynogenetic diploids; all the offspring of a homozygous individual after another cycle of androgenesis or gynogenesis will be genetically identical. Androgenesis might be especially advantageous for this purpose for several reasons: (1) males have shorter generation times than females in some fish species, consequently, androgenesis might make more rapid genetic improvement possible, and (2) inbred lines might be stored in the form of cryopreserved sperm (Stoss, 1983) and recovered through androgenesis.

Androgenesis could potentially be a useful general approach to the recovery of genotypes from cryopreserved sperm (Stoss, 1983; Parsons and Thorgaard, 1985). This is important because egg and embryo cryopreservation have not yet been successful in fish and may be difficult because of the large size of fish eggs and embryos (Stoss, 1983). Present androgenesis approaches have the limitation that only homozygous individuals, with attendant poor survival, are produced. However, other approaches that allow the production of outbred androgenetic individuals and are more suitable for gene-banking applications may be possible.

Another possible application of androgenesis may lie in monitoring the effects of mitochondrial genotype on performance. Mitochondrial DNA in animals is maternally inherited (Avise and Lansman, 1983) and this is also likely to be the case in androgenetic individuals in spite of nuclear genome inactivation. Using androgenesis, it should be possible to compare the performance of fish with identical nuclear genotypes and different mitochondrial genotypes. The approach is analogous to the study of alloplasmic lines in plant breeding (Leggett, 1984); that approach, however, requires many generations of backcrossing while androgenesis might be done in a single generation.

The questions concerning survival and fertility of homozygous individuals and the usefulness of inbred lines in breeding programs will be critical for both gynogenesis and androgenesis. In addition, it will be important to assess whether the egg irradiation treatment used in androgenesis has any long-term detrimental effect on the offspring.

CONCLUSION

Induced polyploidy, gynogenesis and androgenesis are likely to have significant applications in aquaculture as methods for inducing sterility and for rapidly generating inbred lines. It is becoming apparent, however, that other valuable applications such as increased hybrid survival, gene transfer and gene banking may be possible using these chromosome set manipulation techniques.

ACKNOWLEDGMENTS

I thank the Washington State Department of Game and U.S. Fish and Wildlife Service for supplying trout eggs and sperm for our ongoing research. Paul Scheerer, Mike Albert (Washington Department of Game), John Propp and Dr. A.R. Stier provided valuable technical assistance. While this paper was written, I was supported by U.S. Department of Agriculture grant 82-CRSR-2-1058 and Washington Sea Grant project R/A-31.

REFERENCES

Allen, S.K., Jr. and Stanley, J.G., 1981. Polyploidy and gynogenesis in the culture of fish and shellfish. Coop. Res. Rep., Int. Counc. Explor. Sea, Ser. B, 28: 1–18.

Allendorf, F.W. and Leary, R.F., 1984. Heterozygosity in gynogenetic diploids and triploids estimated by gene–centromere recombination rates. Aquaculture, 43: 413–420.

Arai, K., Onozato, H. and Yamazaki, F., 1979. Artificial androgenesis induced with gamma irradiation in masu salmon, *Oncorhynchus masou*. Bull. Fac. Fish. Hokkaido Univ., 30: 181–186.

Avise, J.C. and Lansman, R.A., 1983. Polymorphism of mitochondrial DNA in populations of higher animals. In: M. Nei and R.K. Koehn (Editors), Evolution of Genes and Proteins. Sinauer Associates, Sunderland, MA, pp. 147–164.

Benfey, T.J. and Sutterlin, A.M., 1984. Growth and gonadal development in triploid landlocked Atlantic salmon (*Salmo salar*). Can. J. Fish. Aquat. Sci., 41: 1387–1392.

Bingham, E.T., 1980. Maximizing heterozygosity in autopolyploids. In: W.H. Lewis (Editor), Polyploidy: Biological Relevance. Plenum, New York, NY, pp. 471–489.

Briedis, A. and Elinson, R.P., 1982. Suppression of male pronuclear movement in frog eggs by hydrostatic pressure and deuterium oxide yields androgenetic haploids. J. Exp. Zool., 222: 45–57.

Cherfas, N.B., 1981. Gynogenesis in fishes. In: V.S. Kirpichnikov (Editor), Genetic Bases of Fish Selection. Springer-Verlag, Berlin and New York, NY, pp. 255–273.

Cherfas, N.B. and Truweller, K.A., 1978. Investigation of radiation-induced diploid gynogenesis in carp (*Cyprinus carpio* L.). III. Gynogenetic offspring analysis by biochemical markers. Genetika, 14(4): 599–604.

Chevassus, B., Guyomard, R., Chourrout, D. and Quillet, E., 1983. Production of viable hybrids in salmonids by triploidization. Génet. Sél. Evol., 15: 519–532.

Chourrout, D., 1980. Thermal induction of diploid gynogenesis and triploidy in the eggs of the rainbow trout (*Salmo gairdneri*, Richardson). Reprod. Nutr. Dev., 20: 727–733.

Chourrout, D., 1982a. La gynogénèse chez les vertébrés. Reprod. Nutr. Dev., 22: 713–724.

Chourrout, D., 1982b. Gynogenesis caused by ultraviolet irradiation of salmonid sperm. J. Exp. Zool., 223: 175–181.

Chourrout, D., 1984. Pressure-induced retention of second polar body and suppression of first cleavage in rainbow trout: production of all-triploids, all-tetraploids and heterozygous and homozygous diploid gynogenetics. Aquaculture, 36: 111–126.

Chourrout, D. and Quillet, E., 1982. Induced gynogenesis in the rainbow trout: sex and survival of progenies. Production of all-triploid populations. Theor. Appl. Genet., 63: 201–205.

Donaldson, E.M. and Hunter, G.A., 1982. Sex control in fish with particular reference to salmonids. Can. J. Fish. Aquat. Sci., 39: 99–110.

Fankhauser, G., 1945. The effect of changes in chromosome number on amphibian development. Q. Rev. Biol., 20: 20–78.

Festing, M.F.W., 1979. Inbred Strains in Biomedical Research. MacMillan, London, 483 pp.

Gillespie, L.L. and Armstrong, J.B., 1980. Production of androgenetic diploid axolotls by suppression of first cleavage. J. Exp. Zool., 213: 423–425.

Guyomard, R., 1984. High level of residual heterozygosity in gynogenetic rainbow trout, *Salmo gairdneri*, Richardson. Theor. Appl. Genet., 67: 307–316.

Hunter, G.A. and Donaldson, E.M., 1983. Hormonal sex control and its application to fish culture. In: W.S. Hoar, D.J. Randall and E.M. Donaldson (Editors), Fish Physiology, Vol. 9B. Academic Press, New York, NY, pp. 223–303.

Leary, R.F., Allendorf, F.W., Knudsen, K.L. and Thorgaard, G.H., 1985. Heterozygosity and developmental stability in gynogenetic diploid and triploid rainbow trout. Heredity, 54: 219–225.

Leggett, J.M., 1984. Cytoplasmic substitutions involving six *Avena* species. Can. J. Genet. Cytol., 26: 698–700.

Nagy, A. and Csanyi, V., 1984. A new breeding system using gynogenesis and sex-reversal for fast inbreeding in carp. Theor. Appl. Genet., 67: 485–490.

Nagy, A., Rajki, K., Bakos, J. and Csanyi, V., 1979. Genetic analysis in carp (*Cyprinus carpio*) using gynogenesis. Heredity, 43: 35–40.

Onozato, H., 1982a. The "Hertwig effect" and gynogenesis in chum salmon *Oncorhynchus keta* eggs fertilized with ^{60}Co gamma-ray irradiated milt. Bull. Jpn. Soc. Sci. Fish., 48: 1237–1244.

Onozato, H., 1982b. Artificial induction of androgenetic diploid embryos in salmonids. Abstr. Annu. Meet. Jpn. Soc. Sci. Fish., p. 142 (in Japanese).

Parsons, J.E. and Thorgaard, G.H., 1984. Induced androgenesis in rainbow trout. J. Exp. Zool., 231: 407–412.

Parsons, J.E. and Thorgaard, G.H., 1985. Production of androgenetic diploid rainbow trout. J. Hered., 76: 177–181.

Parsons, J.E., Busch, R.A., Thorgaard, G.H. and Scheerer, P.D., 1986. Increased resistance of triploid rainbow trout × coho salmon hybrids to infectious hematopoietic necrosis virus. Aquaculture, 57: 337–343.

Purdom, C.E., 1969. Radiation-induced gynogenesis and androgenesis in fish. Heredity, 24: 431–444.

Purdom, C.E., 1983. Genetic engineering by the manipulation of chromosomes. Aquaculture, 33: 287–300.

Romashov, D.D. and Belyaeva, V.N., 1964. Cytology of radiation gynogenesis and androgenesis in the loach (*Misgurnus fossilis* L.). Dokl. Akad. Nauk. S.S.S.R., 157(4): 964–967.

Scheerer, P.D. and Thorgaard, G.H., 1983. Increased survival in salmonid hybrids by induced triploidy. Can. J. Fish. Aquat. Sci., 40: 2040–2044.

Solar, I., Donaldson, E.M. and Hunter, G.A., 1984. Induction of triploidy in rainbow trout (*Salmo gairdneri*, Richardson) by heat shock and investigation of early growth. Aquaculture, 42: 57–67.

Stanley, J.G., 1976. Female homogamety in grass carp (*Ctenopharyngodon idella*) determined by gynogenesis. J. Fish. Res. Board Can., 33: 1372–1374.

Stoss, J., 1983. Fish gamete preservation and spermatozoan physiology. In: W.S. Hoar, D.J. Randall and E.M. Donaldson (Editors), Fish Physiology, Vol. 9B, Academic Press, New York, NY, pp. 305–350.

Streisinger, G., Walker, C., Dower, N., Knauber, D. and Singer, F., 1981. Production of clones of homozygous diploid zebra fish (*Brachydanio rerio*). Nature (London), 291: 293–296.

Thompson, D., 1983. The efficiency of induced diploid gynogenesis in inbreeding. Aquaculture, 33: 237–244.

Thorgaard, G.H., 1983. Chromosome set manipulation and sex control in fish. In W.S. Hoar, D.J. Randall and E.M. Donaldson (Editors), Fish Physiology, Vol. 9B, Academic Press, New York, NY, pp. 405–434.

Thorgaard, G.H. and Allen, S.K., Jr., 1986. Chromosome manipulation and markers in fishery management. In: N. Ryman and F.M. Utter (Editors), Population Genetics and its Application to Fishery Management. Washington Sea Grant, Seattle, in press.

Thorgaard, G.H., Jazwin, M.E. and Stier, A.R., 1981. Polyploidy induced by heat shock in rainbow trout. Trans. Am. Fish. Soc., 110: 546–550.

Thorgaard, G.H., Rabinovitch, P.S., Shen, M.W., Gall, G.A.E., Propp, J. and Utter, F.M., 1982. Triploid rainbow trout identified by flow cytometry. Aquaculture, 29: 305–309.

Thorgaard, G.H., Allendorf, F.W. and Knudsen, K.L., 1983. Gene-centromere mapping in rainbow trout: high interference over long map distances. Genetics, 103: 771–783.

Thorgaard, G.H., Scheerer, P.D. and Parsons, J.E., 1985. Residual paternal inheritance in gynogenetic rainbow trout: implications for gene transfer. Theor. Appl. Genet., 71: 119–121.

Wattendorf, R.J. and Anderson, R.S., 1986. *Hydrilla* consumption by triploid grass carp in aquaria. Proc. Annu. Conf. Southeast Assoc. Fish Wildlife Agencies, in press.

Yamazaki, F., 1983. Sex control and manipulation in fish. Aquaculture, 33: 329–354.

Developing a Commercial Breeding Program

F.T. SHULTZ

Animal Breeding Consultants, P.O. Box 313, Sonoma, CA 95476 (U.S.A.)

ABSTRACT

Shultz, F.T., 1986. Developing a commercial breeding program. *Aquaculture*, 57: 65–76.

A commercial breeding program may be partitioned into 10 elements: industry assessment, establish goals, measure goals, initial stocks, estimate parameters, mating system, selection criteria, ancillary studies, monitoring procedures, and periodic reconsideration. These elements are interrelated such that a decision made within one element may affect decisions within others. Illustrations are given of decisions to be made for programs of various levels of sophistication using as type animals abalone, salmon, and macroalgae. Examples of successes and failures in the application of population genetics to breeding are taken from the extensive experience in commercial poultry breeding. The effect of early analysis and evaluation of genetic considerations on the success of new aquaculture projects is discussed.

INTRODUCTION

"Genetics begins at the beginning". Success in an aquaculture enterprise depends on a viable concept, sound management, adequate financing, and an organism genetically suitable for the production system. The general belief of founders of such enterprises is that "we will consider a breeding program after we get all the other problems solved". This attitude is based on the naive assumption that breeding is only for future improvement. In any given production system, some genotypes will make success come easy whereas others will make the job difficult, if not impossible. The choice of initial stocks is crucial.

The majority of people at this symposium constitute what may be called the "support" group for future commercial aquacultural breeding programs. We support in two vital ways, by training the people who will be the future commercial geneticists and by doing the basic research to provide the knowledge which they will draw on for their breeding programs. I will address my remarks to the teachers of the future commercial geneticists.

I will concentrate on non-technical matters. It goes without saying, you will teach them the genetics. But the commercial geneticist must know much besides

TABLE 1

The elements of a breeding program

1. Industry assessment	6. Mating system
2. Establish goals	7. Selection criteria
3. Measure goals	8. Ancillary studies
4. Initial stocks	9. Monitoring procedures
5. Estimate parameters	10. Periodic reconsideration

genetics if he is to have a chance at success in the commercial world. He must be expert in pathology, nutrition, production management, and total economics. The approach, the frame of mind, the attitude, and the utilization of genetic knowledge is quite different in commercial breeding than in institutional research and academic teaching.

My purpose in this paper is to outline the various "elements" of a breeding program (Table 1). Each geneticist might classify these elements differently (e.g., Harris et al., 1984). There is nothing sacred about mine. The important thing is to realize that there are many components and not to leave any of them out in planning a program. For convenience, I have numbered them, but that is not to imply that they are "steps" necessarily to be taken in sequence. They are all interrelated such that a new decision in one may require modification of decisions made previously in others.

I have selected three organisms as "type" organisms to illustrate how these elements might apply to a commercial situation: abalone, salmon, and kelp. I have some experience with each of them. However, what I say about them is not what I would necessarily decide if I were to seriously set up a program. They are for illustration only.

INDUSTRY ASSESSMENT

The geneticist must have a clear picture of the industry within which he is going to work. He must understand the production system, processing methods, and marketing organization and requirements. Most important, he must know the consumers — what they will buy, and what they will not buy.

Unfortunately, it is very difficult for others to satisfactorily enlighten the geneticist in this matter; the geneiticist must see for himself. It is the geneticist who will determine the actual direction of the breeding program. Others may tell him in what direction they think it should go, but there is no way they can determine whether or not the geneticist actually goes in that direction. It is a responsible position and the commercial geneticist stands all alone.

The breeder's present and future position in the industry must be ascertained. He may breed solely for his own production system, or he may sell breeding stock to other producers in the industry. This decision determines how much money should be put into breeding. The goals will be different also.

A crucial question arises! The financial resources available to the program must be incorporated into the general plan. There are different levels of sophistication in breeding programs. The simplest might be nicknamed the "keeping out of trouble" program designed to prevent a loss of quality as the production system reproduces itself. Next comes "improvement within a production system". Breeding procedures are interwoven with production procedures within the production facilities. The next level of complexity is the "small discrete breeding program" from which there is a continuous range up to the "ultimate in sophistication". Complicated genetic programs are not always what is called for — and certainly not always necessary!

ESTABLISH GOALS

This area demands precision as it is easy to go wrong. The goals must be genetically feasible and compatible with the conclusions reached in the industry assessment. There are a couple of points frequently overlooked.

Both short- and long-term goals must be considered if the breeding program is to survive. If the first noticeable improvements are 5 years away, the front office may get tired of waiting. Besides, the company may need something to sell soon. On the other hand, most desirable genetic changes require several generations. The commercial geneticist has to foresee what is coming, and he may be the only one in the whole company who really understands that fact. Others may flamboyantly make great predictions for the future. The salesman can change his mind next year but the geneticist has to commit himself now.

Minimize the number of traits to improve. Selection pressure is limited. A good geneticist can always think of more things to do than there is money and time for. Stay within capabilities.

Never set up a breeding program unless the company is already dedicated to the idea. If you start without dedication, you are going to run out of their money long before you can prove you did anything.

Increasing growth rate will be one goal for all three type organisms. Meat yield will be critical in abalone. It would be very disheartening to select mightily for growth rate only to discover later that all the gains were made in shell weight. Egg number, measured by millions and thousands in abalone and salmon, respectively, should not be a goal in the early stages. What to do genetically about sexual maturity will require soul searching for both animals. Livability is very important in most organisms, especially during later stages of growth. Chemical composition of kelp will probably be a goal whether the kelp is to be used for chemicals, animal feed, or human food.

MEASURE GOALS

The measurement of a goal should consider heritability, cost of measurement, and biological efficacy. Weighing an abalone should be easy, but when you take an animal out of the water some of the water comes with it. The outside can be dried quite easily, but each time you squeeze, some more water comes out. So how many times should you squeeze? To measure three animals is no problem. To measure 300 is a problem, and thousands will need to be weighed in a breeding program.

The measurement must be of the actual goal desired. The most important goal in chicken broiler breeding for many years has been to increase growth rate. The measurement used was the total weight of the chicken at some age. Tremendous gains were made, but the breeders got something the consumers did not want — excessive fat.

This brings up the problem of breeding from dead animals, because the best way to measure fat in a bird is to kill it, then extract and weigh the fat. The search is on for an effective way to measure fat on a live bird. Family selection is one alternative, but the commercial geneticists must use ingenuity.

Ingenuity is also needed for feed conversion in broilers and turkeys — and certainly will be for aquatic animals. One approach is to measure feed conversion directly by comparing the feed eaten with the weight gained for each individual. Dr. Nordskog and his students have elucidated that process (Wing and Nordskog, 1982). At this symposium Dr. Nordskog has discussed substituting an index to avoid the laborious measurement of individual feed conversion.

A new approach entails setting up a nutritional regimen that would penalize fat production and credit production of increased muscle. This method takes into consideration the metabolic pathways of fat, carbohydrate, and protein metabolism and their relative energy costs.

Every commercially important trait is a "compound" trait. Feed conversion can be subdivided into growth rate which lowers maintenance cost and has long been recognized as the principal component, shape of the growth curve, sexual maturity which may have effects at earlier ages, activity, excess appetite, gut retention time, digestive efficiency, biochemical efficiency, and body composition. Some of these components can be selected for directly whereas others may be as hard to measure as feed conversion itself. The direct measurement as used today may actually miss some of these components. Thus, the commercial geneticist must decide which traits to divide into subtraits and select as separate entities and which of seemingly separate traits can be combined into one trait and treated better as a unit.

INITIAL STOCKS

As in agriculture, success or failure of a new aquaculture project may depend on the initial stocks. Wild populations as well as domesticated strains vary tremendously in their genetic composition and adaptability. Start with the best. You can spend years catching up to the best that already existed when you started.

Finding and recognizing the best is a genetic matter. Introductions may not perform up to their true genetic potentials in the initial "screening" tests. Then there is the question, seldom considered, of how well a stock will respond in the future genetic program. The past evolutionary history, wild or domestic, of the stock gives clues to the geneticist. An organism which has been under artificial selection for many generations is quite a different situation from one which has been newly domesticated or not yet domesticated.

Salmon offer exciting possibilities with so many populations, subpopulations, isolates and semi-isolates. The commercial geneticists will be very busy sorting these out, at least until the domestic strains are improved beyond the range of the wild stocks for traits such as growth rate.

Sea ranchers on the Pacific Coast should be very careful where they obtain their initial stocks and how they evaluate them (Ricker, 1972; Withler, 1982). It is highly probable that salmon are genetically programmed to do certain things when they reach the sea with regard to returning home. Imprinting is simply the final page in their information handbook. One genotype is programmed to "turn right", another genotype is programmed to "turn left", so to speak; and the program to return home is not necessarily the reverse of the program to leave home.

Salmon have a powerful attribute. Thousands of eggs are available from each of many females. Sperm is also available from each of many males, hopefully at the same time as the eggs. Crosses between lines, complex diallel designs, and stratified matings are easy to make. The genetic potential is startling. Then comes the hard part, how do you keep all these genetic units identified until they grow up big enough to mark them? The commercial geneticist must compromise between the biological opportunities and the economic limitations.

Kelp promises to have great genetic potential. Clines and semi-isolates and even isolates probably exist in most species. Environments include exposed headlands, open and sheltered bays, open ocean water, sewage outfalls, chemical variables, and temperature variations. The genotypes can be easily manipulated.

ESTIMATE PARAMETERS

A thorough review of the literature should be made before the initial form of the breeding program is finalized. Visit with your friendly scientists because their work may not be published yet. The commercial geneticist must at least be aware of everything that is known about the genetics of the organism to this point. You certainly cannot afford, in a commercial breeding program, to "reinvent the wheel".

If there is no information available and if you understand the life cycle, environment, and biology of your organism, you can make some reasonably safe guesses. If the program is to be a reasonably sophisticated one, it should be designed so as to generate data which can be eventually analyzed.

Parameters for abalone must at present be estimated "by guess and by gosh". The same situation is true for kelp. For salmon, some data have been published and several people at this symposium are developing more.

MATING SYSTEM

The choice of the mating system is a primary tool of the breeder. Inbreeding, outcrossing, population sizes, and choice of mates are but a few areas of decision for the geneticist. Inbreeding is both friend and foe (Nordskog and Hardiman, 1980; Cahaner et al., 1980; Woodard et al., 1983). The commercial geneticist works with finite populations. He must understand what inbreeding means genetically if he is to use it to advantage while at the same time avoiding its serious consequences (Shultz, 1953). Inbreeding as we have had to deal with it in livestock will be a totally different story in many aquatic organism. Self-fertilization, gynogenesis, parthenogenesis, polyploidy, physiological manipulation of sex, diverse genetic mechanisms for sex determination, sequential sex changes, hermaphrodism, and other biological tools are available in various organisms (Nace, 1968; Shultz, 1970; Purdom, 1983).

The tremendous reproductive capability of many aquatic organisms invites error. For example, a strain of pen-reared Atlantic salmon can be reproduced easily from say two males and three females each generation. Selection pressure would be intense with over 5000 adult market fish to choose from. Since there are only six matings total, a slight superiority of one mating could result in the five fish selected to be parents of the next generation, all being full-sibs. Hence, selection pressure in the next generation will go mostly to counteract the effects of inbreeding. Even the simple "keeping out of trouble" breeding program would avoid this mistake and loss in productivity.

SELECTION CRITERIA

Natural selection is not to be ignored, although it usually is. It is occurring in domestic, artificial-selected populations just as in wild populations (Lerner, 1954). As with inbreeding, it can be both friend and foe (Wright, 1931; Doyle, 1983). The geneticist must use it to advantage as well as minimize its adverse effects (Lerner, 1958).

In commercial breeding "no–no traits" can be a problem. These are characteristics which someone in the company, or worse yet a customer, wants you to improve, but it has no economic value. The geneticist must be prepared to counteract such ideas. However, there are also many commercially important traits which should not be included in the selection criteria. Some low-frequency abnormalities will not respond, and some just come and go. If natural selection is taking care of it, don't bother. Some loss in egg number in abalone and salmon can perhaps be tolerated. Monitor such traits but do not waste selection pressure on them. Selection pressure is limited.

A common practice in commercial poultry breeding, when assigning dams to sires, is to avoid mating birds with common grandparents. The assumption is that avoiding common grandparents will reduce the rate of increase in inbreeding in the line. Actually this practice has little, if any, effect. The real value is that the effectiveness of selection is not impaired. Full- and half-sib matings rarely produce offspring with a chance to be selected, whether the selected trait be egg number, livability, growth rate, or meat yield.

Balanced genetic polymorphisms may arise from artificial as well as natural selection (Crow, 1961; Connor and Bellucci, 1979). They may be of such a nature as to overpower additive differences and seriously reduce additive genetic gains. This area needs attention, by both private breeders and institutional researchers. Marker genes will be necessary.

The commercial geneticist must remember he is not running a selection experiment. This is perhaps the toughest transition to be made from academic research to commercial breeding. You do not select to find out what happens. The purpose is to get somewhere, and to know where before you start! The moment a change in your program is indicated, you change.

ANCILLARY STUDIES

Ancillary studies make up a very critical element of a breeding program. The mathematical aspects usually receive adequate attention in sophisticated programs, especially calculation of heritabilities, correlations, and sometimes selection indexes (Lerner, 1950; Nordskog, 1977). However, very effective breeding programs are possible without such calculations. The biological aspects are seriously neglected in most commercial programs. The fat and feed conversion problem in poultry is an example.

Growth rate, a combination of mathematics and biology, is a primary trait in most organisms. Each segment of the growth curve should be studied. It may be desirable to allocate more of the available selection pressure to one segment than to another.

Body shape may be a good indicator of edible meat yield. It can be measured on the live animal in poultry and in most, if not all, fin fish, lobsters, shrimp, abalone, oysters, and certainly macro-algae. After many years of selection for breast-meat yield in turkeys using a subjective measurement of breast shape, it was found that gains had been made primarily in the outer superficial pectoral muscle (F.T. Shultz and F.L. Cherms, in preparation). The inner deep pectoral muscle showed little change from the unimproved controls. Furthermore, the ratio of superficial pectoral to total body weight increased during the growing period, but the rate of increase sharply accelerated as the bird approached market weight. Knowledge of the shape of these curves helps future program development.

The biology of sexual maturity must be understood before the geneticist tries to improve it genetically. In layer chickens, geneticists worked very hard for early sexual maturity and made great genetic progress. Then, it was found that if sexual maturity was delayed by environmental means, total egg production was increased. The geneticists only did half the job! The problem in salmon and abalone of course is the reverse, namely, how to prevent it until it is needed.

A very important rule in commercial breeding programs should be emphasized at this time. "If you can improve a trait or eliminate a problem by management, nutrition, veterinary medicine, physiology, endocrinology, or whatever, it is suicide to attempt to do it by genetics in the commercial world".

Two kinds of chromosome manipulations should be recognized. The first involves whole chromosomes and sets of chromosomes (e.g., polyploidy and gynogenesis). The second involves segments of chromosomes. Breeders have been manipulating segments for years without realizing it. To illustrate, two strains of turkeys, one with high reproduction and slow growth rate and the other with poor reproduction and fast growth rate, are crossed. Beginning with the cross, strong selection for reproduction will tend to duplicate the high reproduction strain with very few genes being incorporated from the other strain. Likewise strong selection for growth will tend to duplicate the high growth strain with very few genes being incorporated from the other strain. The explanation is that blocks of genes (segments of chromosomes) are being selected.

The moral is, to start a new synthetic line or to begin a selection experiment from two or more lines, allow a few generations with very weak selection so the blocks can break up by simple crossing-over; then the best of both strains can be combined, with proper breeding techniques of course.

Marker genes may be defined as "genes which are identifiable such that they can be followed through a line of descent and whose presence or absence in an individual can be detected by immunological, electrophoretic, or other methods". These genes are very useful for studies in evolution, taxonomy, population structure, migration, and chromosome tracking. In breeding, they are useful as pedigree markers and maintaining the identity of pure lines (Moav et al., 1976; Hedgecock, 1977).

About 30 years ago, many poultry breeders thought that blood group genes would be the ultimate tool in commercial breeding. Millions of dollars were spent from private and public funds based on the concept that the proper combination of alleles from the proper loci would result in quantum gains in certain commercially important traits. The breeders failed to understand that these genes were marker genes, not pleiotropic genes (Shultz and Briles, 1953; Neimann-Borensen and Robertson, 1961). By the time thay carried a "superior" combination through the production system into the hands of the final producer, the mystical effects had dissipated by simple crossing-over. The money was not all wasted because the histocompatibility locus, with all of its ramifications, was brought to light in chickens and turkeys (Cherms, 1969; Nordskog, 1983; Briles, 1984; Miller et al., 1984; Abplanalp et al., 1985).

The suggestion was made by I.M. Lerner, (personal communication, 1952) that if 25–30 marker loci were available, the approach to poultry breeding would drastically change. Extending that concept, the breeding unit could become the segment of a chromosome instead of the whole organism. A new mathematical approach would be required. Sufficient loci are still not available in avian species, but electrophoretic techniques take on added importance as aquatic breeding programs develop. Marker genes, where available, will be invaluable in several ways, but they should be used correctly. In only rare cases are pleiotropic effects of commercial importance likely to be found (Vrijenhoek, 1985). Restriction fragment length polymorphisms will also add a new dimension to breeding (Beckmann and Soller, 1983; Soller and Beckmann, 1983; Tanksley, 1983).

Genetic engineering will some day be an important addition to the tools of the breeder. The commercial geneticist needs to know what it can do and what it cannot do, but he does not need to be a molecular geneticist. Advances will come out of institutional research and cooperation between private breeding and genetic engineering companies.

Statements in the public press as "...speed selective breeding into a single generation" can be taken in good humor. Turkeys already exist which grow 10–15% faster than the present commercial turkey, but they are not economically viable products. They are too costly to produce! Any organism, except perhaps certain lower forms, which receives by genetic engineering a gene or genes that makes a substantial change in the phenotype will almost certainly require several generations of "fine tuning" by the commercial geneticist using orthodox population methods to make the new creature commercially useful. There are exciting things in biotechnology on the horizon; most will turn out not so good but there will be some successes. However, we must not overlook the fact that, in the "unimproved" aquatic organisms, there are tremendous genetic gains to be made by orthodox breeding techniques.

Obviously the breeder cannot perform all of the desirable biological studies, but he should know what has been done and how the results affect his program. The private breeder must do some research; for other problems the breeder can stimulate institutional research. In some cases, joint projects will be indicated for the benefit of both.

MONITOR PROCEDURES

Fredeen (1986) used the term "genetic change" rather than the usual term "genetic progress". Genetic change is more appropriate as it includes the negative as well as the positive. It also fits the fact that both natural and artificial selection are acting in domestic populations.

Not all genetic changes are phenotypically obvious even with the best of monitoring system. It is imperative that hidden genetic changes such as balanced genetic polymorphisms be detected before several generations of reduced gains occur. Marker genes can be used for monitoring purposes (Abplanalp, 1984).

A monitoring procedure does not contribute directly to immediate phenotypic gain, but is costs money and facilities. Therein lies a dilemma. Should the breeder increase his chance of survival by doing something positive such as have another line under development? Or should monitoring be done to assure that the program is moving and in the right direction? The advantage of the large breeder is not that he can do something better, but that he has more chances at success. The small breeder has to take his best shot. Monitoring has been an important aspect of poultry breeding programs (Flock, 1980). Breeding programs on newly-domesticated species should not overlook the possibility of using wild populations as controls.

PERIODIC RECONSIDERATION

The commercial geneticist must constantly re-evaluate the program. Even so, a specific time should be set to have a comprehensive re-evaluation of each element at least annually. It is very easy to get so involved in day to day problems that what is happening in the rest of the world is overlooked. This symposium, and others such as the National Breeders Roundtable, provide excellent opportunities to break the routine and discuss new developments with one's peers, commercial and institutional.

REFERENCES

Abplanalp, H., 1984. The genetic control of immunity in chickens. Taiwan Livestock Res., 17: 125–134.
Abplanalp, H., Schat, K.A. and Calnek, B.W., 1985. Resistance to Marek's disease of congenic lines differing in major histocompatibility haplotypes to 3 virus strains. In: B.W. Calnek and J.L. Spencer (Editors), International Symposium on Marek's Disease. American Association of Avian Pathologists, Inc., Univ. Pennsylvania, pp. 347–358.
Beckmann, J.S. and Soller, M., 1983. Restriction fragment length polymorphisms in genetic improvement: methodologies, mapping and costs. Theor. Appl. Genet., 67: 35–43.
Briles, W.E., 1984. Early chicken blood group investigations. Immunogenetics, 20: 217–226.
Cahaner, A., Abplanalp, A. and Shultz, F.T., 1980. Effects of inbreeding on production traits in turkeys. Poult. Sci., 59:1353–1362,
Cherms, F.L., 1969. Influence of the histocomppatibility system on reproductive efficiency of turkeys. J. Reprod. Fertil., 19: 195–197.

Connor, J.L. and Bellucci, M.J., 1979. Natural selection resisting inbreeding depression in captive wild housemice (*Mus musculus*). Evolution, 33: 929-940.

Crow, J.F., 1961. Population genetics. Am. J. Hum. Genet., 13: 137-150.

Doyle, R.W., 1983. An approach to the quantitative analysis of domestication selection in aquaculture. Aquaculture, 33: 167-185.

Flock, D.K., 1980. Genetic improvement of egg production in laying type chickens. In: A. Robertson (Editor), Proc. Symposium on Selection Experiments in Laboratory and Domestic Animals. Farnham Royal, Slough, 1979. Commonwealth Agricultural Bureaux, pp. 214-224.

Fredeen, H., 1986. Monitoring genetic change. Aquaculture, 57: 1-26.

Harris, D.L., Stewart, T.S. and Arboleda, C.R., 1984. Animal breeding programs: a systematic approach to their design. Adv. Agric. Tech. AAT-NC-8, Afric. Res. Ser. U.S. Dept. Agric.

Hedgecock, D., 1977. Biochemical genetic markers for broodstock identification in aquaculture. Proc. World Maricult. Soc., 8: 523-531.

Lerner, I.M., 1950. Population Genetics and Animal Improvement. Cambridge University Press, Cambridge, 342 pp.

Lerner, I.M., 1954. Genetic Homeostatis. John Wiley and Sons, New York, 134 pp.

Lerner, I.M., 1958. The Genetic Basis of Selection. John Wiley and Sons, New York, 298 pp.

Miller, M.M., Goto, R. and Abplanalp, H., 1984. Analysis of the B-G antigens of the chicken MHC by two-dimensional gel electrophoresis. Immunogenetics, 20: 373-385.

Moav, R., Brody, T., Wohlfahrt, G. and Hulata, G., 1976. Application of electrophoretic genetic markers to fish breeding. I. Advantages and methods. Aquaculture, 9: 217-228.

Nace, G.W., 1968. The amphibian facility of the University of Michigan. BioScience, 18: 767-775.

Neimann-Sorensen, A. and Robertson, A., 1961. The association between blood groups and several production characteristics in three Danish cattle breeds. Acta Agric. Scand., 11: 163-196.

Nordskog, A.W., 1977. Success and failure of quantitative genetic theory in poultry. In: E. Pollak, O. Kempthorne and T.B. Bailey, Jr. (Editors), Proc. Int. Conf. Quant. Genet. 16-20 August 1976. ISU Press, Ames, IA, pp. 569-586.

Nordskog, A.W., 1983. Immunogenetics as an aid to selction for disease resistance in the fowl. World's Poult. Sci. J., 39: 199-209.

Nordskog, A.W. and Hardiman, J., 1980. Inbreeding depression and natural selection as factors limiting progress from selection in poultry. In: A. Robertson (Editor), Proc. Symposium on Selection Experiments in Laboratory and Domestic Animals. Farnham Royal, Slough, 1979. Commonwealth Agricultural Bureaux, pp. 91-99.

Purdom, C.E., 1983. Genetic engineering by the manipulation of chromosomes. Aquaculture, 33: 287-300.

Ricker, W.E., 1972. Hereditary and environmental factors affecting certain salmonid populations. In: R.C. Simon and P.A. Larkin (Editors), The Stock Concept in Pacific Salmon. H.R. MacMillan Lectures, Vancouver, pp. 19-160.

Shultz, F.T., 1953. Concurrent inbreeding amd selection in the domestic fowl. Heredity, 7: 1-21.

Shultz, F.T., 1970. Genetic potentials in aquaculture. In: H.W. Youngken, Jr. (Editor), Food-Drugs from the Sea Proc. 1969. Marine Tech. Soc., Washington, DC, pp. 119-134.

Shultz, F.T. and Briles, W.E., 1953. The adaptive value of blood group genes in chickens. Genetics, 38: 34-50.

Soller, M. and Beckmann, J.S., 1983. Genetic polymorphism in varietal identification and genetic improvement. Theor. Appl. Genet., 67: 25-33.

Tanksley, S.D., 1983. Molecular markers in plant breeding. Plant Molecular Biol. Reporter, 1: 3-8.

Vrijenhoek, R.C., 1985. Homozygosity and interstrain variation in the self-fertilizing hermaphroditic fish, *Rivulus marmoratus*. J. Hered., 76: 82-84.

Wing, T.L. and Nordskog, A.W., 1982. Use of individual feed records in a selection program for egg production efficiency. 1. Heritability of the residual component of feed efficiency. Poult. Sci., 61: 226-230.

Withler, F.C., 1982. Transplanting Pacific salmon. Can. Tech. Rep. Fish Aquat. Sci., No. 1079. Dept. Fish. and Oceans, Canada.
Woodard, A.E., Abplanalp, H., Pisenti, J.M. and Snyder, L.R., 1983. Inbreeding effects on reproduction traits in the Ring-Necked pheasant. Poult. Sci., 62: 1725–1730.
Wright, S., 1931. Evolution in Mendelian populations. Genetics, 16: 97–159.

Breeding Plan for Sea Ranching

TRYGVE GJEDREM

The Agricultural Research Council of Norway, Institute of Aquaculture Research, 1432 As-NLH (Norway)

ABSTRACT

Gjedrem, T., 1986. Breeding plan for sea ranching. *Aquaculture*, 57: 77-80.

Cultivation of freshwater as well as marine fishes to increase food production is a great challenge to mankind. Different systems of sea ranching, particularly those using anadromous species, are already in use. However, breeding programs for sea-ranching systems have been infrequently discussed.

In this paper breeding goals are discussed and it is concluded that the most important traits are growth rate, age at maturation, and recapture frequency.

Practical problems in running a sea-ranching program such as fish marking, how to organize and record recapture of the fish, and production of broodstock are discussed. The necessity of utilizing family testing on a large scale is stressed.

The possibility of genetic improvement of growth rate and age at maturation is quite good. Because of the lack of information it is difficult to estimate the possible response to selection of recapture frequency. However, the limited information available indicates that it is a heritable trait.

It is concluded that more research should be carried out to increase our knowledge in this field.

INTRODUCTION

There are several ways in which ocean resources can be utilized better to increase human food production. One alternative applicable to anadromous and marine fish species is sea ranching. Sea ranching is practised in several places and in several ways, but the basic principle of all operations is the same. Part of the population undergoes artificial reproduction, and the offspring are kept in captivity to a particular development stage (eggs/alevins/fry/fingerlings) before release.

The present paper discusses a breeding program for salmon. This species was chosen as an example because more is known about genetic parameters in salmon than in other species used for sea ranching. Large-scale programs using salmon are currently operating in the northern Pacific Ocean.

Cultivation of wild stocks of salmon goes back more than 100 years. During the early years alevins from hatcheries were released into rivers and lakes.

Later the fry were fed for some period before release. Fingerlings and smolts are frequently produced and released into the rivers to compensate for the damage to the fish stocks by dam construction and other habitat alterations.

Dr. Lauren Donaldson did pioneering work in sea ranching when he released chinook smolts of 1949 brood fish from a small pond on the University of Washington campus in Seattle (Donaldson, 1968). The successful return of the first fish in 1953 is described by Hine (1976). Later, many successful sea-ranching programs were started by private companies, river cooperatives and national institutes. The largest programs by far are run in the North Pacific in Japan, the U.S.S.R., Canada, and the U.S.A. Successful programs have been running in Iceland and Sweden for many years. Salmon runs have been established in Chile, and salmon have been introduced into the Great Lakes in the U.S.A. (Donaldson and Joyner, 1983).

The success of a sea-ranching program depends on many factors, but in particular on the production cost of smolts and return percent of fish. Many experiments have been carried out to produce smolts which return in large numbers. Important factors are feeding, age and size of smolts, equipment used, water quality and temperature, and time and place of release. Certainly more suitable methods can be developed to optimize the environmental factors to improve the return from salmon runs, among which site selection is one of the most important. However, to date, few investigations have been carried out to study the possibilities for selective breeding in sea ranching and today no efficient selection program aimed at increasing return of fish is in progress.

BREEDING GOALS

There are three traits, in particular, of economic importance which should be improved through selection: growth rate, age at maturation, and return percent. It is quite evident that growth rate and return percent should be improved. It is not clear, however, what the breeding goal for age at maturation should be. Early-maturing fish will return with higher frequency than late-maturing fish, but there will be large differences in body weight between these two classes. The breeding goal for age at maturation should therefore be determined separately for each project.

Other traits of interest for selection could be early growth rate and date of return from the sea.

POSSIBILITIES FOR SELECTIVE BREEDING

It is now well documented that large genetic variation exists in growth rate and age at maturation in Atlantic salmon (Gjerde, 1984; Gjerde and Gjedrem, 1984). It is also well known that the salmon has a high fecundity which makes intensive selection possible.

We still have only limited knowledge about the magnitude of the genetic variance for return percent. No estimates of heritability for return frequency in fish are known. Carlin (1969) and Ryman (1970) report significant differences in recapture frequency between full-sib groups released as smolts in Indalselva which migrated into the Baltic Sea. Bailey and Saunders (1984) found significant differences between Atlantic salmon strains in return percent. Their return percent was rather low, 0.64% in 1977 and 2.14% in 1978. Ryman (1970) found a significantly lower return percent in inbred vs. non-inbred fish, while Bailey and Saunders (1984) found no evidence of heterosis in strain crosses.

It can be concluded from the literature that there are good possibilities for improvement of growth rate and age at maturation in salmon, and it is likely that return percent can be improved through selection.

In a selection program one should combine these three characters in a selection index. However, our limited knowledge about return percent makes this difficult to practice today. After adjusting each trait for environmental factors one could, for simplification, use kg salmon return/100 smolts released as a selection criterion.

BREEDING PLAN

A breeding plan for sea ranching should be based on family selection because individual selection is not efficient for all-or-none traits such as age at maturation and return percent. Progeny testing is of no interest since it will increase the generation interval considerably, and in most cases the sire and dam will be dead when the test results are known.

At fertilization one should make both full- and half-sib groups. This makes it possible to etimate both additive and non-additive genetic variance. An alternative is to make only sire half-sib groups, which will allow estimation of additive genetic variance only. It is important to test many families each year, at least 100.

Each full-sib group must be reared separately until marking can take place, usually at 10–20 g. After marking, the families should be mixed to reduce environmental differences.

The number of smolts which should be marked and released from each family depends on the return percent. As a guideline, one could on average ask for 20 returning salmon per family; then 200 smolts should be marked if the return percent is 10% and 2000 should be marked if the return percent is 1%. The smolts should be released in the river.

In order to obtain good records on the returning fish a fish trap must be placed in the river. Date of return, family, sex, weight and length should be recorded for all returning fish.

One approach to selection is to keep all returning fish alive until the season

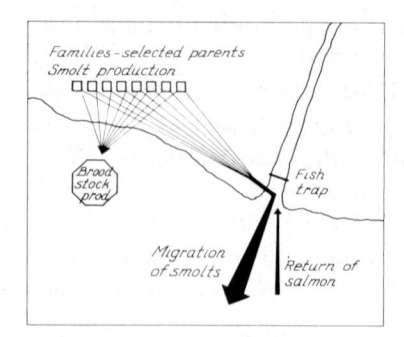

Fig. 1. Breeding plan for salmon, sea ranching.

is ended. Selection can then be carried out based on a ranking of families determined from the collected data. In practice, it is very difficult to keep all fish during this period; many of them will die, making selection inefficient. An additional problem is that there will be too few fish in each family to permit individual selection within families.

An alternative plan is to mark some smolts (e.g., 150–200) from each family and rear them in cages to produce broodstock. This will allow the sea rancher to slaughter all fish at return. When all records have been taken, a ranking of all families can be calculated. Selection of broodstock will then be practised on the cage-reared sibs of the returning fish (Fig. 1).

The selection intensity will depend on the total number of families, number of fish alive in the cage, and number of brood fish required for the next generation. The same principle can also be applied with small modifications to other species. I am convinced that selection programs will be a profitable investment in every sea-ranching system.

REFERENCES

Bailey, J.K. and Saunders, R.L., 1984. Returns of three year-classes of sea-ranched Atlantic salmon of various river strains and strain crosses. Aquaculture, 41:259–270.
Carlin, B., 1969. Salmon tagging experiments. Laxforskningsinstitutet (Swedish Salmon Research Institute). Medd.: 2–4/1969: 8–13.
Donaldson, L.R., 1968. Selective breeding of salmonid fishes. Marine Aquaculture, 23, 24 May 1968. Oregon State University Press, pp. 65–74.
Donaldson, L.R. and Joyner, T., 1983. The salmonid fishes as a natural livestock. Sci. Am., July 1983: 49–56.
Gjerde, B., 1984. Response to individual selection for age at sexual maturity in Atlantic salmon. Aquaculture, 38: 229–240.
Gjerde, B. and Gjedrem, T., 1984. Estimates of phenotypic and genetic parameters for carcass traits in Atlantic salmon and rainbow trout. Aquaculture, 36: 97–110.
Hine, N., 1976. Fish of Rare Breeding, Salmon and Trout of the Donaldson Strains. Smithsonian Institution Press, Washington D.C., 167 pp.
Ryman, N., 1970. A genetic analysis of recapture frequencies of released young salmon (*Salmo salar*). Hereditas, 65: 159–160.

Application of a Genetic Fitness Model to Extensive Aquaculture[1]

JAMES E. LANNAN and ANNE R.D. KAPUSCINSKI*

Department of Fisheries and Wildlife, Oregon State University, Marine Science Center, Newport, OR 97365 (U.S.A.)
*Department of Fisheries and Wildlife, University of Minnesota, 200 Hodson Hall, 1980 Folwell Avenue, St. Paul, MN 55108 (U.S.A.)

ABSTRACT

Lannan, J.E. and Kapuscinski, A.R.D., 1986. Application of a genetic fitness model to extensive aquaculture. *Aquaculture*, 57: 81–87.

Substantial progress has been made in the breeding improvement of a variety of species of fish and shellfish. At the present time the opportunities for genetic improvement of aquaculture broodstock are limited to intensive production systems where the environment remains relatively constant and predictable from one generation to another. This is not the case in extensive aquaculture systems where the environment may be highly variable from one generation to the next.

Stocks of fish and shellfish produced in extensive aquaculture systems may experience reduced genetic diversity as a consequence of inbreeding and genetic drift if only a small proportion of the population is used as broodstock.

This report describes the application of a genetic fitness model developed for natural resource management to the problem of broodstock management in extensive aquaculture. The model describes the probability distribution of stock fitness in terms of population numbers, age structure, and immigration into the broodstock. The output of the model is the size and structure of the broodstock required to maintain the probability distribution of fitness in subsequent generations.

INTRODUCTION

The opportunities to realize economic gains in commercially important fish and shellfish will continue to increase as the information about the genetics of these species increases. However, at the present time the opportunities for substantial genetic improvement of aquaculture broodstocks is limited to species produced in intensive culture systems. The term 'intensity' is used here to denote the relative degree of environmental control in an aquaculture system.

---

[1]Oregon Agricultural Experiment Station Technical Paper No. 7430.

The aquaculture environment is likely to be more constant in intensive than in extensive systems. In the latter, juveniles may be produced in intensive culture systems, but grow-out occurs in the absence of environmental control. Therefore, the environment may be highly variable from one generation to the next. Because the environment cannot be predicted, there is uncertainty in the prediction of performance from year to year. Organisms that have been selected for superior performance in a certain environment may exhibit inferior performance as environmental conditions change. Unless selected strains exhibit superior performance throughout the range of environments encountered by the species over time, selective breeding efforts may actually result in economic losses rather than gains.

To place the problem in perspective, consider the frequency distribution of fitness in a fictitious population. The phenotypic value of a particular individual in this breeding population will vary depending upon the specific sequence of environmental circumstances. Because the sequence of environmental events cannot be predicted before the fact, there is uncertainty associated with predicting individual fitness in the breeding population.

We have proposed a conceptual model for managing fisheries to maintain the diversity of genetic information in exploited stocks and described its application to Pacific salmon fisheries (Kapuscinski and Lannan, 1984). In this paper, we describe how the general fisheries management model can be applied to maintaining genetic diversity in broodstocks for extensive aquaculture operations.

DESCRIPTION OF THE CONCEPTUAL MODEL

The model is based upon the genetic concept of fitness. Fitness is defined as the relative contribution of an individual in a breeding population to the next generation. All populations have a probability distribution of individual fitnesses characterized by a mean ($\bar{W}$) and a variance (V_w).

The model assumes that fitness is a quantitative genetic character with both genotypic and environmental sources of variance. If this assumption is not valid then there is no reason for concern about the maintenance of genetic variability in broodstocks in extensive aquaculture because, in the absence of genetic variation, the population cannot respond to natural or artificial selection. Fitness is the character upon which natural selection acts (Lewontin, 1970; Falconer, 1981).

In extensive aquaculture, as in fisheries management, the maintenance of fitness may be an important goal if genetic diversity is to be conserved. Assuming that fitness reflects genetic variability, and because fitness cannot be predicted in future time, satisfaction of this management goal requires that the probability distribution of fitness for a given population does not decline substantially over time. Thus, the broodstock management goal can be stated as maintaining the probability distribution of fitness.

Unfortunately, the direct measurement of individual and population fitness is technically complicated and impractical in most fish culture situations. However, the goal of maintaining the probability distribution of fitness may be achieved without assigning an absolute value to $\bar{W}$ or V_w.

Let W equal the probability distribution of fitness (with mean $\bar{W}$ and variance V_w) for a population which is considered to have adequate genetic diversity to perpetuate itself throughout the range of environments it may encounter through its life history. Next, let W' equal the probability distribution of fitness of a broodstock for extensive aquaculture. Consistent with the goal of maintaining the probability distribution of fitness of the broodstock, conservative management dictates that $W = W'$. This conceptual model can be made quantitative by deriving the functional relationships

$$W = f \text{ (population variables)} \text{ and } W' = f \text{ (population variables}') \qquad (1)$$

The population variables may be limited to those that influence stock fitness. These include changes in population numbers, immigration into the stock, and the presence or absence of age structure in the reproductive stock. Although mutation and selection also influence broodstock fitness, mutation may be considered negligible in the short term and the maintenance of the probability distribution of fitness through the course of natural selection is implicit in the management goal. It is not necessary to determine genotypic properties of the population because fitness is modeled in phenotypic space.

As an illustrative example of functional relationships, we have adapted standard equations from population genetics to the application of the conceptual fitness model for managing Pacific salmon fisheries (Kapuscinski and Lannan, 1984). Because a probability distribution is (partly) described by its mean and variance, we started with the equation of Kimura and Crow (1963) and Crow and Kimura (1970) which relates the variance effective number, $N_{e(v)}$, to the mean and variance of fitness. In this equation, fitness is represented by a diploid individual's contribution of gametes to the next generation. The equation was weighted for immigration by applying an Island Model (Wright, 1951). The resulting expression is

$$N_{e(v)} = \frac{N_{t-1}\bar{k}}{1 + \frac{V_k}{\bar{k}}(1-M)^2} \qquad (2)$$

where $(1-M)^2$ is the probability that neither of two genes uniting in a zygote has been exchanged for a migrant gene; N_{t-1} is the number of parents at breeding season $t-1$; and $\bar{k}$ and V_k are, respectively, the mean and variance of successful gametes (i.e., which survive to become parents in the next generation) per parent breeding at $t-1$.

This equation was corrected for age structure by incorporating terms expressing the average age of reproducing adults. We have subsequently derived an improved method of dealing with age structure by considering each age class as a cohort and computing a weighted sum over all cohorts to estimate $\bar{k}$ and V_k (Kapuscinski, 1984; A.R.D. Kapuscinski and J.E. Lannan, unpublished data, 1985). Thus

$$\bar{k} = \frac{\sum_x \bar{k}_x N_{x,t-1}}{N_{t-1}} \tag{3}$$

and

$$V_k = \frac{\sum_x s_{k,x}^2 N_{x,t-1}}{\sum_x N_{x,t-1}} = \frac{\sum_x s_{k,x}^2 N_{x,t-1}}{N_{t-1}} \tag{4}$$

where x is the age of reproduction; and s_k^2 is the variance of k.

These terms are now substituted into Eqn. 2. The final form of Eqn. 2 is not equivalent to the classical concept of $N_{e(v)}$ described in the literature of population genetics (Crow and Kimura, 1970). Instead, it provides a functional relationship between population variables and the probability distribution of fitness.

Because a broodstock's probability distribution of fitness is a function of $N_{e(v)}$, the effective number for the reference population and the broodstock of concern are equated to one another. Thus

$$N_{e(v)} = N'_{e(v)} \tag{5}$$

Finally, the expressions derived for $N_{e(v)}$ and $N'_{e(v)}$ can be substituted into Eqn. 5 and the resulting equation rearranged to solve for the required number of individuals in the broodstock in the breeding season of concern. The solution of the resulting equation requires the following data:

(1) estimates of stock abundance in appropriate breeding seasons;
(2) the proportion of reproducing adults in these breeding seasons;
(3) the number of migrants in the stock in the breeding season of concern;
(4) the age composition of reproducing adults in the appropriate breeding seasons; and
(5) estimates of age-specific variance of successful progeny in the breeding season of concern.

APPLICATION OF THE MODEL TO BROODSTOCK MANAGEMENT IN EXTENSIVE AQUACULTURE

Application of the conceptual model to real management situations would involve the following stepwise process.

(1) Estimate the mean and variance of fitness ($\bar{W}$ and V_w) for each reproductive age-class in the reference population. These may be estimates of the theoretical probability distribution inferred from the life-history patterns of the appropriate species or by empirical observation.

(2) Using the estimates of $\bar{W}$ and V_w for each age-class, integrate over all age-classes in the breeding population to generate a functional relationship between fitness and age at reproduction.

(3) From these data and estimates of stock abundance, estimate the numerical values for the input variables in the functional relationships (Eqns. 2–5).

(4) Solve for the number of parents in the breeding season of concern that are required to maintain the probability distribution of fitness.

If the age structure of the selected broodstock is identical to that of the reference population, the linear relationship for the reference population satisfies both sides of Eqn. 5. Thus, it can be used to estimate the broodstock size required to maintain the probability distribution of fitness.

If the age structure of the broodstock differs from that of the reference population, $N'_{e(v)}$ (Eqn. 5) must be recalculated to determine the unique linear relationship between the number of parents in the broodstock and the effective number. For example, consider a broodstock with predominantly young age structure and one with two reproductive ages missing. The unique linear relationship for each broodstock must be used to estimate the broodstock size required to maintain the probability distribution of fitness associated with the reference population.

The genetic fitness model can be applied to species and populations with different life histories. For example, consider a change from a Poisson to a Gamma distribution ($\sigma^2_{k,x} = \beta \bar{k}_x = 3\bar{k}_x$) of successful gametes per parent for each reproductive age of the fictitious population. This change is reflected in new values for the variance of successful gametes for each reproductive age. New values of $N_{e(v)}$ for the reference population and $N'_{e(v)}$ for different broodstock age structures are then used to estimate the broodstock size required to maintain the probability distribution of fitness associated with the reference population.

DISCUSSION

The predominant contemporary practice for determining the size of broodstocks for aquaculture is to control the rate of inbreeding at each generation to maintain heterozygosity and reduce inbreeding depression in cultured fish and shellfish (for example, see Kincaid, 1983). The number of parents required can be estimated from the equation for the rate of inbreeding and the associated inbreeding effective number (Falconer, 1981).

The inbreeding method of determining broodstock size has a number of serious limitations for extensive aquaculture. First, it assumes random pairing of the broodstock and equal fitness of all individuals in the broodstock, i.e., equal numbers of successful progeny (or gametes) per parent. Because of the latter assumption, the inbreeding model is insensitive to changes in population variables that influence the fitness of a broodstock.

Additionally, the breeding history of the broodstock must be known for the inbreeding method; otherwise the level of inbreeding (inbreeding coefficient) in the broodstock cannot be calculated. Finally, the allowable rate of change of inbreeding must be known to solve the inbreeding equation. For most species of fish and shellfish this value is not presently known.

A more important philosophical concern about the inbreeding model is that it is a genotypic model based on the probabilities that the two alleles at any locus are identical (descended from a common ancestor). However, fitness is a phenotype and as implied above, cannot be predicted in future time even if the entire genotype of all breeding individuals is known because of uncertainty about future environments. The maintenance of heterozygosity in the broodstock does not in itself insure the maintenance of fitness in the broodstock.

In contrast, our genetic fitness model is a phenotypic model that addresses the maintenance of phenotypic variation of traits influencing fitness. Therefore, the broodstock management objective for extensive aquaculture is the maintenance of the life-history variation found in the naturally occurring, reference population. Many naturally occurring broodstocks of commercially important species are known to exhibit substantial variation in life-history traits (e.g., Healey and Heard, 1984). Furthermore Saunders and Schom (1985) suggested that the maintenance of tremendous variation in life-history parameters may be the mechanism that allows small, wild broodstocks of Atlantic salmon (*Salmo salar*) to persist.

Application of our current equations for the fitness model to the management of real broodstocks depends upon the availability of estimates for the mean and variance of successful gametes per parent for each reproductive age-class. Because these estimates are unavailable for many extensively cultured species, we recommend research to develop (1) empirical estimates of successful gametes per parent or (2) indirect estimates from existing life-table data.

In the absence of numerical estimates for the equations of the fitness model, the conceptual model (Eqn. 1) can still be applied to extensive aquaculture by following these minimum guidelines for the makeup of the broodstock.

(1) Include all possible reproductive ages in the same relative proportions as they occurred in the reference population.

(2) Include the appropriate proportion of individuals from all portions of the frequency distribution observed in the reference population for any other trait known or believed to have a major influence on fitness. For example, all portions of the size distribution of females should be represented if fecundity is known to be a direct function of size.

(3) Use random mating procedures and equal numbers of males and females.

The conceptual fitness model is intended to insure the maintenance of a broodstock's fitness over a range of future environments and this implies the maintenance of the diversity of genetic information over time. The inbreeding method of broodstock management is based solely upon the maintenance of gene frequencies over time and consequently would present a greater risk of loss of fitness in a dynamic environment than a phenotypic fitness model. The latter provides a conservative approach to the management of broodstocks in extensive aquaculture by focusing on the maintenance of the phenotypic variability of traits influencing fitness.

REFERENCES

Crow, J.F. and Kimura, M., 1970. An Introduction to Population Genetics Theory. Harper and Row, New York, NY, 591 pp.

Falconer, D.S., 1981. An Introduction to Quantitative Genetics. Longman Group Limited, New York, NY, 340 pp.

Healey, M.C. and Heard, W.R., 1984. Inter- and intra-population variation in the fecundity of chinook salmon (*Oncorhynchus tshawytscha*) and its relevance to life history theory. Can. J. Fish. Aquat. Sci., 41: 476-483.

Kapuscinski, A.R.D., 1984. A Genetic Fitness Model for Fisheries Management. Ph.D. Thesis, Oregon State University, 128 pp.

Kapuscinski, A.R.D. and Lannan, J.E., 1984. Application of a conceptual fitness model for managing Pacific salmon fisheries. Aquaculture, 43: 135-146.

Kimura, M. and Crow, J.F., 1963. The measurement of effective population number. Evolution, 17: 279-288.

Kincaid, H.L., 1983. Inbreeding in fish populations used for aquaculture. Aquaculture, 33: 215-227.

Lewontin, R.C., 1970. The units of selection. Annu. Rev. Ecol. Syst., 1: 1-18.

Saunders, R.L. and Schom, C.B., 1985. Importance of the variation in life history parameters of Atlantic salmon (*Salmo salar*). Can. J. Fish. Aquat. Sci., 42: 615-618.

Wright, S., 1951. The genetical structure of populations. Ann. Eugen., 15: 323-354.

Duration of Feeding and Indirect Selection for Growth of Tilapia[1]

C.T. VILLEGAS and R.W. DOYLE*

Aquaculture Department, Southeast Asian Fisheries Development Center (SEAFDEC), Tigbauan (Philippines)
*Department of Biology, Dalhousie University, Halifax, N.S. B3H 4J1 (Canada)

ABSTRACT

Villegas, C.T. and Doyle, R.W., 1986. Duration of feeding and indirect selection for growth of tilapia. *Aquaculture*, 57: 89–92.

Duration of spontaneous feeding was observed at three times each day in a laboratory population of nine juvenile tilapia (*Oreochromis mossambicus*). Growth of the fish was measured as change in length and weight, and also as uptake of ^{14}C-labelled glycine by isolated scales. Duration of the first morning feeding was highly correlated with all measures of growth and was independent of initial size. Later feedings were not correlated with growth. Selection on feeding duration could be used to select indirectly for growth rate; this might be a valuable procedure where individuals in the population are not exactly the same age (i.e., where size-at-age is an inaccurate measure of growth).

INTRODUCTION

Growth in fish is a complex process affected by many behavioral, physiological, nutritional and environmental factors. In principle, it should be possible to achieve genetic improvement in growth indirectly, by selecting for one or more of these component traits or responses. Although indirect selection is usually less efficient than direct selection, the latter can be difficult to manage in practice. An indirect approach may be of great practical value under certain circumstances, as, for example, when the base population contains fish of several ages so a direct "select the biggest" approach cannot work.

In the present paper we show that at least one aspect of feeding behavior in a laboratory population of *Oreochromis mossambicus* is highly correlated with growth rate, but not with size. Indirect selection on this or similar traits might offer possibilities for genetic improvement of populations where the age distribution is under less-than-perfect control.

---

[1]Contribution No. 155 of the SEAFDEC Aquaculture Department.

TABLE 1

Experimental data on nine individually measured *O.mossambicus*. Initial weight (W), length (L), growth in weight (DW) and length (DL), glycine uptake (GLY) and duration of spontaneous morning, noon and evening feeding; glycine uptake expressed in cpm $\times 10^{-3}$

W	L	DW	DL	GLY	Feeding duration (min)		
					Morning	Noon	Night
4.01	49	0.84	3	8.80	11.5	16.7	9.3
4.43	48	0.09	2	10.30	6.1	1.5	2.0
4.49	50	0.24	1	5.09	11.4	18.3	16.4
7.83	59	−0.23	−1	4.10	3.9	3.1	2.6
9.24	62	1.93	5	19.30	15.1	7.7	7.1
9.99	61	1.81	5	9.57	13.1	6.0	4.3
10.65	64	−0.77	−1	3.59	5.2	6.1	7.7
11.15	62	0.47	3	9.13	14.6	10.3	5.0
11.28	68	0.55	3	8.90	10.2	6.6	3.0

METHODS

Oreochromis mossambicus juveniles used in this study came from a laboratory population maintained in the Department of Biology, Dalhousie University. Three fish out of a total of eight in each aquarium were differentially fin-clipped to permit identification of individuals during behavioral observations. The experiment took place in three 55-l aquaria at 27.5°C. Fish were fed pellets three times daily (morning, noon, late afternoon) throughout the 9-day observation period. Duration of spontaneous feeding and number of pellets eaten by each of the nine marked fish were recorded until the time of "satiation", defined as the voluntary cessation of feeding by the individuals under observation.

Growth was measured in three ways: change of length, change of weight, and uptake of ^{14}C-labelled glycine by isolated scales at the end of the experiment. The latter procedure was similar to that of Ottaway and Simkiss (1977). Scales were removed from the posterior part of the body above the lateral line and individually incubated in 0.4 ml fish saline containing 0.4 μCi/ml of ^{14}C-glycine at pH 7.9. After a 2-h incubation, the scales were washed, dried, weighed, and digested in 0.4 ml Protosol for 24 h prior to counting.

RESULTS AND DISCUSSION

Data on growth and feeding duration are given in Table 1. All three growth measures were highly correlated with the duration of feeding in the first feeding episode in the morning (Table 2) ($p<0.05$ for glycine uptake, $p<0.01$ for change in length and weight). The total number of pellets eaten by a fish during

TABLE 2

Simple and partial correlations of growth measures with feeding duration at three times of the day.
Partial correlations in parentheses: first two rows, constant initial weight; third row, constant initial length.

Growth measure	Correlation with Feeding Duration		
	Morning	Noon	Night
Weight change	0.80 (0.80)	0.14 (0.19)	−0.02 (0.03)
Glycine uptake	0.65 (0.64)	−0.09 (−0.05)	−0.16 (−0.14)
Length change	0.83 (0.83)	0.11 (0.17)	−0.13 (−0.08)

the 27 periods of observation was also highly correlated with the growth measures ($p < 0.01$); however, this is not a readily selectable trait. The duration of morning feeding was not correlated with initial length or weight, and the simple correlations approximately equalled the partial correlations with initial length and weight held constant. The non-parametric (Spearman) correlations corresponding to the significant correlations in Table 2 were also significant ($p < 0.05$).

The high correlation between growth and morning feeding may be the result of increased appetite and decreased "distractability" at that time. Several of our marked fish exhibited territorial behavior that interfered somewhat with feeding, but this was only obvious later in the day. Dr. A.D. Munro (personal communication 1985) has recently observed that the hormonal levels as well as the reproduction-related behavior of male tilapia has a 24-h periodicity and is minimal in the early morning. Morning feeding may therefore be relatively unaffected by reproductive behavior. If so, early morning feeding could be used as an indicator – or indirect selector – of capacity for growth after the onset of reproduction, when the actual growth of the faster-growing (and usually earlier-maturing) fish slows down.

Indirect selection on a trait can be useful when the heritability of the primary character (growth rate in this case) is relatively low and the genetic correlation of the traits is high (Falconer, 1981). Growth rate can be expected to have a low heritability in populations where age differences are confounded with growth rate differences (Doyle and Talbot, 1986a), and also where there is variation in maturity status. This will be especially true if there is size-dependent competition for resources among the fish (Moav and Wohlfarth, 1974; Doyle and Talbot, 1986b). The heritability of the duration of morning feeding may be less affected by these confounding effects.

We have not yet tested a feasible way of implementing indirect selection on feeding duration. A lift-net or box-trap under a feeder might work and be simple enough for artisanal use. Methodological experiments of this sort are underway at the Southeast Asian Fisheries Development Center in the Philippines; if they succeed, genetic investigations will follow.

ACKNOWLEDGEMENTS

This work was supported by a training grant to SEAFDEC by the International Development Research Centre (Canada). The assistance of T. Hay and A. Talbot is gratefully acknowledged.

REFERENCES

Doyle, R.W. and Talbot, A.J., 1986a. Effective population size and selection in variable aquaculture stocks. Aquaculture, 57: 27–35.
Doyle, R.W. and Talbot, A.J. 1986b. Artificial selection on growth and correlated selection on competitive behaviour in fish. Can. J. Fish. Aquat. Sci., 43: 1059–1064.
Falconer, D.S., 1981. Introduction to Quantitative Genetics, 2nd edition. Longman, New York, NY, 340 pp.
Moav, R. and Wohlfarth, G.W., 1974. Magnification through competition of genetic differences in yield capacity in carp. Heredity, 33: 181–202.
Ottaway, E.M. and Simkiss, K., 1977. "Instantaneous" growth rates of fish scales and their use in studies of fish production. J. Zool. (London), 181: 407–419.

Replicate Variance and the Choice of Selection Procedures for Tilapia (*Oreochromis niloticus*) Stock Improvement in Thailand

SUPATTRA URAIWAN and ROGER W. DOYLE*

National Inland Fisheries Institute, Bangkok (Thailand)
**Department of Biology, Dalhousie University, Halifax, N.S. B3H 4J1 (Canada)*

ABSTRACT

Uraiwan, S. and Doyle, R.W., 1986. Replicate variance and the choice of selection procedures for tilapia (*Oreochromis niloticus*) stock improvement in Thailand. *Aquaculture*, 57: 93–98.

Mass selection has a higher predicted response than within-family selection unless the intra-family correlation coefficient, t, exceeds 0.75 or selection intensities are higher within families. At the National Inland Fisheries Institute (NIFI) of Thailand, feasible selection intensities are higher using within-family selection. The within- and between-replicate variance of a large number of tilapia grow-out experiments in ponds and cages was analysed to aid the design of future selection experiments. Within-family selection is preferable at NIFI and probably other small-to-medium selection projects in Asia. Inbreeding and other practical considerations reinforce this view.

INTRODUCTION

Three types of selection are available for routine aquaculture stock improvement programs, namely, mass, family, and within-family selection. Mass selection is usually considered most effective when the heritability is reasonably high. When heritability is low either family or within-family selection may be more useful when non-genetic sources of sib covariation are low or high, respectively (Falconer, 1981). A combination of mass and family selection has also been recommended for selection in fish when heritability information is available (Kirpichnikov, 1981; Gall, 1983; Gjedrem, 1983). The objective of the present study was to decide which selection method is best for tilapia stock improvement at the National Inland Fisheries Institute (NIFI) of Thailand. The facilities at NIFI are typical of, or perhaps superior to, those generally available for genetic research in Southeast Asia.

PROCEDURE

Data from experiments on tilapia (*Oreochromis niloticus*) grown under various conditions at NIFI and Bangsai between 1982 and 1985 were available for analysis. The experiments had different patterns of replication, as described below. Analysis of variance was used to estimate intra-class correlation coefficients and replicate variances to be expected during selection. The experiments can be classified into the following three types.

Experiment type A. Members of each family were divided equally between two replicates, either cages or ponds. Each pair of replicates contained only one family. Between-replicate mean squares, within-replicate mean squares, intra-class correlations (t) and F-ratios were calculated separately for each sex in each experiment. The between-replicate variance includes components due to variation in the grow-out environment but not genetic or maternal effects. It is thus a minimal estimate of the spurious among-group variance to be expected in family selection, or in mass selection when more than one pond is involved.

The mean square between groups was estimated from the between-replicate mean squares as

$$\text{MSG} = \bar{n} \left(\frac{\bar{x}_1 - \bar{x}_2}{2} \right)^2$$

where $\bar{x}_1$, $\bar{x}_2$, are the replicate means and $\bar{n}$ is the mean sample size. Sample sizes ranged between 40 and 250 and were similar but not usually identical for replicate pairs.

The added variance component between groups was estimated as

$$\text{MSA} = \frac{\text{MSG} - \text{MSW}}{\bar{n}}$$

where MSW is the within-replicate or error variance. The intra-class correlation coefficient, t, was estimated as

$t = \text{MSA}/(\text{MSA} + \text{MSW})$

Experiment type B. In this type of experiment, full-sib families were grown separately in ponds or cages without replication. This is the procedure most likely to be followed in a between- or within-family selection program. Variance components were calculated by conventional one-way ANOVA. The added variance component among groups contains genetic and maternal as well as common environment effects.

Experiment type C. In these experiments, many families were spawned "simul-

TABLE 1

Probability of the F-ratio for added variance between replicates in 93 Type A experiments; the table contains the number of observations at the indicated probability level

Experimental conditions	Probability level		
	$P > 0.05$	$0.05 > P > 0.001$	$P > 0.001$
Males reared separately	1	1	1
Females reared separately	4	0	0
Males reared mixed sex	8	3	8
Females reared mixed sex	8	1	8
Juveniles (pond)	0	0	
Sexes not distinguished (pond)	3	3	15
Sexes not distinguished (cage)	20	2	5
Totals	44 (47%)	10 (11%)	39 (42%)

taneously" — actually within 1 week — and grown together in a common pond. This is a typical procedure for mass selection programs involving tilapia or carp.

RESULTS

The probability of the F-ratio (between-group/within-group) observed in 93 replicate comparisons in Type A experiments is given in Table 1. Added variance due to cage or pond effects occurs very frequently, with 53% of the ratios having a statistical significance $P > 0.05$. Values for t are also shown in Fig. 1 for those Type A experiments for which the calculation was appropriate. The range of estimates is large, with 30% of the observations exceeding 0.1.

Fewer experiments of Type B were performed (five only) but there was a minimum of five groups in each experiment rather than two as in Type A. The t values (shaded blocks, Fig. 1) fall towards the high end of the range, as would be expected because of the extra maternal and genetic variance among groups. The median value of t was 0.38 for this type of experiment with a high of 0.61.

Only one Type C experiment was available for analysis at the time of writing. The first section of Table 2 contain an ANOVA for this experiment which involved five ponds, each with 25 different, full-sib families spawned within a week. The second section is an ANOVA for a contemporaneous Type B experiment with five ponds containing one family each.

The third section in Table 2 is a "synthetic" ANOVA for a hypothetical experiment on families grown together in a single pond as if the families were individually marked. The ANOVA was constructed by setting the total variance in such a pond equal to the within-group (pond) variance in the first

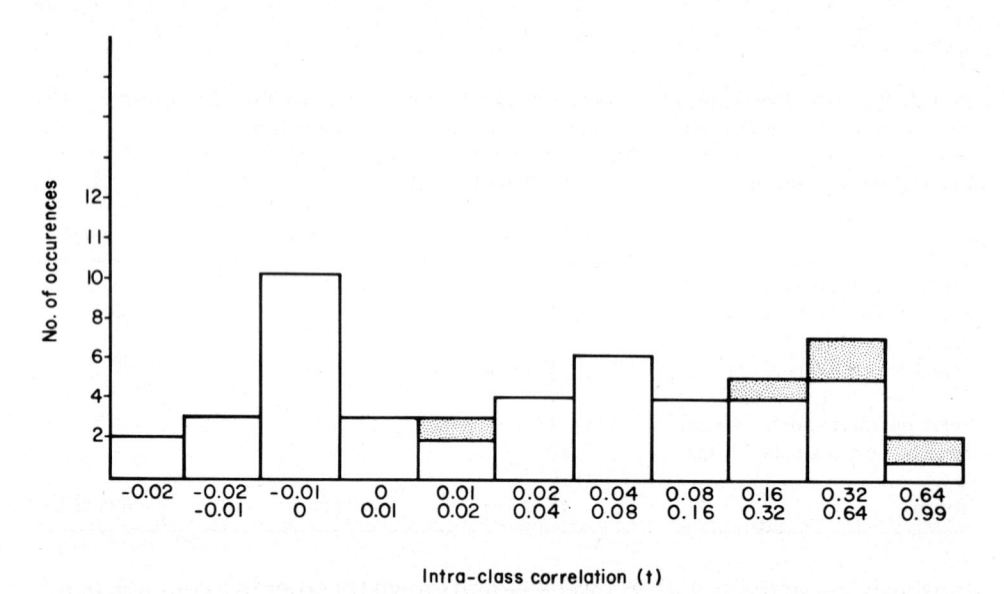

Fig. 1. Estimated intra-class correlation coefficients (t) among replicate ponds and cages. Open blocks: full-sib *O. niloticus* families split among replicates; shaded blocks: different full-sib family in each replicate.

ANOVA, and setting the within-family variance equal to the within-group variance in the second ANOVA. The calculation is outlined in the legend of Table 2. The resulting estimate of t is very large, 0.79.

TABLE 2

Analysis of variance for the Type C and contemporaneous Type B experiment
The values in the third section were calculated as follows, in order of calculation: (a) $99 \times 0.369 = 36.5$; (b) $75 \times 0.0789 = 5.92$; (c) $36.5 - 5.92 = 30.6$; (d) assumes $n = 4$ individuals measured per family

Source	d.f.	SS	MS	t
Among ponds (25 families/pond)	4	5.4	1.37	0.026
Within ponds	495	182.7	0.369	
Among ponds (one family/pond)	4	50.6	12.6	0.61
Within ponds	495	39.1	0.0789	
Among families in one pond	24	30.6(c)	1.28	0.79
Within families	75(d)	5.92(b)	0.0789	
Total	99	36.5(a)	0.369	

DISCUSSION

We have used the above results to decide which of the three principal selection schemes is likely to be most successful at NIFI.

Family selection

When t is less than 0.25 family selection is generally superior unless other factors intervene (Falconer, 1981). A realistic appraisal of the facilities at NIFI suggests that family selection could only be conducted at about 1/3 the intensity of mass selection because of the large number of ponds required, the difficulty of obtaining simultaneous spawning, and a wish to maintain at least 20 families to minimize inbreeding. The corresponding cut-off for t is approximately 0.1. Half of the Type A and four out of five Type B estimates lie above this value. We conclude that family selection is not a promising technique at NIFI with current cage and pond culture techniques.

Mass vs. within-family selection

The relative efficiencies of these two selection procedures is also affected by the relative selection intensities that can be routinely attained. *Oreochromis niloticus* can produce several hundred offspring per brood so selection within families is not limited by fecundity. (We propose to cull no more than 95% of the population to avoid random loss of favorable alleles in the early stages of selection.) A more important limiting factor for mass selection, is the number of families that can be spawned at the same time to produce an equal-aged cohort. A relatively low culling threshold is needed to achieve sufficient spawning redundancy. Within-family selection does not require spawning synchrony and we estimate that at NIFI, selection can be at least 50—100% stronger within families than with mass selection. The corresponding cut-off points for t are 0.44 and zero, respectively.

Two of the five estimates of t in the Type B experiment exceed 0.44, as does the estimate conjectured in the third ANOVA of Table 2. Only the latter estimate includes the added variance caused by competition among families that would be expected in a mass-selection environment. There is reason to believe that non-genetic covariation of full-sibs would be greater when families are reared together than when reared in separate ponds, because of competitive magnification of variation in maternal effects and spawning dates (Moav and Wohlfarth, 1974; Doyle and Talbot, 1986). The high estimate of t in the third section of Table 2 is consistent with the existence of such effects.

Within-family selection has the genetic advantage of minimizing inbreeding and random loss of alleles at any given population size. It also has the practical advantage of equalizing work load and pond utilization throughout the year since synchronous spawning is not required. Our statistical analysis suggests that it is as efficient, and possibly more efficient, than alternative procedures. We conclude that within-family selection is likely to be the most effective selection technique at NIFI and perhaps other research institutions in Southeast Asia that use similar spawning and grow-out procedures.

ACKNOWLEDGEMENTS

This work was supported by the International Development Research Centre (Canada) and by the Royal Thai Department of Fisheries. Thanks are extended to the staff of the Aquaculture Unit at the National Inland Fisheries Institute and in particular to the director of the Institute, Dr. Thiraphan Bhukaswan.

REFERENCES

Doyle, R.W. and Talbot, A.J., 1986. Artificial selection on growth and correlated selection on competitive behaviour in fish. Can. J. Fish. Aquat. Sci., 43: 1059–1064.
Falconer, D.S., 1981. Introduction to Quantitative Genetics, 2nd edition. Longman, New York, NY, 340 pp.
Gall, G.A., 1983. Genetics of fish: a summary of discussion. Aquaculture, 33: 383—394.
Gjedrem, T., 1983. Genetic variation in quantitative traits and selective breeding in fish and shellfish. Aquaculture, 33: 51—71.
Kirpichnikov, V.S., 1981. Genetic Bases of Fish Selection (G. Gause, transl.). Springer-Verlag, New York, NY, 410 pp.
Moav, R. and Wohlfarth, G.W., 1974. Magnification through competition of genetic differences in yield capacity in carp. Heredity, 33: 181–202.

Genetic Aspects of Hatchery Rearing of the Scallop, *Pecten maximus* (L.)

A.R. BEAUMONT

Marine Science Labs., University College of N. Wales, Menai Bridge, Anglesey, Gwynedd LL59 5EH (Great Britain)

ABSTRACT

Beaumont, A.R., 1986. Genetic aspects of hatchery rearing of the scallop, *Pecten maximus* (L.). *Aquaculture*, 57: 99–110.

The scallop, *Pecten maximus*, is a bivalve of commercial importance in Europe with considerable potential for aquaculture. Existing data are reviewed with a view to hatchery culture, and new data on the induction of triploidy in this species are presented. Electrophoretic surveys of scallops show them to be genetically very variable (72% loci polymorphic, mean heterozygosity 0.215), and this natural variation is a valuable feature for hatchery broodstock. Considerable variation in yield and normality of the veliger larval stage has been evident in laboratory rearing trials over a long period and the genetic and environmental causes of these variations are discussed. Self-fertilisation occurs commonly in laboratory spawnings, and although there is considerable evidence that self-fertilised larvae do not grow well, the potential exists for the rapid production of homozygous, inbred lines which would be a valuable source of broodstock for hybridisation. Cytochalasin B was used to induce triploidy in *P. maximus* larvae and the most effective dose was 0.5 mg l^{-1} added 10 min after fertilisation for a period of 20 min. The treatment was observed to interfere with meiosis I spindle formation, and about 30% of embryos were estimated to be triploid. A simple method for assessing the degree of triploidy in embryos based on polar body counts is suggested as an alternative to chromosome counting.

INTRODUCTION

The great scallop, *Pecten maximus*, is a common epi-faunal bivalve, distributed from Northern Norway to the Iberian Peninsula, which lives partly recessed into sandy or gravelly substrates down to 100 m depth. This species can occur in very large numbers in dense 'beds' and these beds have been traditionally fished, by dredging, in European waters. The increasing development of the aquaculture of oysters and clams in Europe has led many workers to consider *P. maximus* as another suitable commercial bivalve for culture. *Patinopecten yessoensis*, the Yezo scallop, has been cultured for many years in Japan by collecting natural spatfall and growing the scallops in hanging cul-

TABLE 1

Estimates of polymorphism and heterozygosity in scallops and other bivalves
P=proportion of loci polymorphic. (A locus is designated polymorphic if the commonest allele is present at a frequency <0.99.) $\bar{H}_0$=mean per-locus heterozygosity

Species	P	$\bar{H}_o \pm$ SE	Author
Argopecten irradians	0.33	0.116±0.069	Wall et al., 1976
Patinopecten yessoensis	0.33	0.094±0.006	Nikiforov and Dolganov, 1982.
Pecten maximus	0.722	0.215±0.064	Beaumont and Beveridge, 1984
Chlamys opercularis	0.591	0.155±0.054	
C. varia	0.723	0.284±0.075	
C. distorta	0.727	0.321±0.081	
Mytilus edulis	0.380	0.095±0.036	Ahmad et al., 1977.
Tridacna maxima	0.79	0.209±0.027	Campbell et al., 1975.
Crassostrea spp.	0.444	0.147±0.018	Buroker, 1980
Saccostrea spp.	0.479	0.184±0.003	
Tiostrea sp.	0.224	0.044±0.023	Buroker et al., 1983.

ture (Ventilla, 1982). However, most spat collecting trials conducted in European waters have failed to produce commercial quantities of *P. maximus* (Brand et al., 1980).

The alternative to the collection of natural spat is, of course, to rear the larval stage in hatcheries and a pilot scale technique for such rearing of *P. maximus* has been successfully developed (Gruffydd and Beaumont, 1972; Beaumont et al., 1982). Rearing has recently been conducted on a commercial scale in France (Gerard, 1985). In the light of these developments and the continuing interest in scallop culture, this paper reviews the relevant genetic data so far published concerning *P. maximus* and presents new data on the induction of triploidy in this species using cytochalasin B.

GENETIC VARIATION IN NATURAL POPULATIONS

The use of electrophoresis to identify and quantify genetic variation at many protein and enzyme loci in animals and plants has enabled geneticists to assess the relative genetic variability of different organisms or groups of organisms. Table 1 lists levels of polymorphism (P) and heterozygosity ($\bar{H}_o$) for a range of bivalves including both commercial and non-commercial species. It can be seen that with 72% of loci polymorphic and a mean observed heterozygosity of 0.215, *P. maximus* ranks genetically as a highly variable bivalve. The relatively high genetic variability of the Pectinidae as a group is also evident. Although the Pectinidae as a whole exhibit high values of P and $\bar{H}_o$, individual species such as *Argopecten irradians* and *Patinopecten yessoensis* have low values (Wall et al., 1976; Nikiforov and Dolganov, 1982). This may be partly due to the inclusion in these studies of a high proportion of non-enzymatic or general protein loci, which, in general, show less variation than enzymatic loci (Smith and Fujio, 1982).

The high natural genetic variation exhibited by *P. maximus* should have immediate value in hatchery production of scallops. When only small numbers of highly variable individuals are spawned and cross-fertilised, the chances are high that the cohorts of larvae derived from such fertilisations will exhibit high genetic variation. This wide variation in genotypes is expected to enable individuals to adapt to, and make the most of, a greater range of environmental conditions during growth. However, in practice there is little hard evidence demonstrating this principle, and certain bivalves, such as mussels which have a very low level of genetic variability (Table 1; Ahmad et al., 1977), are clearly able to colonise and withstand a very wide range of environmental conditions. Several models involving environmental variation have been proposed to account for these variations in heterozygosity both within and between groups of animals but no single model has yet come near to explaining the data fully (Buroker, 1980; Nelson and Hedgecock, 1980; Smith and Fujio, 1982).

GENETIC VARIATION IN LARVAL DEVELOPMENT

In comparison to the straightforward identification of genetic variation in single gene characters such as enzymes, assessment of genetic variability associated with polygenic characters such as larval growth or viability is considerably more difficult. During early years of rearing *P. maximus* larvae, it was noted that there was unexplained variation in the proportion of eggs which developed through the trochophore stage into normal veliger largae ('D' larvae) (Gruffydd and Beaumont, 1970, 1972). Larval growth rates were also very variable . Although larval growth rates have improved slowly and variability reduced over the years, considerable variation is still observed in the percentage of normal larvae produced under laboratory conditions. An opportunity arose, during experiments on self-fertilisation, to investigate the background genetic component of the variance of several features of early larval life in exclusive cross-fertilisations (Beaumont and Budd, 1983). Data were collected on the percentage of eggs fertilised, the percentage of abnormal larvae produced, the shell length of early and late veligers and the mortality of veligers after 18 days. The mean values (± standard error) of these variables are given in Table 2 for the experiment which was a 2×2 factorial mating between wild caught individuals where self-fertilisation was known to be minimal. Eggs were fertilised and larvae were reared in identical conditions in all cultures so that the variation observed should be a measure of genetic rather than environmental variation. The results were analysed using a two-way model II analysis of variance and the F values and their significance are presented in Table 3.

TABLE 2

Pecten maximus, 2×2 factorial mating
Values are means (±SE) of measurements of yield, abnormality, shell length of veligers at day 3 and day 18 and mortality of veligers at day 18 (from Beaumont and Budd, 1983)

Eggs from scallop	1		2	
Sperm from scallop	3	4	3	4
% Eggs fertilised	23.47±0.87	21.13±1.44	18.60±0.70	16.53±0.68
% Larvae abnormal	13.00±1.12	28.57±1.05	8.77±2.03	8.50±0.80
Veliger shell length day 3 (μm)	118.74±0.34	116.81±0.31	117.46±0.28	115.70±0.29
Veliger shell length day 18 (μm)	167.08±1.04	163.35±1.02	168.75±0.81	171.02±0.88
Mortality day 18 (% empty shells)	0.37±0.23	5.33±1.27	1.52±0.48	2.82±0.68

Both eggs and sperm independently affected the percentage of eggs which produced larvae. Eggs from scallop 1 produced more larvae than eggs from scallop 2, but spermatozoa from scallop 3 were slightly more effective than those from scallop 4 in fertilising eggs regardless of the source of the eggs (Table 2). By

TABLE 3

Pecten maximus, 2×2 factorial mating, two-way model II analysis of variance of genetic variation during larval development (Beaumont and Budd, 1983)

Factor measured	Source of variation	F-ratio	Significance
% Eggs fertilised	Eggs	26.25	**
	Sperm	5.67	*
	Interaction	0.02	NS
% Larvae abnormal	Eggs	2.36	NS
	Sperm	0.93	NS
	Interaction	35.20	**
Veliger shell length day 3	Eggs	1.67	NS
	Sperm	5.00	NS
	Interaction	6.28	*
Veliger shell length day 18	Eggs	2.56	NS
	Sperm	0.06	NS
	Interaction	18.79	**
Mortality day 18	Eggs	0.14	NS
	Sperm	2.92	NS
	Interaction	5.76	*

NS = Not significant; * $P < 0.05$; ** $P < 0.01$.

contrast the percentage of abnormal larvae varied considerably between cultures and significant interactive effects were evident. Eggs from scallop 1 produced 13% or 29% abnormal larvae depending on the source of spermatozoa, while eggs from scallop 2 produced only around 9% abnormal larvae irrespective of the source of spermatozoa. Much less extensive, but nevertheless significant, interactive effects are also evident in measurements of larval growth and mortality. These results demonstrate that natural genetic variation between scallops can be sufficient to produce very large variations in the normality of larvae produced in cross-fertilisations and that quite high levels of larval abnormalities may occasionally be inevitable in hatchery production of scallops. Although these data indicate a large genetic component in the production of abnormal larvae, other trials (Beaumont, unpublished data) have demonstrated that low levels of copper (20 ± 5 μg l^{-1}) in the seawater used for egg incubation will cause the failure of all eggs to develop into shelled larvae even though the percentage of eggs fertilised is relatively unaffected by such low levels of copper. Extensive variations in the normality of hatchery produced larvae may, therefore, not be entirely genetic in origin and considerably more research is needed into the abnormalities which occur during this important period of larval development.

SELF-FERTILISATION

P. maximus is hermaphroditic and, in laboratory spawnings, some self-fertilisation is the rule rather than the exception. The principal genetic consequence of self-fertilisation in natural populations is the reduction in heterozygosity which occurs over all genetic loci. Reduction in heterozygosity is expected to reduce viability and growth rate and we have evidence from two sources that growth rate of larvae, at least, is reduced by self-fertilisation in *P. maximus*. Firstly, in a rearing experiment involving eight families of larvae, two derived from self-fertilisations and six from cross-fertilisations, larval growth rate was significantly depressed in the selfed families compared to the crossed families. Data re-analysed from Beaumont and Budd (1983), on mean yield, normality, growth and mortality of larvae from selfed and crossed families are shown in Table 4. Interestingly, it is only the growth of veliger larvae which is affected significantly by self-fertilisation. Larval mortality and abnormalities did not differ significantly between crossed and selfed families. Secondly, data from 72 rearings of *P. maximus* larvae carried out between 1969 and 1977 were divided into those larvae derived from exclusive self-fertilisations and larvae derived from spawnings where only a proportion of the eggs were self-fertilised. These data (Table 5) suggest that larvae from exclusive self-fertilisations grow slower, and metamorphose less readily than other larvae. Although these results must be treated with some caution (Beaumont and

TABLE 4

Comparison between P. maximus larval cultures derived from cross- or self-fertilisation (values are means ± standard error)

	Self-fertilised	Cross-fertilised	Student's 't'
% Fertilisation	20.24 ± 4.77	20.56 ± 1.32	0.09
% Abnormality	14.77 ± 0.21	14.34 ± 3.72	0.06
Veliger shell length day 3 (μm)	117.39 ± 0.21	117.38 ± 0.46	0.01
Veliger shell length day 18 (μm)	152.24 ± 1.77	167.42 ± 1.18	6.56*
Mortality day 18 (% empty shells)	2.85 ± 0.18	2.45 ± 0.86	0.25

*$P < 0.001$.

Budd, 1983) the trend towards reduced growth of larvae derived from self-fertilisations is clear.

There is one possible important use to which self-fertilisation in P. maximus can be put, and that is the production of inbred, homozygous broodstocks (Wilkins, 1981). Crossing between such inbred stocks will produce highly heterozygous, hybrid offspring which should exhibit higher viability and growth rates than non-hybridised offspring. It has yet to be established whether self-fertilised, inbred scallops can be reared to maturity, but as techniques improve this may become a possibility and could provide a very valuable tool in hatchery production of scallops.

HETEROZYGOSITY AND GROWTH

It is well documented that the high overall heterozygosity produced by hybridisation can increase viability and growth rates in domesticated animals and plants (Lerner, 1954). Some recent attempts to demonstrate this principle using

TABLE 5

Pecten maximus Comparison between mean growth rates and success to metamorphosis of larvae derived from exclusive self-fertilisation and all other cultures

Type of fertilisation	Number of cultures	Mean growth rate (μm/day ± SE)	% of cultures reaching metamorphosis
Exclusively self-fertilised	19	2.8 ± 0.3	10
Others	53	3.5 ± 0.2	32

electrophoretic analysis of several polymorphic enzyme loci in cohorts of bivalves have indeed shown a positive correlation between multi-locus heterozygosity and growth rate (Zouros et al., 1980, 1983; Fujio, 1982; Green et al., 1983; Garton et al., 1984; Koehn and Gaffney, 1984). However, other studies, using single families or small numbers of progenitors, have reported no similar correlation (Beaumont et al., 1983, 1985; Adamkewicz et al., 1984; Gaffney and Scott, 1984), possibly due to the reduced overall genetic variation in such studies. Although many of the bivalve species so far studied have shown a positive correlation between growth rate and multi-locus heterozygosity, studies on Pectinids have not found such a relationship (Foltz and Zouros, 1984; Beaumont et al., 1985). Extensive discussion of this topic is outside the scope of this paper and has been covered in other publications (Beaumont et al., 1983, 1985; Gaffney and Scott, 1984; Mitton and Grant, 1984). It is, however, important to point out that failure to find a positive relationship between growth rate and heterozygosity at a small number of enzyme loci in scallops does not, in any significant way, alter the principle that overall polygenic heterozygosity should be positively correlated with growth and viability in these species.

INDUCTION OF TRIPLOIDY

Treatment of newly fertilised eggs with the antibiotic, cytochalasin B, is known to produce triploidy in the American oyster, *Crassostrea virginica* (Stanley et al., 1981), the soft shelled clam, *Mya arenaria* (Allen et al., 1982) and the bay scallop, *Argopecten irradians* (Tabarini, 1984). Cytochalasin B appears to inhibit micro-filament formation in cells (Maclean-Fletcher and Pollard, 1980) and thus, prevents spindle formation while allowing chromosome replication. If this agent is allowed to act upon fertilised eggs at the appropriate time, and at the appropriate concentration, it can prevent the separation of the chromosomes during meiosis I. Subsequent removal of cytochalasin B allows normal development of the meiosis II spindle to proceed, but each chromosomal complement of the nuclei resulting from meiosis II will be diploid instead of haploid. One of these nuclei is the second polar body while the other is the female pronucleus which, when united with a male pronucleus, forms a triploid first cleavage nucleus.

Adult triploid oysters and scallops have been shown to exhibit higher meat weights than their diploid siblings due either to increased heterozygosity (Stanley et al., 1984) or to their failure to develop gonads (Tabarini, 1984). The advantages of producing shellfish which put much of their potential reproductive energy into growth are clear. I report here on a preliminary trial designed to induce triploidy in *P. maximus* using cytochalasin B.

MATERIALS AND METHODS

P. maximus eggs were obtained using standard techniques (Beaumont et al., 1982) and were allowed to stand for 1 h to assess the degree of self-fertilisation by examining for presence of polar bodies. Approximately one-quarter of the eggs proved to be self-fertilised. Spermatozoa from other scallops were then mixed with the eggs to give a ratio of 20 spermatozoa per egg and, 10 min after the addition of spermatozoa, the eggs were divided and placed into eight 1-l crystallising dishes with filtered seawater, giving about 10^5 eggs per dish. Two dishes were each treated with 0.1 mg/l cytochalasin B, introduced in the carrier dimethyl sulphoxide (DMSO) which had a final concentration of 0.01%. Two further dishes were treated with 0.5 mg/l cytochalasin B in 0.02% DMSO, two were treated with 0.01% DMSO alone, and two remained untreated as controls. After 20 min, eggs from the two cytochalasin treatments were drained gently onto a 20-μm sieve and placed in 1 l of filtered seawater containing 0.01% DMSO in order to remove all traces of cytochalasin B. After a further 30 min, eggs from all dishes except the controls were gently sieved out and placed in filtered seawater. Developing embryos were then left undisturbed for 3 days, by which time normal 'D' larvae had developed. Samples were taken for chromosome analysis at the end of the 20-min cytochalasin B treatment period, and again at 3 h after fertilisation. Eggs and embryos were preserved in Carnoy's fixative and were prepared and stained in a similar manner to that described by Beaumont and Gruffydd (1974). Larvae from all dishes were examined during the third day of development and assessments were made of the percentage of eggs which had developed into larvae and the proportion of those larvae that were normal.

RESULTS AND DISCUSSION

Examination of samples taken at the end of the cytochalasin B treatment period revealed that many eggs contained a single cluster of approximately 38 chromosomes and no visible spindle. In control eggs however, meiosis I had just been completed and most eggs contained a first polar body and a spread of approximately 19 chromosomes arranged on the metaphase plate of the meiosis II spindle. The haploid chromosome number of *P. maximus* is 19 (Beaumont and Gruffydd, 1974), thus the early developmental changes expected to be induced by cytochalasin B were apparent.

Estimates of triploidy in cytochalasin B treated eggs were obtained in two ways. Firstly, counts were made of chromosome numbers in spreads where most, or all of the chromosomes were clearly visible. This is a laborious and exacting task and, because of this, an alternative method for measuring triploidy is presented here. Because the cytochalasin B treatment prevents the formation of the first polar body, but not necessarily the production of the second

TABLE 6

Pecten maximus Estimates of triploidy caused by the use of cytochalasin B
Figures in parentheses represent % of triploidy when unclassified embryos are assumed to be diploid

(1) Chromosome counts

Concentration of cytochalasin B (mg/l)	Approx. $2n$	Approx. $3n$	% Triploids
0.1	13	2	13
0.5	12	5	29

(2) Polar body counts

Concentration of cytochalasin B (mg/l)	2 pb	1 pb	Unclassified	% Triploids
0.1	27	8	11	23 (17)
0.5	21	12	7	36 (30)

polar body, estimates of the number of embryos with one, or two polar bodies should indicate the degree of possible triploidy. Certainly, eggs which possess two clear polar bodies cannot have contributed an extra haploid chromosome complement to the first cleavage nucleus. Three hours after fertilisation the polar bodies are still closely associated with the embryo which is at around the 8 to 16 cell stage, and it is generally easy to see whether one or both polar bodies are present. The first polar body is extruded in a small sphere and consists of a tightly packed nucleus, while the second polar body appears direcly below the first and has a much less dense nucleus. Occasional embryos cannot be classified due to the presence of other nuclear material such as undeveloped supernumerary spermatozoa, or other nuclei in the embryo, associated closely with the polar bodies. The data from both methods of assessment of triploidy are given in Table 6, and, if unclassified embryos are included as non-triploids, the estimates agree reasonably well. Between 13% and 17% of the embryos were triploids in the 0.1 mg/l cytochalasin B treatment and 29% to 30% in the 0.5 mg/l cytochalasin B treatment. These values will be an underestimate because 25% of the eggs had already been self-fertilised before the cytochalasin treatments were begun.

Previous workers have used chromosome counting, fluorometric flow cytometry or relative staining intensity after electrophoresis to identify the degree of triploidy (Stanley et al., 1981, 1984; Allen et al., 1982; Tabarini, 1984). Chromosome counting is notoriously time consuming and especially difficult in species with high chromosome numbers, and the other two methods require that putative triploids are reared for several months before estimates can be made. A simple method based on polar body counts could be a valuable technique for rapidly estimating the degree of triploidy within a few hours of treatment and therefore deserves further study.

Analysis of the yield and normality of the larvae produced at day 3 was performed; data for both levels of cytochalasin B treatment were combined because no differences in yield or normality were found between the two treatments. Cytochalasin B significantly reduced the numbers of larvae produced (mean $12.75 \pm 1.00\%$). However, the use of cytochalasin B at either concentration did not produce significantly more abnormal larvae than the controls. Surprisingly, the DMSO treated larvae had significantly higher normality (mean $48.67 \pm 5.25\%$) than the controls (mean $19.87 \pm 3.53\%$) and this may be a result of the altered chemistry of the seawater in the DMSO treated cultures. As pointed out earlier, at least one heavy metal ion, copper, when present at low concentrations, can influence the percentage of normal larvae produced. It is possible that DMSO may have caused a subtle change in the bioavailability of copper or other seawater constituents.

The preliminary data on triploid induction in *P. maximus* indicate that up to one third of the eggs may be induced to form triploid embryos using 0.5 mg/l cytochalasin B for a 20-min period beginning 10 min after fertilisation, and that this treatment, while reducing the total number of larvae produced, does not significantly increase larval abnormalities.

ACKNOWLEDGEMENTS

I am grateful to Miss J.L. Greene for assistance with the triploidy trial and to M.D. Budd and B. Roberts for conditioning scallops.

REFERENCES

Adamkewicz, S.L., Taub, S.R. and Wall, J.R., 1984. Genetics of the clam *Mercenaria mercenaria*. I. Mendelian inheritance of allozyme variation. Biochem. Genet., 22: 215–219.

Ahmad, M., Skibinski, D.O.F. and Beardmore, J.A., 1977. An estimate of the amount of genetic variation in the common mussel *Mytilus edulis*. Biochem. Genet., 15: 833–846.

Allen, S.K., Gagnon, P.S. and Hidu, H., 1982. Induced triploidy in soft-shell clam (*Mya arenaria*): cytogenic and allozymic confirmation. J. Hered., 73: 421–428.

Beaumont, A.R. and Beveridge, C.M., 1984. Electrophoretic survey of genetic variation in *Pecten maximus*, *Chlamys opercularis*, *C. varia* and *C. distorta* from the Irish Sea. Mar. Biol., 81: 299–306.

Beaumont, A.R., Beveridge, C.M. and Budd, M.D., 1983. Selection and heterozygosity within single families of the mussel, *Mytilus edulis* (L.). Mar. Biol. Lett., 4: 151–161.

Beaumont, A.R. and Budd, M.D., 1983. Effects of self-fertilisation and other factors on the early development of the scallop *Pecten maximus*. Mar. Biol., 76: 285–289.

Beaumont, A.R. and Gruffydd, Ll.D., 1974. Studies on the chromosomes of the scallop *Pecten maximus* (L.) and related species. J. Mar. Biol. Assoc. U.K., 54: 713–718.

Beaumont, A.R., Budd, M.D. and Gruffydd, Ll.D., 1982. The culture of scallop larvae. Fish Farming Int., 9: 10–11.

Beaumont, A.R., Gosling, E.M., Beveridge, C.M., Budd, M.D. and Burnell, G.M., 1985. Studies on heterozygosity and growth rate in the scallop, *Pecten maximus* (L.). In: P.E. Gibbs (Editor), Proc. 19th Eur. Mar. Biol. Symp. Plymouth, U.K., 1984. Cambridge University Press, pp. 443–455.

Brand, A.R., Paul, J.D. and Hoogesteger, J.N., 1980. Spat settlement of the scallops *Chlamys opercularis* (L.) and *Pecten maximus* (L.) on artificial collectors. J. Mar. Biol. Assoc. U.K., 60: 379–390.

Buroker, N.E., 1980. An examination of the trophic resource stability theory using oyster species of the family Ostreidae. Evolution, 34: 204–207.

Buroker, N.E., Chanley, P., Cranfield, H.J. and Dinamani, P., 1983. Systematic status of two oyster populations of the genus *Tiostrea* from New Zealand and Chile. Mar. Biol., 77: 191–200.

Campbell, C.A., Valentine, J.W. and Ayala, F.J., 1975. High genetic variability in a population of *Tridacna maxima* from the Great Barrier Reef. Mar. Biol., 33: 341–345.

Foltz, D.W. and Zouros, E., 1984. Enzyme heterozygosity in the scallop *Placopecten magellanicus* (Gmelin) in relation to age and size. Mar. Biol. Lett., 5: 255–263.

Fujio, Y., 1982. A correlation of heterozygosity with growth rate in the Pacific oyster, *Crassostrea gigas.* Tohoku J. Agric. Res., 33:66–75.

Gaffney, P.M. and Scott, T.M., 1984. Genetic heterozygosity and production traits in natural and hatchery populations of bivalves. Aquaculture, 42: 289–302.

Garton, D.W., Koehn, R.K. and Scott, T.M., 1984. Multiple-locus heterozygosity and the physiological energetics of growth in the coot clam, *Mulinia lateralis,* from a natural population. Genetics, 108: 445–455.

Gerard, A., 1985. Observations sur la sensibilité des élevages larvaires et postlarvaires de *Pecten maximus* á la température. 5th Int. Pectinid Workshop, La Coruna, Spain. Mimeo, 5 pp.

Green, R.H., Singh, S.M., Hicks, B. and McCuaig, J., 1983. An intertidal population of *Macoma balthica* (Mollusca, Pelecypoda): genotypic and phenotypic components of population structure. Can. J. Fish. Aquat. Sci., 40: 1360–1371.

Gruffydd, Ll.D. and Beaumont, A.R., 1970. Determination of the optimum concentration of eggs and spermatozoa for the production of normal larvae in *Pecten maximus* (Mollusca, Lamellibranchia). Helgol. Wiss. Meeresunters., 20: 486–497.

Gruffydd, Ll.D. and Beaumont, A.R., 1972. A method for rearing *Pecten maximus* in the laboratory. Mar. Biol., 15: 350–355.

Koehn, R.K. and Gaffney, P.M., 1984. Genetic heterozygosity and growth rate in *Mytilus edulis.* Mar. Biol., 82: 1–7.

Lerner, I.M., 1954. Genetic Homeostasis. Oliver and Boyd, Edinburgh, London, 134 pp.

Maclean-Fletcher, S. and Pollard, T., 1980. Mechanism of action of cytochalasin B on actin. Cell, 20: 329–341.

Mitton, J.B. and Grant, M.C., 1984. Associations among protein heterozygosity, growth rate and developmental homeostasis. Annu. Rev. Ecol. Syst., 15: 479–499.

Nelson, K. and Hedgecock, D., 1980. Enzyme polymorphism and adaptive strategy in the decapod Crustacea. Am. Nat., 116: 238–280.

Nikiforov, S.M. and Dolganov, S.M., 1982. Genetic variation of the Japanese scallop *Patinopecten yessoensis* from the Vostok Bay, Sea of Japan. Biol. Morya Vladivostok, 2: 46–50 (English translation).

Smith, P.J. and Fujio, Y., 1982. Genetic variation in marine teleosts: high variability in habitat specialists and low variability in habitat generalists. Mar. Biol., 69: 7–20.

Stanley, J.G., Allen, S.K. and Hidu, H., 1981. Polyploidy induced in the American oyster, *Crassostrea virginica* with cytochalasin B. Aquaculture, 23: 1–10.

Stanley, J.G., Hidu, H. and Allen, S.K., 1984. Growth of American oysters increased by polyploidy induced by blocking meiosis I but not meiosis II. Aquaculture, 37: 147–155.

Tabarini, C.L., 1984. Induced triploidy in the bay scallop, *Argopecten irradians*, and its effect on growth and gametogenesis. Aquaculture, 42: 151–160.

Ventilla, R.F., 1982. The scallop industry in Japan. Adv. Mar. Biol., 20: 309–382.

Wall, J.R., Wall, S.R. and Castagna, M., 1976. Enzyme polymorphisms and genetic variation in the bay scallop, *Argopecten irradians*. Genetics, 83: (3: part 1) Suppl., p. 81.

Wilkins, N.P., 1981. The rationale and relevance of genetics in aquaculture: an overview. Aquaculture, 22: 209–228.

Zouros, E., Singh, S.M. and Miles, H.E., 1980. Growth rate in oysters—an overdominant phenotype and its possible explanations. Evolution, 34: 856–867.

Zouros, E., Singh, S.M., Foltz, D.W. and Mallet, A.L., 1983. Post-settlement viability in the American oyster (*Crassostrea virginica*): an overdominant phenotype. Genet. Res. Camb., 41: 259–270.

Controlled Mating of the European Oyster, *Ostrea edulis*

GARY F. NEWKIRK

Biology Department, Dalhousie University, Halifax, N.S. B3H 4J1 (Canada)

ABSTRACT

Newkirk, G.F., 1986. Controlled mating of the European oyster, *Ostrea edulis. Aquaculture*, 57: 111–116.

Full-sib families were produced of the European oyster, *Ostrea edulis*. This is the first recorded success at controlled mating in the genus *Ostrea*. Matings were made of selected stock from Nova Scotia and a recently imported stock from Maine producing pure and hybrid families. The spawning rate differed in the 2 years of observation (70% and 40%). There were also significant differences between the stocks in spawning rate. The Maine stock had a higher rate each year. However, there was no apparent difference between the stocks in growth rate to 2 years of age. Thus, there was little evidence of heterosis in the hybrids.

INTRODUCTION

There are no reports in the literature of the controlled spawning of *Ostrea* species although some hatchery managers (e.g., C. Heinig, personal communication, 1982) have used elevated temperatures to stimulate groups of *Ostrea edulis* to initiate spawning and DiSalvo et al. (1983) have used the technique with *Ostrea chilensis*. Unlike *Crassostrea* species, the females of the genus *Ostrea* brood the eggs and resulting larvae and release free-swimming larvae after considerable development has occurred. Without control of spawning the production of single pair matings is difficult. Thus, commercial hatcheries and researchers have used mass spawnings (Walne, 1974, Newkirk and Haley, 1982, 1983). The selection program at Dalhousie University which has been underway since 1977 has been utilizing mass spawning of the oyster, *Ostrea edulis*.

Results of mass-spawned lines produced in 1980 (G.F. Newkirk, unpublished data, 1980) indicated that the stock of *O. edulis* imported from Maine in 1978 was superior to the second-generation selected stock which had been in Nova Scotia since 1973. Aside from the desire to maintain the lines with as much pedigree control as possible, the ability to produce single pair matings is necessary to determine if there would be any heterotic effects.

During the early years of our selection program we experimented with single pair matings by isolating two oysters in separate containers prior to spawning. This was done without biopsy to determine the sex of the individuals. Initial trials gave successful spawnings of almost 50% which is what one would expect as the frequency of pairs containing both sexes. Thus, a program was started to produce single pair matings of *O. edulis*.

The results of two year classes of single pair matings of the oyster, *O. edulis* are presented here. The crosses include pure crosses of the Maine stock and the Nova Scotia selected stock (both first- and second-generation selected individuals) and hybrid crosses between the Maine stock and the Nova Scotia stocks.

MATERIALS AND METHODS

Parental oysters for the pairs produced in 1981 were taken from the 1977 and 1978 selected lines of the Nova Scotia stock (Newkirk and Haley, 1983), and the offspring of the oysters imported from Maine, U.S.A. There were 10 pairs set up to spawn of the Nova Scotia pure crosses, 10 pairs of the Maine pure crosses and 10 pairs of the hybrid Nova Scotia × Maine. The handling of the oysters prior to spawning was the same as in Newkirk and Haley (1983) except that after 4 weeks at an elevated temperature the pairs were isolated in 8-l buckets. The water in the buckets was changed daily. In every case the first spawning was at least 2 weeks after the isolation of the oysters by pairs. Assuming that there was no sperm release prior to pair isolation and no sperm storage (for which there is no known mechanism) each brood of larvae was considered to be a full-sib family.

In 1983 the same procedure was used except the parents of both stocks came from the 1980 year class. The Nova Scotia stock in this case was selected offspring of second-generation selected oysters. The Maine stock was selected for the first time from the offspring of unselected parents. In 1983 there were three sets of spawnings with variable numbers of each cross.

In 1981 the offspring were not put out in the field until the end of July or the beginning of August. The mean size at the end of the second growing season was below expectation because of the late start in 1981. Thus, the data reported here are from the end of the third growing season, November 1983.

The 1983 offspring were out in the field in late June and early July. Consequently, the mean size at the end of the second season, November 1984, was over 20 g and these data are reported here.

Since the spawning in 1983 was in three sets (blocks), there were differences in the time of spawning. In both years there were differences in the location of the nets in which the oysters were held. Thus, a standard score for each individual weight has been calculated as

$SSW_{ij} = (X_{ij} - X_{i.})/S_{i.}$

TABLE 1

Number of pairs of oysters spawning in 1981 and 1983 (percentage in parentheses)

Year		Cross		
		Nova Scotia	Hybrid	Maine
1981	Number of pairs	10	10	10
	Spawn larvae	5(50)	8(80)	8(80)
	Not spawning	5(50)	2(20)	2(20)
1983	Number of pairs	32	41	33
	Spawn eggs (only)	8(25)	4(10)	2(6)
	Spawn larvae	5(16)	20(49)	17(52)
	Not spawning	19(59)	17(41)	14(42)

where SSW_{ij} is the standard score for live weight, X_{ij} is the weight of the jth individual in the ith net and X_i, and S_i, are the mean and standard deviation of the oysters in the ith net.

An adjusted weight has also been calculated as

$$ADJWT = (SSW_{ij} \times S..) + X..$$

where $X..$ and $S..$ are the overall pooled (unweighted) mean and standard deviation.

RESULTS

The reproductive performance of the oyster crosses was evaluated using larval production in 1981 and egg and larval production in 1983. The numbers of pairs that actually produced broods of larvae are broken down by cross type in Table 1. In 1981 the Nova Scotia pure crosses produced fewer broods of larvae but the difference was not significant. The hybrid and pure Maine crosses both produced more broods of larvae than the Nova Scotia crosses in 1983 (G-test, $P<0.01$). Of the 20 hybrid crosses in 1983 producing larvae, only five (25%) were from Nova Scotia female parents.

The correlation between the number of larvae produced and the live weight of the female was very small in the 1983 families ($r = -0.08$, $n = 28$). However, the range in size of the females was not very great.

In 1983, seven pairs had repeat spawning with each individual spawning as a female. Of these seven pairs, five were pure Maine crosses, one was a hybrid and one was pure Nova Scotia.

The number of pairs producing eggs, but not followed by a larval release, is indicated in Table 1. This was most common among the Nova Scotia pure

TABLE 2

Mean weight and mean standard score (SSW) for offspring of the families produced in 1981 and weighed in November 1983

Cross	Family	No.	Mean wt. (g)	SSW
Nova Scotia	202	3	15.5	−0.79
	203	51	13.5	−0.21
	205	2	12.8	−0.38
	208	10	13.7	−0.34
	Total	66	Weighted mean	−0.26
Hybrid	211	61	19.5	0.04
	212	119	18.1	0.09
	213	40	13.2	−0.19
	214	78	29.4	0.38
	216	1	19.8	−0.15
	218	166	20.4	0.00
	219	141	20.7	0.14
	220	65	18.6	−0.02
	Total	671	Weighted mean	0.08
Maine	221	56	15.5	−0.13
	222	132	16.3	−0.32
	224	93	20.5	−0.02
	226	51	16.6	−0.33
	227	96	19.4	−0.15
	228	7	22.9	0.09
	229	90	25.5	0.23
	230	152	22.3	0.08
	Total	677	Weighted mean	−0.07

crosses, less so among the hybrids and rarest in the pure Maine crosses. There was significant heterogeneity ($G=14.0$, $P<0.01$) among crosses.

We were able to compare the growth of the 1981 offspring by the type of cross (Table 2). Of the four pure Nova Scotia crosses that had offspring survive through three growing seasons, two of the families had very few individuals (two and three). The family means of the standard scores for weight for these crosses were consistently negative with a weighted average of -0.263. If the two families with two and three animals are deleted, the weighted mean is -0.234. The mean for the hybrid crosses was slightly higher than that of the pure Maine crosses. The pure Nova Scotia crosses were smaller than the other two crosses but in the analysis of variance of this set of families, the effect of cross type was large but not significant ($F=2.74$, d.f.$=2, 17$, $P=0.09$).

In the 1983 crosses there was only one pure Nova Scotia cross that had animals survive through two growing seasons. There were again a reasonably large number of hybrid (13) and pure Maine (12) crosses with large family sizes. Although the adjusted mean weight of the pure Maine crosses (22.5 g) was slightly larger than the hybrids (21.2 g) the difference was not significant.

DISCUSSION

In this work we have been able to produce a number of single pair matings of the oyster, *Ostrea edulis*. This is of considerable importance for the control of breeding programs and genetic studies of this species. The families produced are full-sib families. In a few cases the individuals of a pair changed sex and successfully produced a second brood. These would be reciprocal crosses between the individuals. There were too few of these cases to warrant a detailed genetic analysis but this does provide a breeding scheme which is somewhat unique among animals in that the sex change occurs within a few weeks.

There was no significant heterosis in growth rate in the hybrid crosses between the two strains. However, this may have gone undetected. In 1981, there were only 10 matings attempted for each cross. Of these only two of the Nova Scotia pure stock matings were really successful in producing surviving offspring. Of the 1983 matings, only one out of 32 attempted Nova Scotia pure stock matings produced offspring surviving to 2 years. Thus, there was very little on which to evaluate growth of this cross type.

The fact that the Nova Scotia pure stock matings produced so few successful broods was in itself an indication of differences between the stocks. In both years, the hybrid crosses were as successful as the Maine pure stock crosses. It is impossible to determine what component of reproductive performance was affected. There were more cases of eggs being produced without being followed by larval release in the Nova Scotia pure crosses. This could be due to the production of inviable eggs, infertile sperm or a failure of the male to spawn. The fact that in most of the successful hybrid matings it was the Maine stock animal which spawned as a female and, of the seven repeat spawnings, five were pairs of Maine stock suggests that the problem with the Nova Scotia matings was with female fertility.

There was a higher rate of spawning in the 1981 set of matings. This could be due to environmental differences associated with different years or due to the use of advanced generations in 1983. The results of Newkirk and Haley (1983) and G.F. Newkirk (unpublished data, 1984) suggest that there is evidence of inbreeding depression in the advanced selected generations of the Nova Scotia stock. The selected Nova Scotia oysters used in 1983 are one more generation advanced from those used in the work just cited. Since matings prior to this work were by mass spawning the inbreeding coefficient cannot be calculated.

In summary, there was little indication of heterosis in growth of *Ostrea edulis* in the full-sib families produced by crossing two stocks which had been reproductively isolated for almost 40 years. However, there is an indication of differences in reproductive output which may be the result of inbreeding in the Nova Scotia stock.

ACKNOWLEDGEMENTS

This work was supported by grants from the National Sciences and Engineering Research Council.

REFERENCES

DiSalvo, L.H., Alarcon, E. and Martinez, E., 1983. Induced spat production from *Ostrea chilensis* Philippi 1845 in mid-winter. Aquaculture, 30: 357—362.

Newkirk, G.F. and Haley, L.E., 1982. Progress in selection for growth rate in the European oyster, *Ostrea edulis*. Mar. Biol. Prog . Ser., 10: 77—79.

Newkirk , G.F. and Haley, L.E., 1983. Selection for growth rate in The European oyster, *Ostrea edulis*: response of second generation groups. Aquaculture, 33 149—155.

Walne, P.R., 1974. Culture of Bivalve Molluscs, 50 Years Experience at Conwy. Fishing News (Books) Ltd., Surrey, Great Britain, 173 pp.

Calluses, Cells, and Protoplasts in Studies Towards Genetic Improvement of Seaweeds

MIRIAM POLNE-FULLER and AHARON GIBOR

Biological Sciences and Marine Science Institute, University of California Santa Barbara, Santa Barbara, CA 93106 (U.S.A.)

ABSTRACT

Polne-Fuller, M. and Gibor, A., 1986. Calluses, cells, and protoplasts in studies towards genetic improvement of seaweeds. *Aquaculture*, 57: 117-123.

We are developing techniques for tissue, cell, and protoplast culture to be used in the conversion of wild seaweeds into cultivated crop plants with desirable properties. Seaweeds have complex life histories and do not have the equivalent of "seeds" capable of extended survival in storage. Our efforts concentrate on three major aspects: (1) development of techniques for vegetative propagation from isolated cells and protoplasts, (2) development of procedures for maintenance of tissues as seed stock, and (3) production of improved strains by selection of mutants and hybridization through protoplast fusion. In this paper we report the state of our advances in the vegetative propagation of representatives of the three groups of seaweeds from isolated cells. We also report on efforts in selection for disease-resistant strains of *Porphyra yesoensis* (Nori), a marine crop plant.

INTRODUCTION

During the last 50 years agricultural productivity in the U.S.A. has more than doubled while the amount of land under cultivation has actually decreased. This increase in productivity is largely attributed to newly developed strains selected and bred from naturally occurring varieties. However, it is predicted that farmers may not be able to obtain much more of a rise in productivity, because the number of varieties obtained by cross-breeding is approaching its limit (Board on Agriculture, NRC, 1984). It is expected that new techniques, specifically the use of tissue and cell culture, in combination with induced mutations, somatic hybridization and genetic engineering will create a promising future for further improvements.

Limited experience with seaweed farming has taught us that ocean plants, just like land plants, express a range of genetically inherited traits such as growth rates and product quality (Chapman and Doyle, 1979; Waaland, 1979;

Hansen, 1984). Recent pioneering research in seaweed genetics has involved the study of whole plants which were repeatedly selected for specific morphologies, fast growth rates and high iodine content (Fang and Dai, 1980; Brinkhuis et al., 1984; Hanisak and Ryther, 1984; Van der Meer et al., 1984). The prospect for continued selection of improved seaweed strains is promising, but it is still limited by the qualitative potential of the combinations of traits which both exist in the natural population and can be selected for.

It is expected that tissue, cell and protoplast technology in combination with methods in genetic engineering, will become a means of enhancing algal strain improvement. Such techniques will produce new mutations, and create new desirable combinations of complementing traits as is presently done in transformations of yeast and bacterial cells. In this paper we summarize the results of recently developed techniques for the isolation and culture of vegetative cells of seaweeds. These techniques allow us to study isolated cells of complex seaweed tissues with and without their walls.

MATERIAL AND METHODS

Plants were collected locally or shipped by air while wrapped in moist newspapers, sealed in plastic bags, and packed in styrofoam containers. Healthy segments were selected and cleaned (Gibor et al., 1981), then incubated in enzyme solutions or cultured for induction of calluses. Resulting dissociated cells, protoplasts or developing calluses were collected and cultured (Polne-Fuller and Gibor, 1984a,b; Saga et al., 1986).

Infected plants were shipped to us from the Nori farming project, in Puget Sound, Washington, by Dr. T.R. Mumford, for the purpose of selecting disease-resistant mutants. Heavily damaged areas of tissue where all but a few cells were dead, were collected. The tissue was: soaked for 7 min in 5% Betadine prepared in fresh water; sonicated for 2 min with rinses in sterile seawater every 0.5 min; and incubated in an antibiotic mixture (Polne et al., 1980; Polne-Fuller and Gibor, 1986) for 5 days. The few living cells were hand-picked with capillary glass tubes, and cultured on agar surfaces to induce callus growth (Polne-Fuller and Gibor, 1984a,b). The resulting calluses were enzymatically treated (as above), and single cells and small clumps of cells cultured on agar surfaces on which the disease-causing fungus, *Pythium*, was cultured. All cultures were kept at $18°C$, $30 \mu E\ m^{-2}\ s^{-1}$, and 12/12 L/D cycle.

RESULTS

We now know that viable calluses, single cells, and protoplasts can be induced and isolated from representatives of all three seaweed groups, although not reliably.

Callus formation

Calluses were induced from the following sources: protoplasts of *Ulva lactuca* and *Enteromorpha linza*, single cells and protoplasts of four species of *Porphyra* (Polne-Fuller and Gibor, 1984a), sections of blades, vesicles and stipes of *Macrocystis pyrifera*, whole germlings, sectioned stipes and vesicles of *Sargassum muticum*, *S. filipendula*, *S. natans*, *S. hystrix* and *S. sinicola*.

In all cases calluses developed from sterile pieces of tissue which were cultured on agar surfaces at the interface between the agar and the air (Fig. 1a,b). In a single case calluses developed in a liquid culture of young germlings of *Sargassum muticum* which had been starved for 4 months at 18°C, 15 μE m^{-2} s^{-1}, 12/12 D/L cycle.

Callus growth and regeneration

The calluses of both of the green algae, all five *Porphyra* species, and *S. muticum* (which developed from germlings) were able to grow and regenerate when they were removed from their support tissue. Under the appropriate culture conditions cells isolated from these calluses regenerated into plants of normal appearance (Fig. 1f). Calluses from other spp. of *Sargassum* grew well on their source tissue but failed to divide when transferred onto various liquid or solidified agar media. The *Macrocystis* calluses divided relatively well as long as they were attached to the source tissue, but developed into a filamentous growth when removed and cultured independently.

Single cells and protoplasts

Single cells and protoplasts were successfully isolated from *Ulva lactuca*, *Enteromorpha linza*, the five species of *Porphyra*, *S. muticum*, *S. hystrix* and *Macrocystis pyrifera*. In all cases the cells were healthy looking, excluded Evans blue dye, and surface cells which contained pigments fluoresced well. However, only single cells and protoplasts of the green algae and the *Porphyra* species regenerated into fully developed plants (Fig. 1e). Protoplasts of the *Sargassum* species divided three to five times producing a small mound of cells which then stopped dividing. *Macrocystis* protoplasts have not divided at the time of writing.

Selection of mutants

Single cells and protoplasts of *Enteromorpha* and *Porphyra* responded to mutagenic agents such as UV radiation and ethyl methyl sulphanate. Clones were selected on high sugar medium, which normal cells cannot survive on,

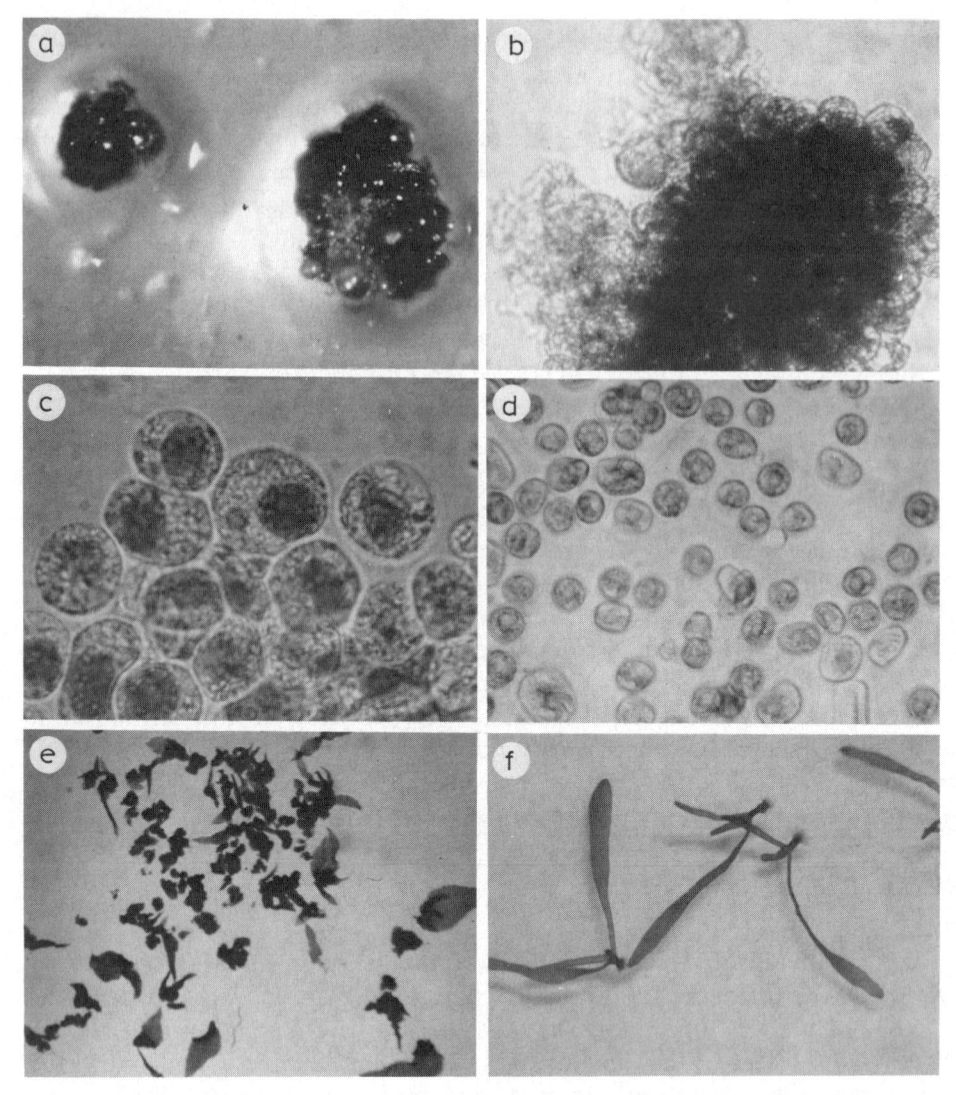

Fig. 1. Calluses, protoplasts and regeneration of plantlets in brown and red seaweeds.
a, Callus of *Porphyra perforata*, developed from an isolated protoplast cultured on agar. Diameter 700 μm.
b, Callus of *Sargassum muticum*, developed from sectioned stipe cultured on agar. Long axis = 500 μm.
c, Protoplasts of *Porphyra lanceolata* after centrifugation. Cell diameter 40—55 μm.
d, Protoplasts of *Sargassum muticum* isolated from surface tissue. Cell diameter 25—30 μm.
e, Young plants of *Porphyra yesoensis* regenerated from protoplasts. Longest plant 2 mm.
f, Young *Sargassum muticum* plants regenerated from cells of calluses which developed in liquid culture on the surface of young plants. Largest plant 3 cm.

and on light and temperature gradients. Survivors are being cultured and studied.

Isolation of disease resistant mutants

The healthy cells isolated from badly damaged tissue of diseased *Porphyra yesoensis* developed into calluses. The calluses were enzymatically dissociated and plated on agar in which an infectious fungus, *Pythium*, was growing. A second generation of healthy, normal looking calluses developed in the mixed culture.

Somatic hybridization

In preliminary tests protoplasts of *Sargassum muticum* were observed to spontaneously merge their cell membranes. Moreover, protoplasts of unrelated seaweed species, *Porphyra yesoensis* and *Enteromorpha linza*, fused somatically during exposure to PEG (Saga et al., 1986).

DISCUSSION

Much work is required to improve the reliability of the presently used methods for the isolation and regeneration of seaweed protoplasts and cells. However, available publications from independent research groups (Millner et al., 1979; Fries, 1980; Tang, 1982; Cheney, 1984; Polne-Fuller and Gibor, 1984a,b; Yan, 1984; Saga et al., 1986) indicate that culturing tissues, calluses, single cells and protoplasts of seaweeds is a feasible task.

We are presently facing several important obstacles which must be addressed methodically.

(1) Insufficient knowledge of the composition and structure of seaweed cell walls and their variability both within the plant and during different seasons hampers the search for specific enzymes which will attack these walls reliably.

(2) The crude enzyme preparations which are presently used with partial success are extracted from guts of abalone or limpit snails. The animals are collected in the field, and their gut enzyme composition may vary depending on the season, the physiological state of the organism, and its diet. A search for micro-organisms which will produce efficient wall-degrading enzyme(s) is in progress (Preston et al., 1985).

The usefulness of dissociated tissues for mutant selection and vegetative propagation seems obvious. We cannot claim positive isolation of disease-resistant clones since the *Pythium* culture used for infection was isolated by Japanese scientists as a pathogen of a Japanese strain of *Porphyra*. The Japanese fungal strain may not be identical with that infecting the Washington fields. It is also possible that callus cells are more resistant to fungal infections than actively dividing thallus cells. Further experiments will be carried out to answer these questions.

ACKNOWLEDGEMENTS

This work was supported in part by a grant from the Gas Research Institute, Chicago, IL, and in part by NOAA—National Sea Grant College Program, grant #NA80AA-D-00120, through the California Sea Grant Office.

REFERENCES

Board on Agriculture, National Research Council, 1984. Genetic Engineering of Plants. National Academy Press, Washington D.C., 83 pp.
Brinkhuis, B.H., Mariani, E.C., Breda, V.A. and Brady-Campbell, M.M., 1984. Cultivation of *Laminaria saccharina* in the New York Marine Biomass Program. Hydrobiologia, 116/117: 266—271.
Chapman, A.R.O. and Doyle, R.W., 1979. Genetic analysis of alginate content in *Laminaria longicrueis*. In: A. Jensen and J.R. Stein (Editors), Proc. 9th Int. Seaweed Symp. 1977, Santa Barbara, CA. Science Press, Princeton, pp. 125-132.
Cheney, D.P., 1984. Genetic modifications in seaweeds: applications to commercial utilization and cultivation. In: R.R. Colwell, E.R. Pariser and A.J. Sinskey (Editors), Biotechnology in the Marine Sciences, Proc. MIT Sea Grant Conference. Wiley, New York, NY, pp. 161-175.
Fang, T.C. and Dai, J.X., 1980. The use of haploid phases in the genetic study of *Laminaria japonica*. Acta Genet. Sin., 7: 19—25 (in Chinese with English abstract).
Fries, L., 1980. Axenic tissue culture from the sporophytes of *Laminaria hyperborea*. J. Phycol., 16: 475—477.
Gibor, A., Polne, M., Biniaminov, M. and Neushul, M., 1981. Exploratory studies of vegetative propagation of marine algae: procedure for obtaining axenic tissue. In: T. Levering (Editor), Proc. 10th Int. Seaweed Symp. 1980, Sweden. Walter de Gruyter Publ., pp. 587-593.
Hanisak, M.D. and Ryther, J.H., 1984. Cultivation biology of *Gracilaria tikvahiae* in the United States. Hydrobiologia, 116/117: 295—298.
Hansen, J.E., 1984. Strain selection and physiology in the development of *Gracilaria* mariculture. Hydrobiologia, 116/117: 89—94.
Millner, P., Callow, M. and Evans, L., 1979. Preparation of protoplasts from the green alga *Enteromorpha intestinalis*. Planta, 14: 174—177.
Polne, M., Gibor, A. and Neushul, M., 1980. The use of ultrasound for the removal of macro-algal epiphytes. Bot. Mar., 23: 731-734.
Polne-Fuller, M. and Gibor, A., 1984a. Developmental studies in *Porpyra*. I. Blade differentiation in *Porphyra perforata* as expressed by morphology, enzymatic digestion, and protoplast regeneration. J. Phycol., 20: 609—616.
Polne-Fuller, M. and Gibor, A., 1984b. Vegetative propagation of *Porphyra perforata*. Hydrobiologia, 116/117: 308—313.
Polne-Fuller, M. and Gibor, A., 1986. Tissue culture of seaweeds. In: K. Bird and P.H. Benson (Editors), Seaweed Cultivation for Renewable Resources. Gas Research Inst., Chicago (in press).
Preston, F.J., Romeo, T., Bromely, J.C., Robi, R.W. and Aldrich, H.C., 1985. Alginate lyase secreting bacteria associated with the algal genus *Sargassum*. Dev. Ind. Microbiol., 26: 727—740.

Saga, N., Polne-Fuller, M. and Gibor, A., 1986. Protoplasts from seaweeds: production and fusion. In: W.R. Barclay and R.P. McIntosh (Editors), Algal Biomass Technologies. Nova Hedwigia, 83: 37—43.

Tang, Yanlin, 1982. Isolation and cultivation of the vegetative cells and protoplasts of *Porphyra suborbiculata*. J. Shandong College of Oceano, 22 (4).

Van der Meer, J.P., Patwary, M.U. and Bird, C.J., 1984. Genetics of *Gracilaria tikvahiae* (Rhodophyceae). X. Studies on a bisexual clone. J. Phycol., 20: 42—46.

Waaland, J.R., 1979. Growth and strain selection in *Gigartina exasperata*. In: A. Jensen and J.R. Stein (Editors), Proc. 9th Int. Seaweed Symp. 1977, Santa Barbara, CA. Science Press, Princeton, pp. 241-247.

Yan, Zuo-mei, 1984. Studies on tissue culture of *Laminaria japonica* and *Undaria pinnatifida*. Hydrobiologia, 116/117: 324—316.

Genetic and Environmental Components of Variation for Growth of Juvenile Atlantic Salmon (*Salmo salar*)

JOHN K. BAILEY and ERIC J. LOUDENSLAGER[1]

Salmon Genetics Research Program, Atlantic Salmon Federation, St. Andrews, N.B. E0G 2X0 (Canada)

[1]Present address: Humboldt State University, Department of Fisheries, College of Natural Resources, Arcata, CA 95521 (U.S.A.)

ABSTRACT

Bailey, J.K. and Loudenslager, E.J., 1986. Genetic and environmental components of variation for growth of juvenile Atlantic salmon *(Salmo salar). Aquaculture*, 57: 125-132.

Full- and half-sib families were produced by hierarchical matings within three stocks in 1978 and two stocks in 1979 of Atlantic salmon, *Salmo salar*. There were no significant differences in length or weight at 3, 6 or 15 months post-hatch among the stocks examined under standardized hatchery conditions. Half-sib heritability estimates for length and weight were low to intermediate (0.1 - 0.4) in 1978 and intermediate to high (0.4 - 0.9) in 1979. Differences among stocks and levels of domestication between years are suggested as possible explanations for such variation. Genetic, environmental and phenotypic correlations between length and weight at the three sampling times were positive and high (0.8 - 1.0). The results suggest that relatively rapid genetic gain for these traits is possible if selection intensities are high.

INTRODUCTION

Atlantic salmon are grown for both aquaculture and enhancement purposes. Regardless of their ultimate use, cultured salmon spend some time during development in a hatchery. The use of selection to improve traits which affect hatchery performance is a way of increasing efficiency (Saunders and Bailey, 1978). To design a breeding program the heritabilities, phenotypic variances, economic values and maternal influences for all traits under selection and the genetic and phenotypic correlations between traits should be known. However, few such parameters are available and comparisons among studies are difficult because there has been a lack of standardization among methods used in their determination (Gjedrem, 1983). Kirpichnikov (1981) provides a recent summary of parameter estimates for Atlantic salmon and other fish species. In this paper, the genetic and environmental sources of variation affecting juvenile growth of Atlantic salmon are estimated.

MATERIALS AND METHODS

Atlantic salmon from five New Brunswick stocks were spawned in 1978 and 1979. Three stocks were involved in the 1978 spawning, wild Saint John River stock and two stocks, Waweig and SRC_1, which were recaptures from earlier sea ranching studies. In 1979 two stocks were used, wild Big Salmon River and SRC_2 stocks. The SRC stocks are synthetics derived from complete 4×4 diallel sets involving four stocks (Friars et al., 1979). Subscripts indicate that SRC_1 and SRC_2 were produced in different years. The SRC stocks were dominated by grilse of pure Big Salmon parentage or crosses which involved one Big Salmon parent (Bailey and Saunders, 1984). In both years approximately 70% of the SRC gene pools descended from Big Salmon stock. Spawning was performed during late autumn. All matings were within stocks and some spawning days were cross-classified with stocks (Table 1). Within a year and stock, the mating design was hierarchical and each full-sib family was produced from approximately 1500 fertilized eggs.

In 1978 each full-sib family was divided among two or more hatchery tanks (1.1 m). In 1979 each family was assigned to a single tank. Apart from fertilizing a constant number of eggs (1500), no attempt was made to standardize the number of fish per group. At about 6 months post-hatch the largest 1000 parr (selected by visual inspection) in each full-sib family were batch-marked with coded micro-tags. The 1979 families were also given unique brands. The parr were then transferred to the rearing tanks (8 m), five families to a single

TABLE 1

Summary of numbers of spawning days, sires and dams used in each stock and year

Year	Strain	Spawn days	Sires	Dams
1978	Waweig	4	14	36
	Saint John	4	12	21
	SRC_1	6	17	29
	Pooled	13[a]	43	86
1979	SRC_2	9	54	133
	Big Salmon	2	10	21
	Pooled	9[a]	64	154

[a]Not equal to column totals because some spawning days were cross-classified with stocks.

tank. Family identity was lost in the 1978 group at transfer and that year's experiment was terminated.

Random samples of 25–40 fish per family were taken at 12 weeks, and 6 months post-hatch. A 15-month sample was included for 1979. All groups were sampled, tagged and transferred in the same order that they were fertilized. Each fish was weighed and its fork length measured, except that in the 1978 twelve-week sample, fry were weighed only. The same individuals were not measured during each sample. The 6-month sample actually spanned a two and 1 half month period because each family was measured just before it was marked.

STATISTICAL MODELS

Separate analyses of variance (ANOVAs) were performed for each year. The following model was used for the analysis of the 1978 data

$$Y_{ijklmn} = u + r_i + t_j + rt_{ij} + s_{ijk} + d_{ijkl} + z_{ijklm} + e_{ijklmn}$$

where, Y_{ijklmn} is length or weight of an individual; u is the population mean; r_i is the fixed effect of the ith stock; t_j is the fixed effect of the jth spawning day; rt_{ij} is the fixed effect of the ij^{th} stock by spawning day interaction; s_{ijk} is the random effect of the kth sire nested within the ith stock and spawning day; d_{ijkl} is the random effect of the lth dam nested within the ijk^{th} sire, stock and spawning day; z_{ijklm} is the random effect of the mth replicate nested within the $ijkl^{th}$ dam, sire, stock and spawning day; and e_{ijklmn} is the random error among individual fish in a replicate ($e \sim N(0, I\sigma_e^2)$).

The 1979 statistical model was similar to that of 1978, but the replication term was eliminated and effects due to a common environment were confounded with dam effects. While there were differences among measurement days for the 6-month sample, any variability was confounded with the spawn day effect. In addition, a rearing tank effect was omitted for the 15-month sample because it was partially confounded with spawn day and its inclusion in the model would have resulted in severely disconnected data.

Variance components were determined for all ANOVAs using maximum likelihood procedures (Hemmerle and Hartley, 1973), and half- and full-sib heritabilities were estimated for each year. Length—weight covariance components were estimated (Searle and Rounsaville, 1974) in order to determine phenotypic, genetic and environmental correlations between traits (Becker, 1984). To estimate maternal effects the following general equation was used

$$m^2 = (\sigma_d^2 - \sigma_s^2)/\sigma_t^2$$

where $\sigma_t^2 = \sigma_s^2 + \sigma_d^2 + \sigma_z^2 + \sigma_e^2$ (in 1978) or $\sigma_s^2 + \sigma_d^2 + \sigma_e^2$ (in 1979).

RESULTS AND DISCUSSION

The fixed effects (stock, spawning day, stock × spawning day) were not significant sources of variation ($P > 0.05$) for length or weight in any sample period in either year. This indicates that the growth rates were not significantly different in the hatchery environment (Table 2). Both grilse and larger salmon of these same stocks have shown size differences at later ages (Bailey and Saunders, 1984). However, the older salmon had existed as wild fish for a minimum of 1 year and standardized conditions could not be imposed following smolt release.

The percent contributions of the different sources of variation within years, were similar at all ages (Table 3). Although the number of culled parr was small, selection of the largest 1000 parr at 6 months of age would eliminate a proportion of the fish which would have comprised the lower mode (Bailey et al., 1980). While this would tend to reduce the within-family variance, the 1979 error terms were remarkably constant between 6 and 15 months. Perhaps the 11-month interval was sufficient for the within-family variability to regain its pre-selection level.

Although variance components were similar within years, they were quite different between years, most notably in the error terms (Table 3). A reduction in error variance suggests some type of husbandry effect. However, management practices, and environmental conditions were similar between years and the nature of any husbandry effect is unknown. An explanation may be that different stocks and/or different parents from the same stock were used in each

TABLE 2

Mean fork lengths and weights of the various Atlantic salmon stocks at 12 weeks, 6 months and 15 months post-hatch (standard errors are given in parentheses)

Year	Strain	Fork length (cm)			Weight (g)		
		12 Weeks	6 Months	15 Months	12 Weeks	6 Months	15 Months
1978	Waweig		7.4			0.7	4.8
	Saint John		7.9			0.8	5.7
	SRC_1		8.8			0.6	8.4
	Pooled		8.0			0.7	6.3
			(1.2)			(0.4)	(3.4)
1979	SRC_2	3.3	7.0	14.6	0.4	4.1	34.4
	Big Salmon	3.8	7.7	14.4	0.6	5.4	33.3
	Pooled	3.4	7.1	14.4	0.4	4.3	34.2
		(0.4)	(0.7)	(1.3)	(0.2)	(1.5)	(8.3)

TABLE 3

Maximum likelihood variance components for length and weight of two year-classes of juvenile Atlantic salmon at different ages (percentage of total variance is given in parentheses)

Year	Age	Source	Length (mm)	Weight (g)
1978	12 Weeks	Sire		0.0101 (8.7)
		Dam		0.0154 (13.3)
		Replication		0.0081 (7.0)
		Error		0.0824 (71.0)
1978	6 Months	Sire	0.0278 (2.2)	0.3023 (2.6)
		Dam	0.1500 (11.8)	1.0350 (9.0)
		Replication	0.1345 (10.6)	1.1354 (9.8)
		Error	0.9620 (75.5)	9.0580 (78.6)
1979	12 Weeks	Sire	0.0408 (19.5)	0.0081 (22.4)
		Dam	0.0532 (25.3)	0.0087 (24.0)
		Error	0.1153 (55.1)	0.0194 (53.6)
1979	6 Months	Sire	0.0616 (14.4)	0.1803 (10.0)
		Dam	0.1029 (24.0)	0.5018 (27.8)
		Error	0.2643 (61.6)	1.1224 (62.2)
1979	15 Months	Sire	0.2507 (16.7)	10.9792 (16.8)
		Dam	0.2623 (19.5)	13.6465 (20.9)
		Error	0.8588 (63.8)	40.6805 (62.3)

year. The 1978 Saint John stock was descended largely from older parents, whereas all others were derived primarily from grilse. Furthermore, while the three stocks used in 1978 were completely different, the SRC$_2$ stock contained approximately 70% Big Salmon genes and it is likely that many of the genes in both 1979 stocks originated from common ancestors.

Domestication causes genetic changes in behaviour, morphology or physiology by eliminating genotypes which are unsuited to hatchery environments (Doyle, 1983). In 1978, approximately 25% of the full-sib groups were derived from wild broodstock and had not been exposed to any prior domestication pressure. Perhaps the reduction in error variance was partially due to the fact that only 14% of the full-sib groups in 1979 were derived from wild broodstock.

The replication term (in 1978) accounted for 7–11% of the total varaibility (Table 3). This source of variation was confounded with the dam effect in 1979. In order to compare dam components between years, the 1978 replication and dam components were combined. The pooled values were similar, though somewhat smaller, than the 1979 dam components. The sire components were also larger in 1979. The observed differences among variance components between years can possibly be attributed to the greater similarity between stocks and the increased level of domestication in the 1979 population.

The 1978 replication term was included in the total phenotypic variability. In general, heritability estimates were higher in 1979 than in 1978 (Table 4). The 1978 half-sib estimates ranged from 0.1 to 0.4, and are similar to those reported by Refstie and Steine (1978). The 1979 half-sib estimates were considerably larger (0.4 – 0.9) and are consistent with those of Lindroth (1972). As expected, full-sib estimates were somewhat higher in all cases (Table 4). The results illustrate that heritability estimates for the same traits can be markedly variable when they are determined in different years and in different populations, despite efforts to standardize environmental conditions. Such differences may be due in part to husbandry effects, levels of domestication, differences among stocks and/or sampling.

TABLE 4

Half-sib (h^2_{hs}) and full-sib (h^2_{fs}) heritability and maternal effects (m^2) estimates for length and weight of juvenile Atlantic salmon at different ages (standard errors are given in parentheses)

Age	Parameter	1978 Matings		1979 Matings	
		Length (cm)	Weight (g)	Length (cm)	Weight (g)
12 Weeks	h^2_{hs}		0.35 (0.29)	0.79 (0.31)	0.89 (0.32)
	h^2_{fs}		0.44 (0.14)	0.90 (0.24)	0.93 (0.10)
	m^2		0.05 (0.04)	0.06 (0.04)	0.02 (0.05)
6 Months	h^2_{hs}	0.09 (0.20)	0.10 (0.17)	0.57 (0.28)	0.40 (0.26)
	h^2_{fs}	0.28 (0.10)	0.23 (0.08)	0.77 (0.14)	0.76 (0.13)
	m^2	0.10 (0.04)	0.06 (0.02)	0.10 (0.02)	0.18 (0.07)
15 Months	h^2_{hs}			0.73 (0.32)	0.67 (0.32)
	h^2_{fs}			0.75 (0.16)	0.75 (0.16)
	m^2			0.01 (0.06)	0.04 (0.05)

TABLE 5

Phenotypic (r_p), sire genetic (r_{gs}), dam genetic (r_{gd}) and environmental (r_e) correlations between length and weight of juvenile Atlantic salmon at different ages

Age	Year	r_p	r_{gs}	r_{gd}	r_e
12 Weeks	1979	0.928	0.982	0.912	0.972
6 Months	1978	0.935	0.807	0.992	0.917
	1979	0.920	0.998	0.910	1.026
15 Months	1979	0.950	0.999	0.968	0.820

The m^2 term (Table 4) is only an approximation of the maternal effects because it also includes common environmental plus some non-additive genetic effects. Throughout the study period, m^2 increased slightly between 12 weeks and 6 months and then decreased to negligible levels by 15 months. Similar trends have been observed in rainbow trout (Gall and Gross, 1978; McKay, 1983).

Additive genetic correlations between length and weight were positive and high in all cases (Table 5). This suggests pleiotropy among additive genes for length and weight and that selection for either trait would produce a correlated response in the other. Similar, high correlations have been reported in rainbow trout (McKay, 1983) and channel catfish (Reagan et al., 1976). Generally the environmental correlations (Table 5) were lower than genetic and phenotypic correlations. The higher correlations in 1978 suggest that the replication term was effective in removing common environmental variation.

In summary, there were no differences in length or weight at any sampled age among the salmon stocks examined under common environmental conditions. Although husbandry practices were relatively constant between years, heritability estimates for length and weight differed substantially between years. Genetic correlations between length and weight were positive and high in both years. In general, these results support Gjedrem's (1983) contention that selection for length and weight in juvenile Atlantic salmon should produce relatively rapid genetic gain for these traits.

ACKNOWLEDGEMENTS

We thank R.L. Saunders, G.W. Friars, C.B. Schom, B.W. Kennedy, and I. MacMillan for their constructive reviews of earlier drafts of this manuscript. Computer facilities were made available by the University of New Brunswick and the Canadian National Sportsmen's Foundation via research grants to C.B. Schom and G.W. Friars, respectively.

REFERENCES

Bailey, J.K. and Saunders, R.L., 1984. Returns of three year-classes of sea-ranched Atlantic salmon of various river strains and strain crosses. Aquaculture, 41: 259–270.

Bailey, J.K., Saunders, R.L. and Buzeta, M.-I., 1980. Influence of parental smolt age and sea age on growth and smolting of hatchery-reared Atlantic salmon (*Salmo salar*). Can. J. Fish. Aquat. Sci., 37: 1379–1386.

Becker, W.A., 1984. Manual of Quantitative Genetics, 4th edition. Academic Enterprises, Pullman, WA, 186 pp.

Doyle, R.W., 1983. An approach to the quantitative analysis of domestication selection in aquaculture. Aquaculture, 33: 167–185.

Friars, G.W., Bailey, J.K. and Saunders, R.L., 1979. Considerations of a method of analyzing diallel crosses of Atlantic salmon. Can. J. Genet. Cytol., 21: 121–128.

Gall, G.A.E. and Gross, S.J., 1978. Genetic studies of growth in domesticated rainbow trout. Aquaculture, 13: 225-234.

Gjedrem, T., 1983. Genetic variation in quantitative traits and selective breeding in fish and shellfish. Aquaculture, 33: 51-72.

Hemmerle, W.J. and Hartley, H.O., 1973. Computing maximum likelihood estimates for the mixed AOV model using the W-transformation. Technometrics, 15: 819-831.

Kirpichnikov, V.S., 1981. Genetic Bases of Fish Selection. Springer-Verlag, Berlin, Heidelberg, 410 pp.

Lindroth, A., 1972. Heritability estimates of growth in fish. Aquilo Ser Zool., 13: 77-88.

McKay, L.R., 1983. A Genetic Analysis of Growth in Rainbow Trout (*Salmo gairdneri*). Ph.D. Thesis, Univ. Guelph, Guelph, Ontario, 200 pp.

Reagan, R.E., Pardue, G.B. and Eisen, E.J., 1976. Predicting selection response for growth of channel catfish. J. Hered., 67: 49-53.

Refstie, T. and Steine, T.A., 1978. Selection experiments with salmon. III. Genetic and Environmental sources of variation in length and weight of Atlantic salmon in freshwater phase. Aquaculture, 14: 221-234.

Saunders, R.L. and Bailey, J.K., 1978. The role of genetics in Atlantic salmon management. In: A.E.J. Went (Editor), Atlantic Salmon: Its Future. Fishing News Books, Ltd., Farnham, Great Britain, pp. 182-200.

Searle, S.R. and Rounsaville, T.R., 1974. A note on estimating covariance components. Am. Stat., 28: 67-68.

The Genetics of Production Characters in the Blue Mussel *Mytilus edulis*. I. A Preliminary Analysis

A.L. MALLET, K.R. FREEMAN and L.M. DICKIE

Department of Fisheries and Oceans, Marine Ecology Laboratory, Bedford Institute of Oceanography, Dartmouth, N.S. B2Y 4A2 (Canada)

ABSTRACT

Mallet, A.L., Freeman, K.R. and Dickie, L.M., 1986. The genetics of production characters in the blue mussel *Mytilus edulis*. I. A preliminary analysis. *Aquaculture*, 57: 133–140.

Estimates of genetic parameters in bivalves have typically been based on few families and are limited to larval and juvenile characters. Using mussels from a natural population, 30 replicated families were reared to an average size of 8 mm in the laboratory and subsequently in two natural environments. The effects of rearing full-sibs apart from fertilization and from day 55 have been estimated for adult shell length. Heritabilities were calculated from the sire and dam variance components for mussels up to an average size of 30 mm. In general, the heritabilities for growth were moderate to large, suggesting the possibilities of rapid genetic changes under directional selection. For survival, the additive genetic variance was zero or non-significant. However, the dam component was very large which suggests the presence of either important maternal or non-additive gene effects.

INTRODUCTION

Understanding the role of genotypic variation in relation to diverse production patterns characteristic of natural populations is an important issue for marine systems. Natural selection acts on the phenotypic variation to produce either characters with low variability (homeostasis) or characters with high variability (plasticity). Within these limits, short-term responses to variable environmental conditions occur as changes in one or more components of production such as meat yield, shell weight or length, percent metamorphosis and survival. These characters are assumed to be polygenic, that is, their expression is governed by many genes, and the appropriate analytical methods are those of quantitative genetics (see Falconer, 1981). The accurate assessment of the variability in production should therefore be made from measurements on several characters evaluated in different environments.

Attempts at describing the phenotypic and genotypic variation in marine bivalves have centered on two approaches. Animals of diverse geographic origins have been grown in common environments (Newkirk, 1978; Mallet and Haley, 1983; Dickie et al., 1984) and to a lesser extent, the evaluation of the genetic variance by parent-offsping regressions (Newkirk and Haley, 1982) or by resemblance among sibs (Lannan, 1972; Losee, 1979). In general, several relatively high estimates of heritability have been reported (Lannan, 1972; Newkirk et al., 1977; Innes and Haley, 1977; Losee, 1979). Estimates of additive genetic variance and correlations in these studies have, however, been calculated from a small number of parents. Also, no estimates of heritability are available on characters at market size, and the estimates are further confounded by other genetic or non-genetic factors.

In this paper we describe the variation in production components in half-sib groups produced from a natural population of the blue mussel *Mytilus edulis*. Specifically, we report estimates of genetic variance for growth and survival in larval, juvenile, and mature animals from a population not previously subjected to artificial selection.

MATERIAL AND METHODS

Fertilization and larval rearing procedure

Two groups, each of 150 adult mussels, were collected from the estuary of Bideford River near Ellerslie, P.E.I., Canada, in October and November 1982. Adult mortality in this broodstock during the conditioning period was less than 3%. Gonadal maturation of adults was achieved by maintaining the animals in a continuous flow of unfiltered sea-water at 7–8°C, and by feeding them a mixed diet of *Chaetoceros gracilis* and *Isochrysis galbana*. Each animal was individually spawned in a beaker containing 500 ml of 20°C, 1-μm filtered sea-water. Upon spawning, the sex of each individual was identified and fertilizations were done by mixing eggs and sperm from single individuals. Sperm from one male (sire) was used to fertilize separately eggs from three different females (dams). Thirty matings were done in the first day of spawning (day 0) using 30 dams (one per mating) and 10 sires (one per three matings). Each of the 30 matings was replicated; thus, a total of 60 "groups" of sibs was established. The arrangement of the larval rearing containers in our hatchery imposed the following procedure: one of the two family replicates was placed on an upper shelf (referred subsequently as Position 1) and the other replicate on a lower shelf (Position 2). Details of the fertilization, initial culture density and larval rearing have been described previously (Mallet et al., 1985).

Juvenile rearing

Prior to the initiation of metamorphosis, around day 20 of the larval phase, the content of a given funnel was collected on a 110-μm nitex screen and transferred into a 4-l polyethylene container. The water was changed every third day and the juveniles were fed a mixed diet of C. gracilis and I. galbana. The length of the metamorphosis period (visual assessment) varied greatly among cultures from a few days to 2 weeks. Also, the total number of juveniles at day 45 varied from less than 100 to 40000. At day 55 when the smallest individuals were greater than 1200 μm, two replicates were made from every 4-l polyethylene bucket, each with an initial count of 450 animals. Replicates were then transferred into solid cylindrical ABS cages with 500 μm mesh at both ends and covering two large openings on each side. Each cage was labelled with a colour-coded, numbered tag for the appropriate experiment, family, and position. After a few weeks of growth in these cages, the animals were transferred into a similar cage of larger nitex mesh size (750 μm) and then, after a few more weeks, into a bag made with a 1.5-mm insect screen. In April 1984 (day 300), two replicates were made from each bag, placed in 2-mm mesh Bemis cages having a semi-rigid tetrahedronic shape. One set of replicates was transferred to St Margaret's Bay, N.S. (ENV1), and the other set to Bedford Basis, N.S. (ENV2). The final sampling was done 15 March 1985 with 700-day-old animals.

Data and analysis

The following characters were selected for this analysis: larval shell length and viability at day 16, juvenile shell length at day 300, and adult length and viability at day 700 in two environments. Let N denote the number of observed individuals. Let Y_i represent a continuous response variable (shell length) associated with the ith individual. The basic model, a random model for unbalanced data, can be generally described as

$$Y_i = \mu + Z\beta + e$$

where μ is the population mean, common to all observations, Z is a matrix of dimension $N \times r$, β is an $r \times 1$ vector of unknown variance components, and e has the following distribution $(0, \sigma_e^2)$. The variance components were obtained from the general mixed model analysis using maximum likelihood estimation, BMDP-3V (Jenrich and Sampson, 1979). The family means were calculated from the model

$$Y = \mu + X\alpha + Z\beta + e$$

where μ is the population common to all observations, X is a matrix of $N \times r$, α is an $r \times 1$ vector of unknown parameters, and $Z\beta$ and e are the same as above.

Standard errors of heritabilities were estimated by using the variance of the variance components and substituting them in the variance of the h^2 equation (Becker, 1984). Estimates of genetic change were computed as

$$\Delta G/\text{generation} = i\, h^2\, CV$$

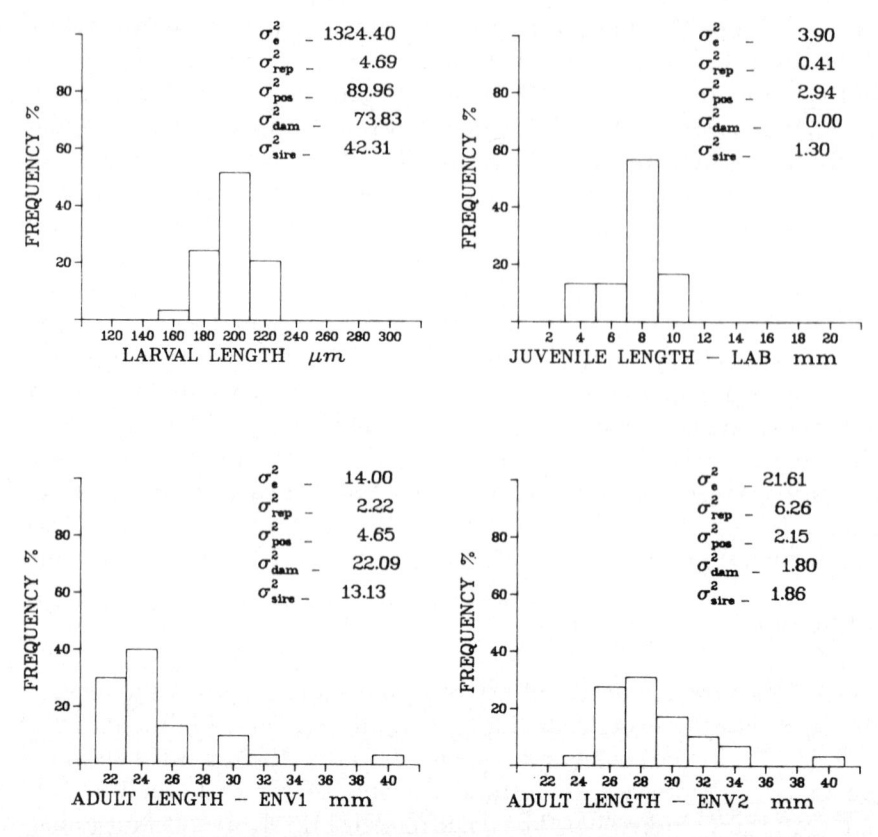

Fig. 1. Histogram of family means, in frequency classes, and the variance components for larval shell length at day 16, juvenile length at day 300, and adult length on day 700. See text for the description of the variance components.

which provides an estimate of percent mean change per generation; i is the selection differential ($=2.063$ for our prediction), h^2 the heritability, and CV is the coefficient of variation (the phenotypic standard deviation divided by the mean).

RESULTS

The mussels for this study were exposed to both laboratory conditions (up to and including day 300) and natural environments (day 300 onwards). The variability in family means and the variance components for all characters are illustrated in Figs. 1 and 2. The means, phenotypic standard deviations, heritabilities, coefficients of variation and predicted percent change per generation assuming a 5% selection intensity are shown in Table 1.

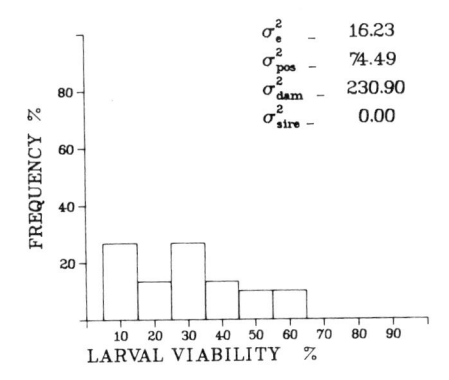

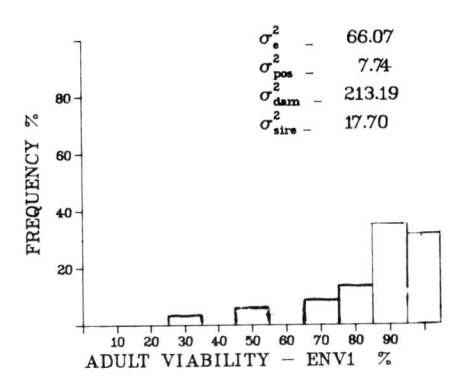

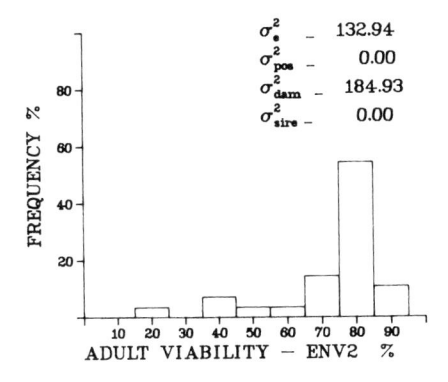

Fig. 2. Histograms of family means, in frequency classes, and the variance components for larval (day 16) and adult (day 700) viability.

TABLE 1

Means ($\bar{x}$), phenotypic standard deviations (SD), coefficients of variation (CV), sire heritabilities (h^2_s) with standard errors and percent expected mean changes (EMC) for all characters

Traits	N	$\bar{x}$	SD	CV	h^2_s	EMC
Length						
Larval	1265	189.8	39.2	20.6	0.11±0.02	4.7
Juvenile	2007	8.2	2.9	35.2	0.62±0.06	45.36
Adult-env1	1931	24.3	7.5	30.9	0.92±0.27	58.6
Adult-env2	1789	29.8	5.8	19.5	0.22±0.07	8.9
Survival						
Larval	120	24.0	3.1	12.9	0.0	—
Adult-env1	99	81.0	17.5	21.6	0.15±0.44	—
Adult-env2	87	68.3	17.8	26.1	0.0	—

Larval characters

The Position (σ^2_{pos}) component reflects the effect of rearing full-sibs apart from the time of fertilization and the Replicate (σ^2_{rep}) component evaluates the experimental sampling error. The sampling error is negligible but position has an important effect on larval characters. The heritability of shell length is significant which shows that the sire groups were significantly different. The heritability based on the dam component ($h^2_{dam} = 0.19 \pm 0.04$) is approximately twice as large as the h^2 based on the sire component. The dam estimate contains four times the maternal effect and all the dominance variance. The percent mean change in this character is predicted to be 4.7% per generation based on a selection intensity of 5% which is probably an underestimate for the larval stage. For survival, a significant non-additive or maternal variance (0.0231±0.007) was found but additive variance was non-significant. The dam component for larval survival accounted for 72% of the total variance.

Juvenile and adult stages

The Position component is as desribed above but the Replicate term reflects the effect of growing full-sibs apart from day 55 onwards. Position explained between 6 and 8% of the total variance for shell length while Replicate accounted for 4–18%, the range being dependent on the environment. The Position effect may have resulted from differential larval densities which provided an initial advantage in some replicates which was maintained. The coefficient of variation for length in ENV1 is almost twice as large as ENV2. Heritabilities estimated from the two environments were both significant; however, compared to the heritability prior to splitting the families in two environments, one h^2 increased while the other h^2 decreased in magnitude.

Only in one environment, ENV1, do maternal or non-additive genetic effects appear to be important. The correlation among family means between environments was positive but non-significant. For survival, overall viability to the adult stage was very high but the distributions of family viability were skewed to the left. All additive variance estimates were zero or non-significant. However, the dam component was very large: in analogy with larval viability, the dam component accounted for 70% and 58% of the total variance in ENV1 and ENV2, respectively. The correlation among family means between environments was positive (Spearman correlation, $r=0.58$, $P=0.05$). The correlation between larval viability and viability to the adult stage in either environment was non-significant. Environment appears to have an effect on mean viability although its effect relative to the genetic effect has not been estimated.

DISCUSSION

Each of the characters analyzed typically has its own level of variation but some general statements can be made, especially between viability versus growth characters. First, there exists considerable variability in all components of production examined; however, the magnitude of the genetic and non-genetic variation differs among components. For survival, the additive genetic variance was absent but the non-additive genetic variance or maternal variance was substantial. Theoretically, stabilizing selection tends to reduce the genetic variance by favoring the less variable animals which, if they are not heterozygous, will drive the gene frequencies to fixation (Robertson, 1956). Electrophoretic studies have shown that heterozygotes have a better viability than homozygotes between the juvenile to adult stages (Zouros et al., 1983). If this observation can be generalized to the type of gene action controlling viability, then the large dam component could be an expression of over-dominant or epistatic gene combinations. Mallet and Haley (1984) have also reported a large dam component in a study on *Crassostrea virginica* larvae, although in that study a significant sire component was found.

A recent electrophoretic survey of certain families revealed that 4% of the progeny had wrong pedigrees (Mallet et al., 1985). Van Vleck (1970) has argued that mistaken pedigrees will not affect genetic progress as long as the fraction of false pedigrees is not large; for example, 20% error in progeny groups of 50 animals will reduce the correlation from 0.88 to 0.82. Nonetheless, we did take some preventative steps for the elimination of contamination.

Shell length, our indicator of growth rate, showed consistently significant heritability estimates. The discrepancy in the magnitude of the h^2 estimates between environments suggests that the difference among sire groups is more acute in ENV1 than in ENV2. This may be a reflection of the environmental variability favoring specific genotypes over others. The position and replicate variance components allowed identification of a pre- and post-metamorphic environmental influence on shell length but further investigations are needed to understand the nature of this variation. If one assumes an intensity of selection of 5%, substantial phenotypic changes could be accomplished in a few generations. For juvenile and adult growth, individual selection would appear to be justified. For other characters such as those for larval viability, the genotype of the parent can be seen only through the performance of the progeny. Selection in this case would need to be based on family means but it is predicted that substantial change could be achieved.

REFERENCES

Becker, W.A., 1984. Manual of Quantitative Genetics, 4th Edition. Academic Enterprises, Pullman, WA.
Dickie, L.M., Boudreau, P. and Freeman, K.F., 1984. Influences of stocks and sites on growth and mortality in the blue mussel *Mytilus edulis*. Can. J. Aquat. Sci., 41: 134–140.
Falconer, D.S., 1981. Introduction to Quantitative Genetics, 2nd edition. Longman Group, London.
Innes, D.J. and Haley, L.E., 1977. Genetic aspects of larval growth under reduced salinity in *Mytilus edulis*. Biol. Bull., 153: 312–321.
Jenrich, R. and Sampson, P., 1979. General mixed model analysis of variance. In: W.J. Dixon and M.B. Brown (Editors), Biomedical Computer Programs P-Series. University of California Press, Berkeley, CA, pp. 581–598.
Lannan, J.E., 1972. Estimating heritability and predicting response to selection for the Pacific oyster, *Crassostrea gigas*. Proc. Natl. Shellfish. Assoc., 62: 62–66.
Losee, E., 1979. Relationship between larval and spat growth rates in the oyster *Crassostrea virginica*. Aquaculture, 16: 123–126.
Mallet, A.L. and Haley, L.E., 1983. Growth rate and survival in pure population matings and crosses of the oyster *Crassostrea virgica*. Can. J. Fish. Aquat. Sci., 40: 948–954.
Mallet, A.L. and Haley, L.E., 1984. General and specific combining abilities for larval and juvenile growth and viability estimated from natural oyster populations. Mar. Biol., 81: 53–59.
Mallet, A.L., Zouros, E., Gartner-Kepkay, K.E., Freeman, K.R. and Dickie, L.M., 1985. Larval viability and heterozygote deficiency in populations of marine bivalves: evidence from pair matings of mussels. Mar. Biol., 87: 165–172.
Newkirk, G.F., 1978. Interaction of genotype and salinity in larvae of the oyster *Crassostrea virginica*. Mar. Biol., 41: 49–52.
Newkirk, G.F. and Haley, L.E., 1982. Progress in selection for growth in the European oyster, *Ostrea edulis*. Mar. Biol. Prog. Ser., 10: 77–79.
Newkirk, G.F., Haley, L.E., Waugh, D.L. and Doyle, R.W., 1977. Genetics of larvae and spat growth rate in the oyster *Crassostrea virginica*. Mar. Biol., 41: 49–52.
Robertson, A., 1956. The effect of selection against extreme deviants based on deviation or on homozygosis. J. Genet., 54: 236–248.
Van Vleck, L.D., 1970. Misidentification and sire evaluation. J. Dairy Sci., 55: 218–226.
Zouros, E., Singh, S.M., Foltz, D. and Mallet, A.L., 1983. Post-settlement viability in the American oyster (*Crassostrea virginica*): an overdominant phenotype. Genet. Res. Camb., 41: 259–270.

The Effect of Strain Crossing on the Production Performance in Rainbow Trout

GABRIELE HÖRSTGEN-SCHWARK, H. FRICKE and H.-J. LANGHOLZ

Institut für Tierzucht und Haustiergenetik der Universität Göttingen, Albrecht-Thaer-Weg 1, 3400 Göttingen (Federal Republic of Germany)

ABSTRACT

Hörstgen-Schwark, G., Fricke, H. and Langholz, H.-J., 1986. The effect of strain crossing on the production performance in rainbow trout. *Aquaculture*, 57: 141–152.

A diallel cross with four strains of spring-spawning rainbow trout (*Salmo gairdneri*) was carried out in two replicates to estimate average direct genetic effects, maternal genetic effects, general combining ability, heterosis, and reciprocal effects on growth and carcass performance. All matings within each replicate were performed the same day. Separate hatching and rearing were used for different genetic groups up to a size of 45 g, with common testing after freeze-branding up to a size of 220 g in two production environments (pond and silo).

The average hatching rate of all mating groups was as high as 68.9%, the hatching rate of the crossbreds being slightly higher (+2.5%). For some reciprocal crosses, hatching rate was 15% higher than the midparent rate. During the rearing and fattening period the purebred strains showed significant differences in growth performance, with average line effects being more important than maternal effects. Differences in the general combining ability were also significant. For some strains, this was primarily caused by high average line effects while others showed a high general line heterosis. The mean growth performance of all crossbreds was not significantly different from that of the purebreds. Significant line differences were found for the percentage of fillet, head and fins, and offal, and for the percentage of fat and dry matter in the carcass. The crossbreds had a significantly higher percentage of fillet than the purebreds. However, ranking of the genetic groups according to their absolute weight of fillet was not different from a ranking by total weight. Significantly different average line effects observed for all traits indicated that a distinct strain differentiation existed. Consequently, selection of superior strains with regard to specific traits is of considerable importance.

INTRODUCTION

Although differences in production performance between different rainbow trout populations have been reported (Gall and Gross, 1978a; Reinitz et al., 1978; Reichle and Bergler, 1980; Ayles and Baker, 1983; Morkramer et al., 1985), less information is available about the efficiency of crossbreeding in trout production (Klupp and Pirchner, 1975; Gall and Gross, 1978b; Hershberger, 1978; Ayles and Baker, 1983; Linder et al., 1983). A diallel cross with four spring-spawning trout populations was therefore carried out to evaluate the genetic differences among a fixed set of rainbow trout strains and their reciprocal crosses. The diallel cross was used as experimental design, because it has the advantage of providing simultaneous estimates of main genetic effects including average direct and maternal effects, general combining ability, general line and specific heterosis and reciprocal effects.

MATERIALS AND METHODS

Design of experiment

A complete diallel cross with four strains of spring-spawning rainbow trout (*Salmo gairdneri*) was carried out in two replicates. The following traits were considered:
hatching rate;
body weight of individuals after rearing and fattening;
weight of net carcass, weight of head and fins, weight of offal;
fat, protein and dry matter content (Kim, 1984) and water-holding capacity of carcass (Meyer et al., 1984).

All matings within each replicate were performed on the same day. The date of mating depended on the simultaneous maturity of a representative sample of spawners from each strain. In the first replicate one hundred 3-year-old spawners (50♀ and 50♂) from a population of four hundred 3-year-old spawners were sampled according to their stage of maturity, which had been checked daily by stripping of fish to ensure maximum viability of gametes (Craik and Harvey, 1984). The number of spawners within a strain averaged 25. The same spawners (then 4 years old) and the same procedure were used for the second replicate 1 year later.

The spawners were weighed before and after stripping. Eggs of females within a strain were mixed and then the average 100-egg-weight was determined. Milt was collected in test tubes, the volume taken and the motility and density measured according to Holtz et al. (1977) and Büyükhatipoglu and Holtz (1984) before the milt of all males within a strain was mixed for fertilization of the eggs.

Separate hatching and rearing were used for the different genetic groups up to a size of 45 g, with common testing after freeze-branding up to a size of 220 g in two production environments (pond and silo) according to the Relliehausen standard testing procedure (Morkramer et al., 1985). The only deviation from this procedure consisted in slaughtering all tested fish instead of a sample out of each test group as usual. Total numbers of observations for both replicates after rearing and fattening and for carcass evaluation are given in Table 1. Each replicate lasted 18 months.

Statistical analysis

All individual data were analysed by the least square procedure (Harvey, 1977) using an extension of the diallel analysis proposed by Gardner and Eberhart (1966) to include maternal effects (Eisen et al., 1983). For rearing weight the statistical model was

$$Y_{ijfk} = \bar{y}_a + (l_i + l_j)/2 + m_j + \delta(h_{ij} + r_{ij}^{**}) + \psi_f + e_{ijfk}$$

where

Y_{ijfk} = rearing weight of the kth individual of sire line i and dam line j in replicate f [$i, j = 1, ... 4, \delta = 0$ for parental line progeny ($i = j$), $\delta = 1$ for crossbred progeny ($i \neq j$)]
$\bar{y}_a$ = mean of parental lines
l_i = $\bar{y}_{ii} - \bar{y}_a - m_i$ = average direct effect of line i
m_j = $\bar{y}_{.j} - \bar{y}_{j.}$ = average maternal (genetic) effect of line j, where
 $\bar{y}_{j.}$ ($\bar{y}_{.j}$) = mean of sire (dam) line j including the parental line
h_{ij} = $(\bar{y}_{ij} + \bar{y}_{ji})/2 - (\bar{y}_{ii} + \bar{y}_{jj})/2$ = direct heterosis for the (ij)th cross
r_{ij}^{**} = $r_{ij} - (m_j - m_i)/2$ = specific reciprocal difference between lines i and j, where
 r_{ij} = reciprocal effects = $(\bar{y}_{ij} - \bar{y}_{ji})/2$
ψ_f = effect of replicate f ($f = 1,2$)

TABLE 1

Total number of observations for rearing weight (upper rows) and fattening weight (lower rows)

Females	Strain			
	2	3	4	5
Males				
2	426	257	280	291
	201	124	104	120
3	269	311	278	274
	132	229	125	121
4	219	206	408	272
	95	97	202	110
5	143	213	287	415
	65	128	110	161
	Sum 4549			
	2124			

e_{ijfk} = random error, $k = 1, \ldots n_{ijf}$, where
n_{ijf} is the number of individuals in the (ijf)th subclass.

The following additional genetic effects were estimated according to Eisen et al. (1983):

g_i = general combining ability of line i;
h_i = direct heterosis of line i;
s_{ij} = specific combining ability of the cross;
z_i = an alternative definition of line (direct) heterosis (Eisen et al., 1983);
w_{ij} = an alternative definition of specific (direct) heterosis (Eisen et al., 1983).

Estimates of g_i provide information on the average performance of a line in crosses, disregarding maternal genetic effects. Gardner and Eberhart (1966) showed that $g_i = 1/2 l_i + h_i$, where h_i is line direct heterosis. Direct heterosis of the ith line is the average of all direct heterosis values of crosses involving line i. The terms of line direct heterosis h_i and z_i are exactly linearly related (Eisen et al., 1983). Line direct heterosis is most conveniently partitioned as a component of both general and specific combining ability (Gardner and Eberhart, 1966). The other heterosis term, z_i, has an analytical genetic interpretation (Vencovsky, 1970), in that it detects the degree of departure of the frequencies of favorable alleles from the mean gene frequency at loci exhibiting directional dominance for the trait studied.

Overall direct heterosis ($\bar{h}$) measures mean heterotic effects for all crosses, i.e., $\bar{h} = \bar{y}_c - \bar{y}_a$ where $\bar{y}_c$ is the overall crossbred mean and $\bar{y}_a$ represents the overall purebred mean.

Individual heterosis is defined as the difference between the mean of the two reciprocal crosses and the midparent mean. Two ways of partitioning individual heterosis are

$$h_{ij} = \bar{h} + h_i + h_j + s_{ij} \text{ (Gardner and Eberhart, 1966)}$$

$$= z_i + z_j - 2w_{ij} \text{ (Vencovsky, 1970)}$$

where s_{ij} is specific combining ability, w_{ij} is a measure of the direction of divergence in gene frequencies between the two lines and the degree of directional dominance affecting the trait, and the other terms are as defined previously. Numerically, $w_{ij} < 0$ if gene frequencies of each line deviate in opposite directions from mean gene frequency and $w_{ij} > 0$ if the deviations have the same sign.

Estimates of each effect could be obtained by forming contrasts among the appropriate least square means. Formulas to estimate the genetic effects, their interrelationship and their genetic interpretation were given by Gardner and Eberhart (1966), Vencovsky (1970) and Eisen et al. (1983).

As stocking densities varied for separate rearing between 80 to 150 fish per tank, rearing weight had to be adjusted for linear and quadratic effects of stocking densities. The appropriate regression coefficients were calculated from a growth study done simultaneously to the above mentioned, where fish having the same genetic make-up were reared in four different stocking densities (75, 100, 125, 150 individuals per tank) in four replicates using the same rearing system and with the same management. The model used to analyse fattening weight, carcass- and meat quality included the additional term for the two fattening systems, and the appropriate interaction. For hatching rate only two observations per genetic group were available and therefore only mean percentages were used.

RESULTS

Reproductive performance

The average weight of female spawners before stripping ranged from 959 to 1244 g with an average of 1080 g in the first replicate, and from 1150 to 1435 g with an average of 1270 g in the second replicate. Average weight of male spawners was less, averaging 880 g (770–1080 g) in the first and 1106 g (995–1370 g) in the second replicate. The 100-egg-weight of females varied from 8 to 9 g in the first replicate, with lighter spawners having the higher egg weights. In the second replicate this relation was not observed; egg weights ranged from 7.4 to 8.2 g. Sperm volume of all males was at least 12 ml with high motility and density.

The average overall hatching rate was 68.9%. The average hatching rate of crossbreds (69.5%) was 2.5% higher than for purebreds (67%). As Table 2 shows, distinct effects of mating partners could be observed for population 5. If females belonging the population 5 were used, hatching rates were lower than the corresponding means of parental populations. The males from this population, however, caused higher hatching rates in all crosses than the parental populations. Individual heterosis varied from −17% to 15% (Table 3).

TABLE 2

Deviations of hatching rates of crossbreds from the corresponding mean of parents

Females	Strain			
	2	3	4	5
Males				
2	63.1	+ 7.3	+12.8	− 8.8
3	+ 3.0	68.0	+12.7	− 3.0
4	−24.9	+ 8.9	78.5	−27.5
5	+22.7	+22.5	+ 3.7	58.5

Growth and carcass evaluation

The mean rearing and fattening weights of fish were 43.8±0.36 g and 222.6±1.43 g, respectively. For all genetic groups the percentage of fillet averaged 68.5±0.07%, the percentage of head and fins 20.6±0.05% and the percentage of offal 10.7±0.05%. The mean percentages of fat, protein and dry matter were 6.91±0.04%, 19.44±0.02% and 28.41±0.06%. Water-holding capacity, defined as the ratio of meat area to total area determined by filter absorption method, ranged from 65.8 to 67.6%, with an average of 66.8%. As no significant differences between mating groups could be demonstrated for percentage protein and water-holding capacity, these traits were omitted from Tables 4 and 5.

During the rearing and fattening period, purebreds showed significant ($P<0.05$) differences in growth (Table 4). Ranking of strains according to average direct line effects was 3, 4, 2, 5 for rearing weight and 3, 2, 4, 5 for fattening weight. For the percentage of fillet only strain 5 showed significant positive average line effects due to negative line effects for the percentage of head and fins and percentage offal. Line effect values for fat percentage were also highest for strain 5, whereas strain 4 showed the greatest negative effect for fat and dry matter percentages.

In general, maternal genetic effects were of less importance than direct line effects. Significant maternal effects for rearing weight could be demonstrated for strain 5 only. For the percentages of offal and fat, there were significant negative maternal effects ($P<0.05$) in strains 2 and 5.

Significant differences in general combining ability (g_i) between strains were found for all traits. Strains 2 and 3 showed the highest positive g_i effects for both weights, whereas negative g_i values were found for strains 4 and 5. Population 5 was the only one with a positive g_i for the percentage of fillet, due to negative g_i for the percentage of head and fins and percentage offal. For the percentage of fat and dry matter, g_i was negative for strains 3 and 4, whereas strains 2 and 5 showed positive g_i for both traits.

In general, differences in g_i between strains for the examined traits were due primarily to direct line effects. The contributions of h_i to g_i were relatively

TABLE 3

Individual heterosis (h_{ij}) for hatching rate

	Cross (including reciprocal)					
	2×3	2×4	2×5	3×4	3×5	4×5
h_{ij}	+5.2	−6.1	+ 7.0	+10.8	+ 9.8	−11.9
$h_{ij}\%$	+7.9	−8.6	+11.5	+14.7	+15.5	−17.4

TABLE 4

Mean purebred performance ($\bar{y}$), average direct line effects (l_i), maternal effects (m_i), general combining ability (g_i) and heterosis (h_i, z_i, and $\bar{h}$)

		Rearing weight (g)	Fattening weight (g)	Fillet (%)	Head and fins (%)	Offal (%)	Fat (%)	Dry matter (%)
$\bar{y}$	2	41.2	221.0	68.2	20.9	10.7	6.9	28.31
	3	52.9	235.1	68.0	20.8	11.0	7.0	28.47
	4	43.0	211.9	68.1	20.6	11.1	6.5	27.71
	5	41.9	211.6	68.8	20.4	10.7	7.0	28.29
	SE[d]	1.35	5.89	0.26	0.15	0.16	0.1	0.20
l_i	2	−3.23[ac]*	4.70[ac]	−0.20[a]	0.07[ab]	0.11[a]	0.17[b]	0.43[a]
	3	10.42[b]*	16.54[a]*	−0.21[a]	0.20[b]	0.02[a]	−0.03[ab]	−0.04[ac]
	4	−1.64[c]	−9.77[bc]	−0.23[a]	0.12[ab]	0.12[a]	−0.48[a]*	−0.63[bc]*
	5	−5.55[a]*	−11.47[b]*	0.64[b]*	−0.39[a]*	−0.24[a]	0.34[b]*	0.24[a]
	SE	1.36	5.44	0.26	0.19	0.20	0.16	0.24
m_i	2	−0.33[a]	−3.57[a]	0.15[a]	0.14[a]	−0.29[a]*	−0.09[ac]	−0.32[b]
	3	−2.26[a]	−1.36[a]	−0.06[a]	−0.07[a]	0.13[b]	0.22[bc]	0.32[a]
	4	−0.15[a]	1.75[a]	0.05[a]	−0.18[a]	0.13[b]	0.09[bc]	0.15[ac]
	5	2.75[b]*	3.17[a]	−0.14[a]	0.11[a]	0.03[ab]	−0.23[a]*	−0.15[bc]
	SE	0.89	3.69	0.18	0.13	0.13	0.11	0.17
g_i	2	1.62[a]*	7.68[a]*	−0.02[a]	−0.01[ab]	0.03[ab]	0.33[a]*	0.45[a]*
	3	2.69[a]*	4.54[ac]	−0.14[a]	0.14[b]	0.00[ab]	−0.20[b]*	−0.15[bc]
	4	−1.07[b]	−4.30[bc]	−0.32[a]*	0.09[b]	0.23[a]	−0.29[b]*	−0.43[b]*
	5	−3.24[c]*	−7.92[b]*	0.47[b]*	−0.22[a]	−0.26[b]*	0.16[a]	0.13[ac]
	SE	0.78	3.26	0.16	0.11	0.12	0.10	0.15
h_i	2	3.23[a]*	5.33[a]	0.09[a]	−0.05[a]	−0.02[a]	0.25[a]*	0.24[a]
	3	−2.52[b]*	−3.73[bc]	−0.04[a]	0.03[a]	−0.01[a]	−0.18[b]	−0.13[a]
	4	−0.25[c]	0.59[ac]	−0.20[a]	0.03[a]	0.17[a]	−0.05[b]	−0.11[a]
	5	−0.46[bc]	−2.19[ac]	0.15[a]	−0.02[a]	−0.13[a]	−0.01[ab]	0.01[a]
	SE	0.81	3.24	0.16	0.11	0.12	0.10	0.15
z_i	2	1.10[a]	4.04[a]*	0.16[a]	−0.07[a]	−0.08[a]	0.15[a]*	0.23[a]*
	3	−1.77[b]*	−0.49[a]	0.10[a]	−0.03[a]	−0.07[a]	−0.06[b]	0.04[a]
	4	−0.64[b]	1.67[a]	0.01[a]	−0.03[a]	−0.01[a]	0.00[ab]	0.05[a]
	5	−0.74[b]	0.28[a]	0.19[a]	−0.05[a]	−0.14[a]	0.03[ab]	0.11[a]
	SE	0.48	1.92	0.10	0.07	0.07	0.06	0.09
$\bar{h}$		−1.34	3.66	0.31*	−0.11	−0.19	0.09	0.29*

[a,b,c]Values with no letter in common are significantly different at $P<0.05$
[d]Approximate standard error of least square estimates.
*Significantly different from 0, $P<0.05$.

small compared to those of l_i, except in strain 2. The corresponding z_i estimates were also highest for strain 2, indicating a high divergence in gene frequency

TABLE 5

Mean performance of crossbreds ($\bar{y}_{ij}$, $\bar{y}_{ji}$) direct heterosis (h_{ij}) and percent direct heterosis (h_{ij} %), specific combining ability (s_{ij}, w_{ij}) and specific reciprocal effects (r_{ij}**)

Effect	Cross (sire strain × dam strain)						SE[e]
	2×3	2×4	2×5	3×4	3×5	4×5	
	Rearing weight (g)						
$\bar{y}_{ij}$	47.2	42.7	43.9	45.2	47.2	40.9	1.40
$\bar{y}_{ji}$	45.5	46.1	40.9	41.3	40.4	39.8	1.40
h_{ij}	−0.72[bc]	2.26[b]	0.83[b]	−4.72[a]*	−3.70[ac]*	−2.15[ac]	1.37
h_{ij} %	−1.5	5.4	2.0	−9.8	−7.8	−5.1	—
s_{ij}	−0.07[a]	0.64[a]	−0.58[a]	−0.58[a]	0.64[a]	0.07[a]	0.59
w_{ij}	0.03[bc]	−0.90[b]*	−220.23[bd]	1.15[a]*	0.59[acd]	0.38[acd]	0.39
r_{ij}**	1.84[a]*	−1.79[bc]*	−0.04[ac]	0.92[a]	0.91[a]	−0.87[bc]	0.72
	Fattening weight (g)						
$\bar{y}_{ij}$	231.5	227.1	230.9	226.3	223.6	209.8	5.89
$\bar{y}_{ji}$	231.2	222.8	221.3	227.7	216.4	213.8	5.89
h_{ij}	3.29[a]	8.52[a]	0.85[a]	3.56[a]	−3.32[a]	0.09[a]	5.41
h_{ij} %	1.4	3.9	4.6	1.6	−1.5	0.04	—
s_{ij}	−1.98[a]	−1.07[a]	3.04[a]	3.04[a]	−1.07[a]	−1.98[a]	2.44
w_{ij}	0.13[a]	−1.41[a]	−2.76[a]	−1.19[a]	−1.55[a]	0.93[a]	1.58
	Fillet percentage						
$\bar{y}_{ij}$	68.7	68.2	68.6	68.2	68.7	68.9	0.26
$\bar{y}_{ji}$	68.6	68.1	69.4	68.0	68.5	68.9	0.26
h_{ij}	0.55[a]*	0.01[a]	0.52[a]*	0.04[a]	0.25[a]	0.45[a]	0.25
h_{ij} %	0.81	0.01	0.76	0.06	0.37	0.66	—
s_{ij}	0.20[a]	−0.17[b]	−0.02[ab]	−0.02[ab]	−0.17[b]	0.20[a]	0.12
w_{ij}	−0.15[a]	0.08[b]	−0.09[ab]	0.03[ab]	0.02[ab]	−0.12[ab]	0.08
	Fat percentage						
$\bar{y}_{ij}$	7.10	7.06	7.36	6.57	6.57	6.52	0.17
$\bar{y}_{ji}$	7.04	6.57	7.59	7.04	6.91	6.87	0.17
h_{ij}	0.10[abc]	0.12[abc]	0.53[a]*	0.05[bc]	−0.26[c]	−0.02[bc]	0.16
h_{ij} %	1.43	1.79	7.63	0.74	−3.71	−0.30	—
s_{ij}	−0.05[b]	−0.16[b]*	0.20[a]*	0.20[a]*	−0.16[b]*	−0.05[b]	0.07
w_{ij}	0.00[cd]	0.02[cd]	−0.17[b]*	−0.05[bc]	0.12[ad]*	0.03[cd]	0.05

[a, b, c, d] Values with no letter in common are significantly different at $P<0.05$.
[e] Approximate standard error of least square estimates.
*Significantly different from 0, $P<0.05$.

from the mean frequencies of the other lines.

Overall heterosis ($\bar{h}$) indicates that crossbreds had significantly higher percentages of fillet and dry matter.

Individual heterosis was small for rearing weight, ranging from -10 to $+5\%$. For fattening weight, individual heterosis effects were insignificant (Table 5).

Significant heterosis effects could be demonstrated for percentage of fillet and fat. For all crosses showing the highest positive h_{ij}, the highest negative w_{ij} effects were also found, which indicates that gene frequencies of each line deviate in opposite directions for these two traits. On the contrary, the highest negative h_{ij} effects were accompanied by the highest positive w_{ij} terms, indicating that the participating lines had gene frequencies of the same sign.

Specific reciprocal effets (r_{ij}**) have an expectation of zero for maternal effects. Significant r_{ij}** terms for rearing weight found for crosses 2×3 and 2×4 may represent specific cytoplasmic effects.

DISCUSSION AND CONCLUSIONS

The main aim of this experiment was to quantify differences in production performance of European rainbow trout populations and their reciprocal crosses. The diallel cross was chosen as experimental design, since it provides simultaneous estimates of many key genetic effects.

After rearing and fattening in the two replicates, ranking of purebreds was nearly the same, confirming the results obtained by Morkramer et al. (1985). For crossbreds this stability was not so marked, which is in agreement with the findings obtained in crossbreeding experiments with carp by Wohlfarth et al. (1984).

Hatching rate could be increased by 8–15% over the midparent hatching rate by use of crossbreeding. However, as long as fertility does not decrease excessively, the hatching rate is of less economic importance for trout production than the growth rate, because of the large quantity of eggs produced by a single female. Hatching rate was influenced considerably by a position of a strain as the male or female mating partner. All crosses with females of strain 5 showed a lower hatching rate (-13%), whereas crosses with males of this strain had higher hatching rates ($+16\%$) than the midparents. Assuming equal viability of germ cells, these results may indicate an incompatibility of gametes as Hershberger (1978) suggested for similar results found in strain crosses of coho salmon.

Significant line differences (l_i) between strains exist for all traits examined. Strains with high line effects for rearing and fattening weights (strains 2 and 3) showed negative line effects for percentage of fillet. As nothing is known about the selection history of these strains, differences in line effects may be due to different selection strategies and intensities as well as to different foundation stocks taken from the wild some 25 generations ago.

In general, maternal effects were of less importance than line effects, although significant differences between strains might indicate a potential for further genetic gain by using special dam and sire lines for crossbreeding (i.e., strain 5, rearing weight).

The observed significant differences in general combining ability (g_i) between strains for the traits studied had different genetic backgrounds. G_i effects based on high line effects indicate the suitability of strains for improving the performance of strains with lower efficiency by a single introduction of favorable genes. This is especially true for strain 3. Rearing and fattening weight could be increased up to 6.5% and 5.3%, respectively, for all strain-3 crosses over the purebred performance of the other strains, which is within the range of predicted gain by individual selection purebreeding schemes (Gjedrem, 1975; Morkramer et al., 1985). Strains with g_i effects depending more on high h_i effects, such as strain 2, would be suitable for commercial crossbreeding strategies, as they reflect a higher influence of non-additive genetic effects.

Crossbreds did not show higher rearing and fattening weights than purebreds. Heritabilities estimated for these traits indicate that both weights are influenced primarily by additive genetic variance, and purebreeding based on family selection is a powerful tool for improving these traits (Gjedrem, 1983). For percentage of fillet, crossbreds out-performed purebreds. In Atlantic salmon, no heterosis for dressing percentage in strain crosses has been demonstrated (Gjerde and Refstie, 1984), and heritabilities for this trait are very low (Gjerde and Gjedrem, 1984). Similar heritabilities have been reported for dressing percentage in rainbow trout (Gjerde and Gjedrem, 1984). The results obtained here indicate that crossbreeding may improve performance for this trait. In general, strains with the highest growth rate had the poorest fillet percentage, which agrees with earlier studies by Morkramer et al. (1985). Ranking of genetic groups according to their absolute weight of fillet, however, was not different from a ranking by total weight.

Although significant line and heterosis effects for percentage of fat have been found according to the results of Ayles et al. (1979), there seems to be no important effect on carcass quality and processing of fish, as long as the fat percentage remains in the observed range of 6.5 to 7.5. How important the significant differences in the percentage of dry matter between strains are depends on the future processing of fish.

Finally, the following conclusions can be drawn from this study.

Significantly different average line effects for all traits examined indicate distinct strain differences. Consequently, selection of superior strains with regard to specific traits is of considerable importance.

Systematic crossbreeding programs for improving growth are not advisable at present because of insignificant heterosis effects. However, strategies to upgrade slow growing strains by introduction of genes from fast growing strains are recommended.

Long selection for growth may result in antagonistic development of carcass and meat quality. In such cases combination of both traits might be more efficiently obtained by systematic crossbreeding strategies with specialized sire and dam lines.

ACKNOWLEDGEMENT

The authors are grateful to the Deutsche Forschungsgemeinschaft for supporting this study.

REFERENCES

Ayles, G.B. and Baker, R.F., 1983. Genetic differences in growth and survival between strains and hybrids of rainbow trout (*Salmo gairdneri*) stocked in aquaculture lakes in the Canadian Prairies. Aquaculture, 33: 269–280.
Ayles, G.B., Bernard, D. and Henzel, M., 1979. Genetic differences in lipid and dry matter content between strains of rainbow trout (*Salmo gairdneri*) and their hybrids. Aquaculture, 18: 253–262.
Büyükhatipoglu, S. and Holtz, W., 1984. Sperm output in rainbow trout (*Salmo gairdneri*) — effect of age, timing and frequency of stripping and presence of females. Aquaculture, 37: 63–71.
Craik, G.C.A. and Harvey, S.M., 1984. Egg quality in rainbow trout. The relation between egg viability, selected aspects of egg composition, and time of stripping. Aquaculture, 40: 115–134.
Eisen, E.J., Hörstgen-Schwark, G., Saxton, A.M. and Bandy, T.R., 1983. Genetic interpretation and analysis of diallel crosses with animals. Theor. Appl. Genet., 65: 17–23.
Gall, A.E. and Gross, S.J., 1978a. A genetic analysis of the performance of three rainbow trout broodstocks. Aquaculture, 15: 113–127.
Gall, G.A.E. and Gross, S.J., 1978b. Genetic studies of growth in domesticated rainbow trout. Aquaculture, 13: 225–234.
Gardner, C.O. and Eberhart, S.H., 1966. Analysis and interpretation of the variety cross diallel and related populations. Biometrics, 22: 439–452.
Gjedrem, T., 1975. Possibilities for genetic gain in salmonids. Aquaculture, 6: 23–29.
Gjedrem, T., 1983. Genetic variation in quantitative traits and selection breeding in fish and shellfish. Aquaculture, 33: 51–72.
Gjerde, B. and Gjedrem, T., 1984. Estimates of phenotypic and genetic parameters for carcass traits in Atlantic salmon and rainbow trout. Aquaculture, 36: 97–110.
Gjerde, B. and Refstie, T., 1984. Complete diallel cross between five strains of Atlantic salmon. Livestock Prod. Sci., 11: 207–226.
Harvey, N.R., 1977. User's Guide for LSML 76 (Mixed Model Least Squares and Maximum Likelihood Computer Program). Ohio State University, Polykopie, 80 pp.
Hershberger, W.R., 1978. Use of interpopulation hybridization in development of coho salmon stocks for aquaculture. In: J.W. Avault (Editor), Proceedings of the 9th Annual Meeting, World Mariculture Society. Louisiana State University Press, Atlanta, GA, pp. 147–156.
Holtz, W., Stoss, J. and Büyükhatipoglu, S., 1977. Beobachtungen zur Aktivierbarkeit von Forellenspermatozoen mit Fruchtwasser, Bachwasser und destilliertem Wasser. Zuchthygiene, 12: 82–88.
Kim, B.C., 1984. Fleischqualitätsuntersuchungen bei Regenbogenforellen. Diss., Göttingen, 138 pp.
Klupp, R. and Pirchner, F., 1975. Zuchtmassnahmen bei der Regenbogenforelle. AFZ-Fischwaid, 100: 511–514.

Linder, A., Sumari, O., Nyholm, K. and Sirkomaa, S., 1983. Genetic and phenotypic variation in production traits in rainbow trout strains and strain crosses in Finland. Aquaculture, 33: 129–134.

Meyer, J.-N., Hörstgen-Schwark, G., Kim, B.C. and Fricke, H., 1984. Wasserbindungsvermögen im Forellenmuskel. Z. Tierz. Zübiol., 101: 389–393.

Morkramer, S., Hörstgen-Schwark, G. and Langholz, H.-J., 1985. Comparison of different European rainbow trout populations under intensive production conditions. Aquaculture, 44: 303–320.

Reichle, G. and Bergler, H., 1980. Herkunftsvergleich mit 2-jährigen Regenbogenforellen. Fischer Teichwirt, 2: 105–108.

Reinitz, G.L., Orme, L.E., Lehm, C.A. and Hitzel, F.N., 1978. Differential performance of four strains of rainbow trout reared under standardized conditions. Prog. Fish Cult., 40: 21–23.

Vencovsky, R., 1970. Alguns aspectos teóricos e aplicados relativos a cruzamentos dialelicos de variedades. Tese de Livre-Docencia. ESALQ-USP. Piracicaba, Brazil, 150 pp.

Wohlfarth, G.W., Feneis, F., Hulata, G. and Lukowicz, v. M., 1984. Application of Selective Breeding of the Common Carp to European Aquaculture. Meeting 6 March 1984, GKSS, Forschungszentrum Geesthacht, 13 pp.

Genotype—Environment Interactions for Growth of Rainbow Trout, *Salmo gairdneri*[1]

R.N. IWAMOTO*, J.M. MYERS and W.K. HERSHBERGER

School of Fisheries WH-10, University of Washington, Seattle, WA 98195 (U.S.A.)

*Present address: Domsea Farms, Inc., Bremerton, WA 98312 (U.S.A.)

[1]The project was supported by U.S. Department of Agriculture Grant # 82-CRSR-2-1064.

ABSTRACT

Iwamoto, R.N., Myers, J.M. and Hershberger, W.K., 1986. Genotype—environment interactions for growth of rainbow trout, *Salmo gairdneri*. *Aquaculture*, 57: 153-161.

Three strains of rainbow trout (*Salmo gairdneri*) and the six possible crosses among those strains were reared under 12 different treatments for 6 months. The three strains originated from: (1) a commercial broodstock operation, and (2) the U.S. Fish and Wildlife Genetics Station at Leetown, WV. The 12 rearing treatments were formed by cross-classifying three feeding levels (commercial feed chart, 1.5 times feed chart, and demand feeding) and four static loading density levels (8.0, 16.0, 24.0, and 32.0 kg/m^3).

Genotype—environment (GE) interactions were not major contributors to the total variance of growth, although they were significant factors in the analyses of variance. Response equations for growth rate indicated that a majority of rank changes would occur at growth rates below those measured in this study. This suggested that major GE interactions for the strains and strain crosses would not be expressed fully under the ration levels and densities evaluated. However, the results clearly indicated that there were significant differences in production potential among the strains and strain crosses. Judicious selection of strains and/or strain crosses adapted for aquacultural applications may therefore lead to increased productivity.

INTRODUCTION

The development of genetic strains with desirable performance attributes is one of the basic foundations for the efficient production of any plant or animal species. The farming of rainbow trout (*Salmo gairdneri*) has been based on this tenet, and currently, numerous strains have been developed and are available for commercial culture. Through selection and breeding, generally for performance in specific locations and environments, reproducible differences among the population gene pools have arisen (Kincaid, 1981). The characterization of these differences has been the subject of many studies (e.g., Gall and Gross, 1978; Reinitz et al., 1978; Kincaid, 1981).

The variation in response of specific genotypes under different rearing conditions [genotype—environment (GE) interactions] is another important element in successful strain development. Depending on the magnitude of such interactions, strains can be characterized as "general purpose" broodstocks that perform moderately well under a range of culture conditions or "special purpose" broodstocks that have desirable traits for specific conditions.

As pointed out by Ayles and Baker (1983), observations of the importance of GE interactions with fish have been variable. Chapman (1968) indicated earlier that this may be the norm for other agricultural species. Furthermore, the detection of important GE interactions is largely unpredictable a priori and may require testing on a case-by-case basis.

This study was consequently designed to provide initial estimates of the importance of several density and ration levels on growth in rainbow trout. Three genetically distinct strains and the reciprocal strain crosses were reared under each possible density—ration level combination. Analyses of the data allowed us to determine the nature and extent of strain—density—ration interactions for the conditions of the present study.

METHODS

Strains

Gametes were obtained from two sources: Trout Lodge, Inc. (McMillin, WA), a commercial broodstock operation, and the Fish Genetics Station (Leetown, WV) of the U.S. Fish and Wildlife Service. Rearing performance and history of the strains are available only for the two Fish Genetics Station strains (FGS-22 and FGS-23) and have been summarized by Kincaid (1981). Gametes from each strain were divided into three equal aliquots and mixed to produce the three strains and six strain crosses.

Incubation and initial rearing

Following fertilization, individual strains and strain crosses were placed in vertical flow incubation trays (Heath-Techna, Inc., Kent, WA) at the University of Washington, School of Fisheries hatchery for incubation. Water source for this and all other phases of the experiment was Lake Washington. After 45 days of incubation (611.2 degree-days), three replicates of 1000 individuals from each cross were placed in separate 200-l rearing circulars. Metabolic loading densities were established initially at 0.5 kg l^{-1} min^{-1} (static loading density $=5.3$ kg/m^3) and never exceeded 1.0 kg l^{-1} min^{-1}. Rearing water was heated initially and later used at ambient temperature. Mean water temperature was 13.1°C with a minimum of 11.6°C and a maximum of 19.4°C. Diet for swim-up fry consisted of vacuum-dried salmon meal prepared at the School of Fisheries. Once active feeding had commenced, salmon meal was replaced with a commercial diet (Cenex Inc., Tacoma, WA). Daily rations were set at 120% of the manufacturer's recommendations for fish size and water temperature.

Final rearing

At approximately 10 g body weight (98 days post-incubation), fish of each cross received either freeze brands (eight crosses) or adipose fin excision (one cross) for subsequent identification. Equal weights of each cross were combined and reared under each of the ration—density level combinations for the evaluation of strain by ration level by density interactions. Three ration levels [commercial feed chart, 1.5×feed chart, and demand feeding by use of self-feeders (Babington Enterprises, Inc., Buhl, ID)] and four static loading density levels (8.0, 16.0, 24.0, 32.0 kg/m^3) were evaluated. Fish of the different genotypes were approximately the same weight at the beginning of this rearing phase, minimizing size-related biases in subsequent performance. However, because of the mixed genotype rearing situation, the possibility of intergenotypic competition could not be eliminated nor did the experimental design allow its evaluation.

Fish were reared in six compartmentalized raceways at the School of Fisheries hatchery. Initial densities were established at one-half of the desired test densities by manipulating water depth and length of each raceway section. Densities were periodically adjusted at subsequent sampling points. Water flow per raceway averaged 76 l/min. After 117 days, fish of each genotype-treatment group were transferred en masse to the School of Fisheries satellite hatchery at Seward Park, Seattle, WA. There, each group was placed in one of 12 wedge-shaped compartments of a 13-m diameter pool. Each compartment received an inflow of 38 l/min. At 288 days post-incubation, fish were sampled, and genotype-treatment groups that had reached an average fish weight of 120 g were terminated. All remaining groups were terminated and final data taken at 337 days post-incubation. Facility limitations prohibited further rearing of genotype-treatment groups once a maximum average weight of 120 g/fish was attained.

Response variables

Individual lengths and wet weights of fish from each of the 108 genotype-treatment groups were measured at the beginning, mid-point, and termination of the experiment. Growth rates were evaluated by determination of mean daily weight gains over sampling intervals.

Although the differences in inital weight among the crosses stocked in the rearing treatments were small (Table 1), the potential bias in correlation of

TABLE 1

Weight (g) of fish at initial ponding for the different density (D; kg/m^{-1}) and ration treatments (R; FC=feed chart; DF=demand feeder)

Treatments		Crosses								
D	R	TL × TL	TL × 22	TL × 23	22 × TL	22 × 22	22 × 23	23 × TL	23 × 22	23 × 23
8.0	FC	8.36	8.83	9.35	10.25	9.93	8.36	9.08	9.08	9.08
	1.5FC	10.25	9.35	8.36	10.25	9.08	8.83	9.63	9.63	8.59
	DF	7.95	9.08	9.08	9.35	10.25	9.08	11.35	9.93	9.63
16.0	FC	8.59	9.21	9.49	9.21	9.93	9.35	10.42	9.49	9.63
	1.5FC	8.71	9.49	9.21	8.59	8.95	9.21	9.93	9.63	9.35
	DF	8.47	8.47	9.49	10.09	8.71	8.83	9.78	8.36	9.78
24.0	FC	8.59	9.21	8.83	8.83	9.35	9.08	8.71	8.71	8.05
	1.5FC	9.21	9.63	9.08	8.36	9.21	9.08	9.49	9.08	9.21
	DF	8.71	9.21	8.95	9.49	8.95	8.47	9.78	8.47	9.08
32.0	FC	8.44	8.67	9.44	9.26	9.17	8.51	9.44	9.73	9.35
	1.5FC	8.83	8.99	9.44	8.83	9.26	8.44	9.53	9.73	8.83
	DF	8.99	9.53	8.99	9.35	9.35	8.29	10.04	9.53	8.83

subsequent weight gain with initial weight justified the transformation of observed weight gains, using procedures of Moav et al. (1975). The corrected weight gain values were used to determine corrected final weights (CRWT) and corrected mean daily weight gains (CDWTG).

Statistical analyses

Ration level and density were analyzed as cross-classified factors with genotypes nested within each treatment combination. All factors were considered as fixed effects. Analyses of variance of individual observations were performed to detect statistical significance and also to determine the relative importance of each component of variance. Genotype—environment interactions were also evaluated by determining the conditions under which changes in rank among the genotypes would be expected. For these analyses, corrected mean daily weight gains for each genotype were regressed with the mean daily weight gain of all genotypes for each density—ration combination. The slopes of the different regression lines were used to indicate the degree to which each genotype performed as environmental conditions improved. The regression equations were also used to predict the performance of the various crosses at conditions under which mean daily weight gains of 0.1000, 0.3276, 0.5726, 0.8146, and 2.000 g/day could be achieved. The lowest value represents situations where environmental quality is poor, the three intermediate values represent fairly typical grow-out situations, and the latter value is indicative of a situation under which growth is optimal (unpublished data).

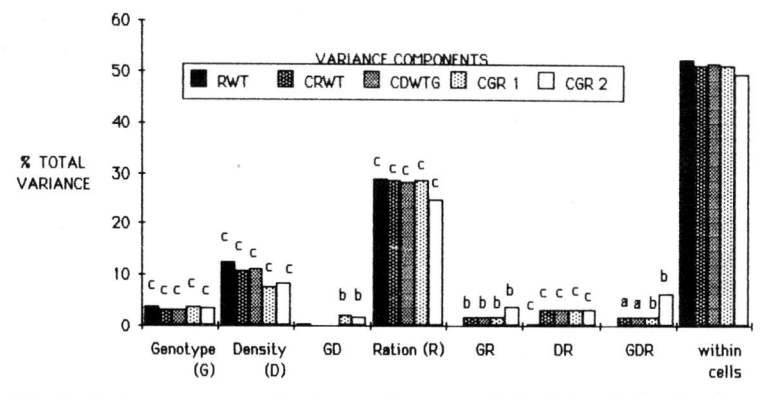

Fig. 1. Variance components expressed as percent of total variation for uncorrected final weights (RWT), corrected final weights (CRWT), corrected mean daily weight gain (CDWTG), corrected instantaneous growth rate per day (CGR1) and per degree-day (CGR2). Statistical significance for each component is indicated by: $a = P \leq 0.05$; $b = P \leq 0.01$; $c = P \leq 0.001$.

RESULTS

The analyses of variance and variance component partitioning for uncorrected final weights indicated that while the main effects including genotype, density, and ration level were significant, interactions were relatively minor. For example, the combined contributions of density and ration levels were responsible for greater than 40% of the total variation for uncorrected final weight while two-way and three-way GE interactions did not exceed 2.0% (Fig. 1).

When final weights were corrected for initial differences in weight, the importance of GE interactions became evident. The interaction terms involving strains and ration levels in particular were significant. The dominant effects, however, continued to be the primary treatments of density and ration levels. As Fig. 1 indicates, none of the interaction terms exceeded 10% of the total variation.

Because two final samples at different times (288 and 337 days post-incubation) were made, the growth rate analyses permitted evaluation of a larger sample size by correcting for differences in number of days (CGR1) and degree-days reared (CGR2). As the results for the different growth rate variables indicate (Fig. 1), genotype, ration, and density were extremely important variables. In addition, two-way and three-way interactions involving the main effects (GR and GDR) were significant, indicating variability in cross responses to the various ration and density treatments. The variability in strain response was evident once response surfaces were plotted for the various strain—density—ration level combinations (Fig. 2). Fish fed on a feed chart basis displayed very similar mean daily weight gains. As the level of feeding increased to 1.5× feed chart levels, variability among the different strains,

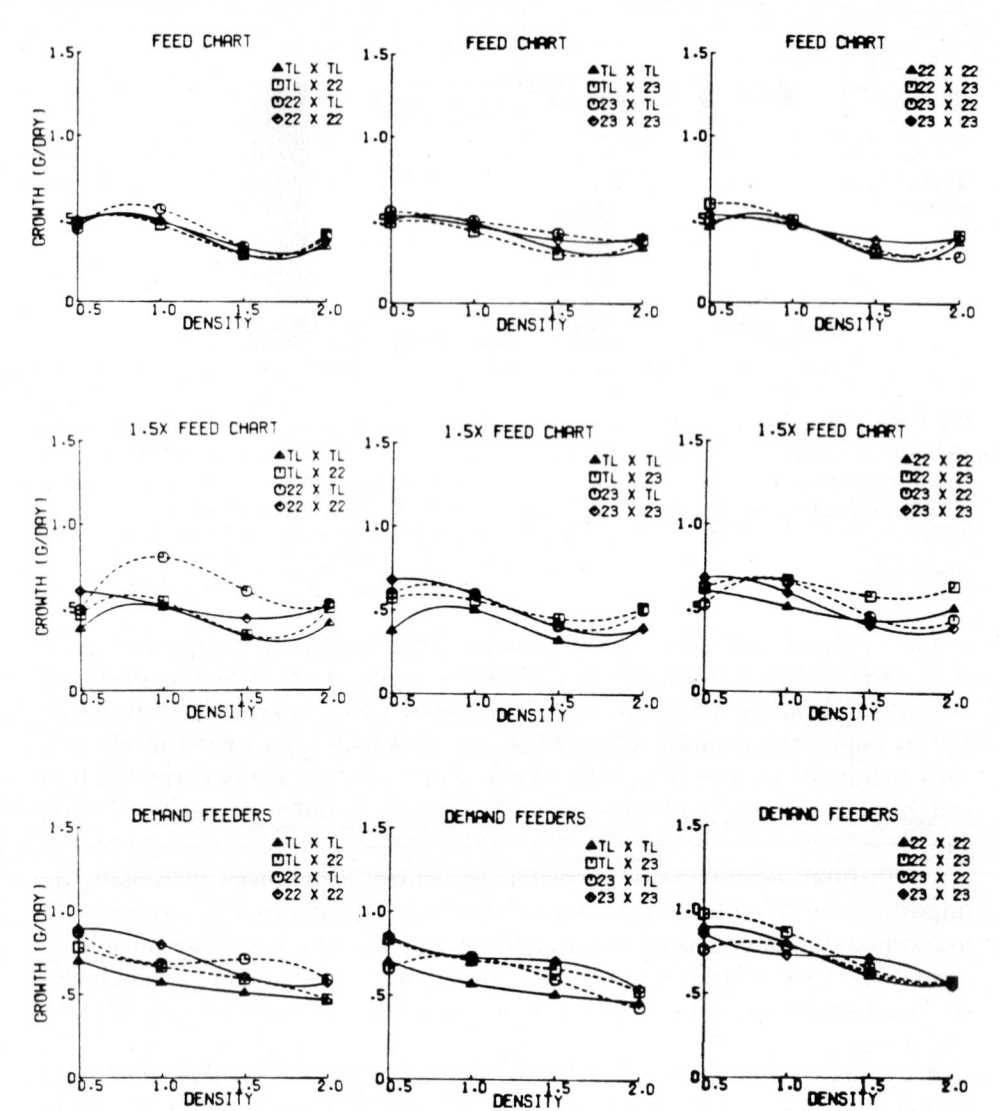

Fig. 2. Corrected mean daily weight gain values for different genotype—density combinations at given ration levels.

particularly that of 22×TL, increased. When food was presented on a demand basis, the differences among the genotypes decreased.

The results from the regression of the mean daily weight gain of each genotype with the mean daily weight gain of all genotypes per density—ration combination are presented in Table 2. The slopes of the different regression lines (each of which was highly significant; $R^2 \geq 0.82$) indicate the degree to which each genotype performs as environmental conditions improve. The growth rate

TABLE 2

Slopes (b), intercepts (a), and R^2 values for response equations from regressing corrected daily weight gain for each cross on the mean daily weight gain for each treatment

Cross		Parameter estimate		
Female	Male	b	a	R^2
TL	TL	1.1206	0.0262	0.827
	22	0.9777	0.0543	0.922
	23	0.9209	0.0449	0.962
22	TL	0.7454	0.1072	0.766
	22	0.7891	0.1086	0.925
	23	0.7450	0.0813	0.922
23	TL	1.0864	−0.0368	0.775
	22	0.8641	0.0788	0.925
	23	0.8201	0.0805	0.864

results for the different genotypes at the different mean daily weight gains are shown in Fig. 3. It became apparent that one strain, TL×TL, and one strain cross, 23×TL, clearly surpassed the other crosses with environmental improvements. The results also indicated that several important rank changes among the genotypes would occur with improvements in rearing conditions, particularly at the lower environmental quality levels. For example, under lower environmental quality levels, the TL×TL strain would be expected to perform relatively poorly. However, with improved conditions, this strain would outperform the other genotypes. Similar performance would be predicted of the

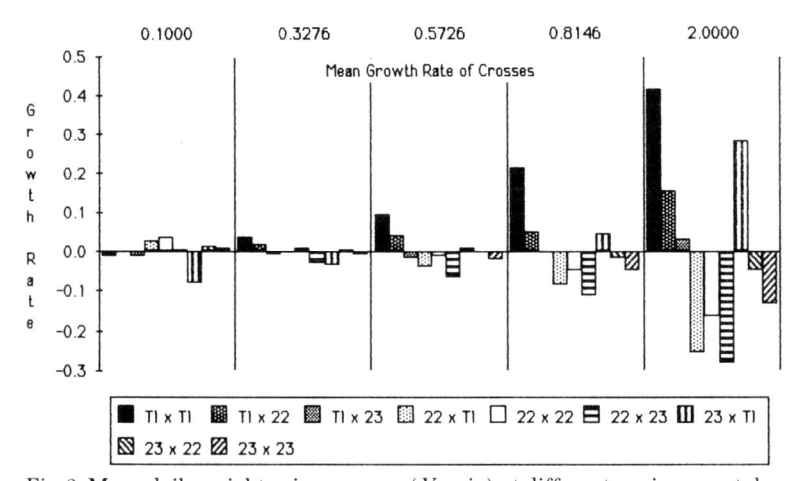

Fig. 3. Mean daily weight gain per cross (Y-axis) at different environmental conditions (X-axis).

TL×22 and 23×TL strain crosses. Clearly, genotype—environment interactions involving both changes in rank and in magnitude would be expected under different ration and feeding levels.

Additionally, when the derived equations for mean daily weight gain were surveyed pairwise for intercepts, 31 intercept points were found, while five genotype-treatment combinations never intercepted to the right of the X axis. Of the 31 intercepts, 17 (54.8%) fell between a mean daily weight gain of 0 and 0.25 g/day, 10 (32.2%) occurred between 0.25 and 0.50 g/day, and only four (12.9%) occurred above the 0.50 g/day mean daily weight gain line.

DISCUSSION

As main effects, environmental treatments were considerably larger contributors to the total variation than the strain or genetic differences. Results of the different density and ration levels confirmed our initial expectations of the importance of those two rearing variables on growth. The magnitude and importance of environmental factors on rainbow trout growth have been documented by other investigators. Donald and Anderson (1982) found that 72% of the total variation in weight of 2-year-old trout stocked in mountain lakes could be attributed to stocking density and overall productivity of food organisms. In another study, Ayles and Baker (1983) determined that growth differences among rainbow trout strains stocked in central Canadian prairie lakes were primarily due to lake-to-lake variability.

Strain differences accounted for significantly less of the total variation. Gjerde and Gjedrem (1984) found that the strain effect in Atlantic salmon crosses accounted for 6.4%, 7.0%, and 0.2% of the total variance in body weight, length, and dressed yield, respectively. Results presented by Webster and Flick (1981), Fraser (1981), and Ayles and Baker (1983), however, indicated that the choice of strains evaluated may dictate the magnitude of strain differences.

Rank changes among genotypes under varying environments, a major indicator of GE interactions, were quite evident in this study. The rapid improvement in rank of the TL×TL strain from the seventh best to the best genotype over a relatively small change in environmental conditions exemplified the importance of GE effects. The performance of the TL×TL strain relative to the other genotypes may be due in part to the past selection of the Trout Lodge strain for improved growth. This genetic advantage also appears to be expressed in the TL×22, TL×23, and 23×TL strain crosses which ranked sixth, eighth, and ninth under slow growth conditions (CDWTG <0.1) but under intermediate conditions (CDWTG=0.9), their ranks improved to third, fourth, and second, respectively.

GE interactions were not significant components of the total variance, although they were significant factors in the analyses of variance for most growth traits during the final rearing phase. It should be noted that in the response equations, 58.1% of the rank changes occurred at growth rates below those found in this study. This would suggest that the major GE interactions for these strains and strain crosses would not be fully expressed under the ration levels and densities used. The environmental parameters evaluated correspond with those normally found in production situations. Thus the importance of ration—density interactions with genotype may in practical terms be limited. Clearly, however, the results did indicate that there are significant differences in production potential among the strains and strain crosses evaluated. Judicious selection of strains and/or strain crosses adapted for specific aquacultural applications may eventually lead to increased productivity.

REFERENCES

Ayles, G.B. and Baker, R.F., 1983. Genetic differences in growth and survival between strains and hybrids of rainbow trout (*Salmo gairdneri*) stocked in aquaculture lakes in the Canadian prairies. Aquaculture, 33: 269-280.

Chapman, A.B., 1968. Genetic—environmental interactions in laboratory and farm mammals. Proc. 17th Natl. Poultry Breeders Roundtable, Kansas City, MO, 25 pp.

Donald, D.B. and Anderson, .S., 1982. Importance of environment and stocking density for growth of rainbow trout in mountain lakes. Trans. Am. Fish. Soc., 111: 675-680.

Fraser, J.M., 1981. Comparative survival and growth of planted wild, hybrid, and domestic strains of brook trout (*Salvelinus fontinalis*) in Ontario lakes. Can. J. Fish. Aquat. Sci., 38: 1672-1684.

Gall, G.A.E. and Gross, S.J., 1978. A genetics analysis of the performance of three rainbow trout broodstocks. Aquaculture, 15: 113-127.

Gjerde, B. and Gjedrem, T., 1984. Estimates of phenotypic and genetic parameters for carcass traits in Atlantic salmon and rainbow trout. Aquaculture, 36: 97-110.

Kincaid, H.L., 1981. Trout Strain Registry. U.S. Fish and Wildl. Serv., National Fisheries Center Leetown, Kearneysville, WV, 118 pp.

Moav, R., Hulata, G. and Wohlfarth, G., 1975. Genetic differences between the Chinese and European races of the common carp. 1. Analysis of genotype—environment interactions for growth rate. heredity, 34: 323-340.

Reinitz, G.L., Orme, L.E., Lemm, C.A. and Hitzel, F.N., 1978. Differential performance of four strains of rainbow trout reared under standardized conditions. Prog. Fish Cult., 40: 21-23.

Webster, D.A. and Flick, W.A., 1981. Performance of indigenous, exotic, and hybrid strains of brook trout (*Salvelinus fontinalis*) in waters of the Adirondack mountains, New York. Can. J. Fish. Aquat. Sci., 38: 1701-1707.

Response of Channel Catfish to Multi-Factor and Divergent Selection of Economic Traits[1]

K. BONDARI

Coastal Plain Experiment Station, University of Georgia, Tifton, GA 31793 (U.S.A.)

ABSTRACT

Bondari, K., 1986. Response of channel catfish to multi-factor and divergent selection of economic traits. *Aquaculture*, 57:163—170.

Three experiments were conducted to evaluate responses of channel catfish (*Ictalurus punctatus*) to divergent selection for commercially important traits. Experiment 1 included lines selected upward (WT+) and downward ((WT−) for 40-week body weight. Catfish in Experiment 2 were selected in two opposite directions (CV+ and CV− lines) for body weight heterogeneity determined by coefficient of variation. Experiment 3 included catfish selected simultaneously in opposite directions (PR+ and PR− lines) for four traits. All three experiments included a control line propagated by random breeding each generation. Results of two generations of selection for Experiments 1 and 2 and three generations for Experiment 3 indicated that: (1) responses to divergent selection for body weight were asymmetrical and varied between generations; (2) genetic and environmental differences among lines replicated in different years strongly influenced selection response for body weight; (3) WT+ and WT− lines diverged in body weight after one generation of selection and remained significantly different in the second generation; (4) channel catfish did not respond to two generations of selection for body weight heterogeneity; (5) after three generations of selection, PR+ catfish were heavier, longer, and more efficient in feed conversion than the control fish; and (6) correlated responses for total length generally agreed with observations made on direct responses for body weight.

INTRODUCTION

The goals of catfish breeding programs usually include production of rapidly growing, uniformly sized, feed-efficient, and disease-resistant catfish. A review of the literature, however, reveals that very little research has been carried out with channel catfish to realize these goals.

Rapid growth rate in channel catfish results in reduced time to market size, which can result in substantial increase in profits for commercial producers. Bondari (1983a) reported that one generation of between- and within- family

---

[1]Supported by State and Hatch funds allocated to the Georgia Agricultural Experiment Station.

selection applied to channel catfish for increased and decreased body weight changed the 40-week body weight by 20% in both directions. Response to selection for improved growth rate is directly related to the heritability of this trait. According to Bondari (1983a) the realized heritability estimate for body weight is 0.10, which is considerably lower than heritability values estimated from the sire or dam components of variance. Estimates of realized heritability reported by Dunham and Smitherman (1983) for body weight of three strains of channel catfish ranged from 0.24 to 0.50, which are considerably higher than the 0.10 reported by Bondari (1983a). The two studies, however, differed in selection procedure since Bondari's (1983a) experiment included sets of full-sibs reared together but the experiment of Dunham and Smitherman did not. No other estimate of realized heritability for growth rate in channel catfish has been reported to date and the estimate given by Bondari (1983a) is not large enough to permit rapid improvement through mass selection. Thus, as recommended by Bondari (1983a,b) and Bondari et al.(1985), a combination of between- and within-family selection should be applied to improve the growth rate in channel catfish.

Genetic improvement of the efficiency of feed conversion in channel catfish has received no attention in selection programs. One of the major reasons is the difficulty in accurately measuring individual feed consumption or weight gain. Genetic progress made from selection for improved feed efficiency, however, cannot be reliably predicted unless the heritability estimate for this trait is determined. Since feed efficiency is expressed as the ratio of feed to gain, direct selection for feed efficiency may not be the most effective method of improving this trait. It is the objective of this study to determine if direct selection for body weight will result in correlated responses for total length or feed efficiency.

This study was also designed to determine the response of channel catfish to (1) divergent selection of body weight and size uniformity; and (2) multifactor selection of economic traits including body weight, total length, weight heterogeneity, and survival. A combination of between- and within-family selection was applied to investigate these objectives. The experimental fish were selected based on family means and the within-family selection was then applied to reduce the inbreeding rate.

MATERIALS AND METHODS

Three experiments, comprised of three lines each, were conducted using the fish from a base population established in 1974. Base population fish were derived from six stocks obtained from five different hatcheries as described by Bondari (1983a). Experiment 1 included lines selected upward (WT+) and downward (WT−) for 40-week body weight. Catfish in Experiment 2 were selected in two opposite directions (CV+ and CV− lines) for body weight heterogeneity determined by the family coefficient of variation. WT+ and WT− line fish were selected on the basis of both family means and individual measurements within selected families, whereas CV+ and CV− lines were selected based only on the family coefficient of variation for 40-week body weight. Experiment 3 included catfish selected simultaneously in opposite directions (PR+ and PR− lines) for four traits (40-week body weight, total length, 8–16-week survival, and weight heterogeneity). All three experiments included an unselected control line propagated by random-breeding each generation.

Fish from the PR− line were kept for only one generation, whereas the PR+ line was maintained for three generations with a generation interval of 3 years. Experiments 1 and 2 were continued for two generations. Whenever possible, a repeat mating trial was conducted in a following year to determine the repeatability of response to selection. Full-sib families were separately raised in indoor tanks (two tanks per family) up to 40 weeks of age and then were transferred to outdoor ponds or raceways. Individual spawning pens, rearing techniques, indoor tanks, out door facilities, feeding and management have been described previously by Bondari (1982, 1983a, 1984) and Bondari et al.(1985).

At 40 weeks of age, fish were selected primarily by their family means. Whenever possible, at least four full sib families were included in each selected line. A representative random sample of all available families was included in the control line. Within-family selection was then applied to reduce the number of fish in each full-sib family (except for the control line) to 20 males and 20 females. These fish were kept under similar culture conditions to become the parents of the next generation. All fish were heat-branded to avoid full-sib mating. In generations 1 and 3, progeny from control and selected lines were performance tested in raceways and tanks, respectively.

Statistical analyses were conducted on body weight, total length, and family CV percent for body weight using the GLM procedure (SAS Institute, 1982). The statistical model included the effects of line, family nested within line, and tank nested within family and line. Sex of fish and sex × line interaction were included in the model for the 1979 performance test data. When significant, the family within line mean square was used to test the significance of line effect. Orthogonal contrasts were used to test the divergence (upwards vs. downward line) and asymmetry of response (average performance of up and vdown lines vs. control). Variance components were also computed to estimate heritability for growth traits as twice the among-family component of variance.

RESULTS

Means ± SD for body weight and total length, respectively, computed from 28 full-sib families in the base population were 8.3 ± 2.2 g and 98.7 ± 7.9 mm at 16 weeks of age and 172.4 ± 57.4 g and 265.1 ± 25.3 mm at 40 weeks of age.

TABLE 1

Weight (g) and total length (mm) measurements and percentage coefficient of variation (CV%) for body weight of control (CTL) and divergently selected lines of channel catfish (+ and − indicate upward and downward directions) over three generations (years combined)

Trait	Weeks of age	Exp.1 — 40-week weight			Exp.2 — weight heterogeneity			Exp.3 — Multiple traits		
		CTL	WT+	WT−	CTL	CV+	CV−	CTL	PR+	PR−
	Generation 1									
Body weight	16	5.2^b	6.5^a	4.6^c	5.2^a	5.7^a	5.4^a	4.9^b	7.2^a	4.5^b
	40	59.1^b	105.1^a	62.6^b	59.1^b	82.0^a	77.9^a	95.7^b	164.9^a	87.3^b
Total length	16	81.9^b	89.7^a	80.8^b	81.9^b	85.6^a	85.0^a	85.0^b	97.2^a	82.7^b
	40	178.8^b	210.1^a	182.4^b	178.8^b	197.3^a	195.9^a	217.6^b	256.5^a	211.6^b
Weight (CV%)	16	29.7^a	30.3^a	28.7^a	29.7^a	32.4^a	26.6^b	29.9^a	29.3^a	34.1^a
	40	43.5^a	37.6^b	41.3^{ab}	43.5^a	45.0^a	39.3^b	41.5^a	33.8^a	42.3^a
	Generation 2									
Body weight	16	5.3^b	6.1^a	4.2^c	5.3^a	5.1^a	5.3^a	7.2^b	8.7^a	
	40	44.6^b	52.2^a	24.9^c	44.6^a	38.3^a	39.7^a	62.8^b	72.7^a	
Total length	16	89.4^b	93.3^a	83.5^c	89.4^a	88.4^a	90.6^a	96.8^b	105.0^a	
	40	174.9^b	182.1^a	146.3^c	174.9^a	167.5^a	170.6^a	192.4^b	202.9^a	
Weight (CV%)	16	28.4^a	26.0^a	26.5^a	28.3^a	27.4^a	26.3^a	31.7^a	34.0^a	
	40	42.0^a	36.3^a	44.0^a	42.0^a	48.5^a	43.0^a	40.9^a	41.5^a	
	Generation 3									
Body weight	16							13.1^b	19.7^a	
	40							65.5^b	95.2^a	
Total length	16							113.7^b	132.2^a	
	40							189.8^b	224.0^a	

^{a,b,c} Means in each experiment, within a row, with different superscripts differ ($P<0.05$).

Heritability estimates from the sire component of variance (one male mated to two females) were 0.56 for body weight and 0.69 for total length at 40 weeks of age.

Least-squares means by line and experiment are given in Table 1 for generations 1–3. Divergence in body weight and total length of the fish included in Experiment 1 occured at 16 weeks and continued to 40 weeks of age in both generations. Catfish in the WT+ line (Experiment 1) were heavier and longer ($P<0.05$) at 16 and 40 weeks of age than those in CTL and WT− lines in both generations (Table 1). The CTL catfish were comparable in weight and length to WT− catfish in generation 1 but were significantly ($P<0.05$) heavier and longer in generation 2. Selected and control catfish did not differ in CV% for body weight at 16 weeks but the WT+ catfish were more homogeneous ($P<0.05$) in weight at 40 weeks of age than other lines.

CV+ and CV− catfish (Experiment 2) did not diverge in body weight or total length in either generation (Table 1). Catfish from both lines were, however, heavier and longer ($P<0.05$) than CTL line catfish in generation 1 but no among-line differences were observed in generation 2. Selection for lower CV% was effective in generation 1 since the CV− catfish were more homogeneous ($P<0.05$) in weight than both CV+ and CTL line catfish. No response to selection was observed in generation 2.

Divergence in body weight and total length of catfish included in Experi-

TABLE 2

Standard deviations (SD), heritability estimates from full-sib analysis (h^2), and response to selection (R) for body weight (BW) and total length (TL) of channel catfish in three experiments

Hatch-year	Age (weeks)	SD		h^2		R(Exp. 1)				R(Exp.3)	
		BW	TL	BW	TL	Up		Down		Up	
						BW	TL	BW	TL	BW	TL
	Generation 1										
1977	16	1.8	8.7	0.31	0.38	2.2	2.1	No[a]	No	1.3	1.4
	40	49.4	27.4	0.27	0.23	2.3	2.0	No	No	1.4	1.4
1978	16	0.4	4.4	0.47	0.50	1.0	1.2	No	No		
	40	9.0	16.6	0.47	0.62	0.9	1.1	No	No		
1979	16	3.3	11.8	0.34	0.37	No	No	0.7	0.6		
	40	29.1	24.5	0.20	0.25	0.5	0.7	0.4	0.3		
	Generation 2										
1979	16	3.6	12.5	0.40	0.45					No	No
	40	36.0	25.7	0.37	0.42					No	No
1981	16	1.5	7.7	0.48	0.51	0.5	0.5	0.7	0.8	No	0.8
	40	18.2	21.2	0.32	0.33	0.4	0.4	1.1	1.3	No	No
	Generation 3										
1982	16	8.6	19.0							0.8	1.0
	40	35.6	27.0							0.8	1.3
	Female performance test (generation 1)										
1979	46	33.5	22.5	0.05	0.05	2.6	2.5	1.1	1.5	2.3	2.0
	60	83.2	27.6	0.24	0.17	2.5	2.7	0.9	1.2	1.9	1.8
	85	106.2	31.8	0.17	0.15	2.4	2.1	1.1	1.3	1.2	1.2
	120	287.1	36.0	0.32	0.29	1.5	1.4	1.1	1.2	1.3	1.2
	Male performance test (generation 1)										
1979	46	47.6	26.5	0.05	0.04	2.2	2.3	1.3	1.8	2.3	1.8
	60	117.4	32.9	0.11	0.12	2.3	2.4	1.0	1.6	2.0	1.8
	85	140.4	36.6	0.12	0.15	2.1	2.0	1.3	1.6	1.3	1.1
	120	351.4	36.3	0.28	0.30	1.2	1.4	1.6	1.8	1.2	1.2

[a]Indicates no response to selection. Also, no response was observed in the downward direction for Exp. 3.

ment 3 occurred at 16 weeks and continued to 40 weeks of age in generation 1. The PR+ line catfish were heavier and longer ($P<0.05$) than CTL catfish in all three generations (Table 1). No line differences were observed for CV%. The PR− line was included in generation 1 only but these catfish did not differ in any respect from those in the CTL line and therefore selection for PR− traits was discontinued. Standard deviations, heritability estimates, and standardized response to selection for each hatch-year, generation, and experiment are shown in Table 2.

Results of the performance test (Experiment 1) indicated that both female and male catfish from the WT+ line remained heavier and longer ($P<0.05$) than CTL catfish, which in turn were heavier and longer than WT− catfish throughout the 85-week test period. As indicated by CV%, the WT+ catfish of both sexes were also more homogeneous ($P<0.05$) in body weight at the conclusion of the test (120 weeks) than were CTL and WT− catfish. In Experiment 2, the CV− catfish were comparable in weight to both CTL and CV+ catfish at 120 weeks but were significantly ($P<0.05$) longer. The CV− catfish of both sexes were also more homogeneous ($P<0.05$) in body weight than CTL catfish. Female and male catfish from the PR+ line (Experiment 3) remained significantly ($P<0.05$) heavier and longer than CTL catfish throughout the test and in all three generations. The two lines did not differ in CV% for body weight. Results of a 30-week performance test conducted on the third generation catfish from Experiment 3 indicated that the PR+ catfish gained 45% more weight, consumed 8% more feed, and were 29% more efficient in feed utilization than CTL catfish.

DISCUSSION

In both Experiments 1 and 3 and in all generations, mean body weight and total length of catfish selected upward were greater ($P<0.05$) than those selected downward or the unselected control line (Table 1). Persistence of genetic progress over three generations of selection and throughout the performance test coupled with the lack of evidence of a reducing trend for additive genetic variance (Table 2) confirms that (1) the application of among- and within-family selection affords considerable scope for further genetic change for growth, and (2) genetic change for body weight is associated with correlated changes in total length (Experiment 1), feed utilization efficiency, and survival (Experiment 3). Such correlated responses may reflect the multiple effects of closely linked groups of genes controlling body weight in channel catfish. This has previously been documented by Bondari (1983a). Records of performance results should be employed in identifying female and male parents for the next generation.

The increased number of traits comprising the aggregate phenotype (Experiment 3) reduced the magnitude of response to selection (Table 2), which is in agreement with the general conclusion of constructing selection indices (Hazel, 1943). Bondari (1983c) has also demonstrated that joint selection for body weight and spawn weight is not as effective as selection for body weight alone on improving growth in channel catfish. Selection indices including body weight and length measurements have been constructed by Reagan et al. (1976) to improve growth traits in channel catfish.

The application of multi-factor selection to channel catfish, however, confirms that a large reservoir of genetic diversity exists for economically important traits which could be successfully exploited. Since feed represents a major part of the cost of production, improved feed efficiency will be of special interest to catfish growers. Although direct selection for feed efficiency has not been practiced in channel catfish, the results of this study provide good evidence that selection for improved growth and survival leads to a significant correlated improvement in feed conversion.

Heritability estimates computed among full-sib families at 16 and 40 weeks of age fall mainly within the range 0.20 — 0.48 for body weight and 0.23 — 0.62 for total length (Table 2). Similar values have also been reported by El-Ibiary and Joyce (1978). Estimates obtained during performance test were somewhat lower (Table 2) but still suggest that selection for body weight and total length will be effective. Bondari (1983a), however, reported that realized heritability estimates were considerably lower, and ranged from 0.10 to 0.14 for body weight and 0.16 to 0.35 for total length of channel catfish at 40 weeks of age. The realized heritability estimates computed from the ratio of response to selection differential for the upward and downward lines, respectively, were 0.34 and 0.03 for body weight in generation 1 and 0.16 and 0.20 in generation 2. Corresponding values for total length were 0.14 and 0.07 in generation 1 and 0.23 and 0.56 in generation 2. These estimates are also considerably lower than those computed from the full-sib components of variance.

Asymmetry of selection responses was observed for 40-week body weight and total length in Experiment 1 (both generations) and Experiment 3 (generation 1). In both experiments (generation 1), the downward line did not respond to selection but the response in the upward direction was 72 — 78% greater than control for body weight and 17 — 18% for total length. In generation 2 (Experiment 1), observed responses were 17.0 and 4.2% greater than control in the upward direction and 44.2 and 16.3% in the downward direction for body weight and total length, respectively. These results confirm the findings of Bondari (1983a) that divergent selection does not yield equal rates of progress in opposite directions.

REFERENCES

Bondari, K., 1982. Interactions of sex and culture conditions on growth, survival, and quality traits of blue tilapia. Growth, 46: 238-246.
Bondari, K., 1983a. Response to bidirectional selection for body weight in channel catfish. Aquaculture, 33: 43-81.
Bondari, K., 1983b. Genetic and environmental control of fingerling size in channel catfish. Aquaculture, 34: 171-176.
Bondari, K., 1983c. Selection of body weight and spawn weight in channel catfish. Proc. Annu. Conf. Southeast. Assoc. Fish. Wildl. Agencies, 37: 348-351.
Bondari, K., 1984. Reproductive performance, growth, and survival of selected and wild × selected channel catfish. Theor. Appl. Genet., 68: 391-395.
Bondari, K., Washburn, K.W. and Ware, G.O., 1985. Effect of initial size on subsequent growth

and carcass characteristics of divergently selected channel catfish. Theor. Appl. Genet., 71: 153–158.

Dunham, R.A. and Smitherman, R.O., 1983. Response to selection and realized heritability for body weight in three strains of channel catfish, *Ictalurus punctatus,* grown in earthen ponds. Aquaculture, 33: 89–96.

El-Ibiary, H.M. and Joyce, J.A., 1978. Heritability of body size traits, dressing weight and lipid content in channel catfish. J. Anim. Sci., 47: 82–88.

Hazel L.N., 1943. The genetic basis for constructing selection indexes. Genetics, 28: 476–490.

Reagan, R.E., Pardue, G.B. and Eisen, E.J., 1976. Predicting selection response for growth of channel catfish. J. Hered. 67: 49–53.

SAS Institute Inc., 1982. SAS User's Guide: Statistics. Cary, NC, 584 pp.

Genetic Selection for Shell Traits in the Japanese Pearl Oyster, *Pinctada fucata martensii*

KATSUHIKO T. WADA

National Research Institute of Aquaculture, Nansei, Mie 516-01 (Japan)

ABSTRACT

Wada, K.T., 1986. Genetic selection for shell traits in the Japanese pearl oyster, *Pinctada fucata martensii. Aquaculture*, 57: 171-176.

Selection experiments are currently being conducted on 3-year-old Japanese pearl oysters, *Pinctada fucata martensii* for shell width and shell convexity, defined as shell width/(shell height + hinge line length + shell width). The third generation was produced in 1981, and offspring were evaluated in 1983 and 1984. Mean shell width was larger for offspring of parents selected for large shell width than in those from parents selected for large shell convexity. Realized heritabilities estimated from regression of the selection response on cumulated selection differential were also larger when selection was based on shell width than when based on shell convexity. Response based on both traits varied every generation and some lines were lost by accident. The variation in selection response is attributed to sampling error and genetic drift since small numbers of parents (6—22 individuals) and offspring were used in each generation.

INTRODUCTION

Selective breeding is a classical, fundamental approach for producing strains which are genetically improved for certain economic traits. Many attempts have been made to estimate the quantitative genetic variance of important traits in bivalve molluscs, such as growth rate, for the purpose of predicting the genetic gain by selection. Selection experiments have shown genetic progress being made in quantitative traits by selective breeding. These kinds of studies have been reported mostly in oysters and clams (Chanley, 1961; Haskin and Ford, 1978; Beatie et al., 1980; Newkirk, 1980; Newkirk and Haley, 1983; Hershberger et al., 1984).

Wada (1984) reported on first and second generation responses to selection for shell width and shell convexity at 3 years of age in the Japanese pearl oyster, *Pinctada fucata martensii*. Shell convexity was calculated as shell width/(shell height + hinge line length + shell width). Realized heritability estimates for these traits were 0.310 and 0.320, respectively. This paper deals with the evaluation of the third generation response.

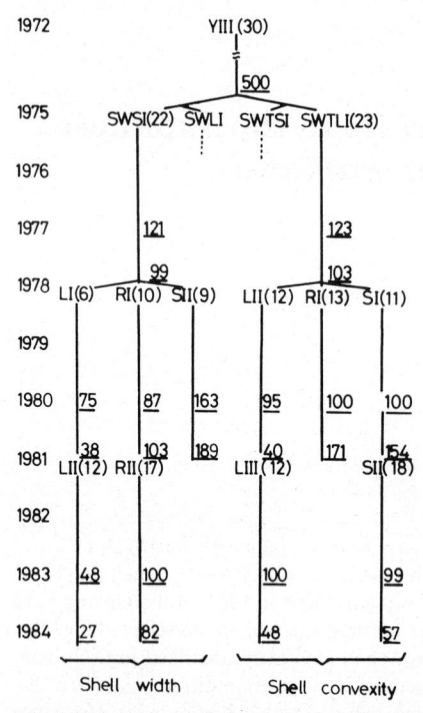

```
1972                    YIII (30)
                           |
                          500
1975        SWSI(22)  SWLI  SWTSI  SWTLI(23)
                :          :
1976
1977           |121               |123
               | 99               |103
1978    LI(6)  RI(10) SI(9)  LII(12) RI(13) SI(11)
1979
1980    |75    |87    163    |95    |100    |100
        |38   |103    189    |40    |171    |154
1981   LII(12) RII(17)       LIII(12)       SII(18)
1982
1983    |48   |100           |100           |99
1984    |27   |82            |48            |57
        _____/  _____/
           Shell width          Shell convexity
```

Fig. 1. Selection experiments on shell width and shell convexity. Figures in parentheses and those underlined refer, respectively, to number of selected parents which spawned and number of offspring measured. In the line designations Y stands for selected for yellow prismatic layer; SW, shell width; SWT, shell convexity; L, large size; R, random control line; S, small size; and I, II and III refer to generations 1, 2 and 3, respectively.

MATERIALS AND METHODS

The base population for these selection experiments was a group of cultured pearl oysters which had been selected for yellow coloration in the prismatic layer of the shell for three generations in Ago Bay, Mie Prefecture, Japan. The frequency of the yellow pearl coloration was 91% in this population. The initial source of the selected animals was a cultured population in Ehime Prefecture, Japan. About 30 parents were used in every generation.

The selection protocol for shell width and shell convexity is shown in Fig. 1. In 1975, five hundred pearl oysters from the 1972 year-class of the Ago Bay population (YIII) were numbered, and their shell height, hinge line length and shell width measured (Wada, 1984). Spawners for the first generation of selection were derived from two groups of oysters: (1) those having largest (SWLI) or smallest (SWSI) shell width, and (2) those having the largest (SWTLI) or smallest (SWTSI) shell convexity. The second generation (1978) of the small line for shell width (SII) was produced by mating, using the same methods as in 1975. From the "SWSI" line, a large selected line (LI) and a random control line (RI) were also produced in 1978. The parents for the random control line (RI) were selected by using the table of random numbers. Subsequent generations were obtained in the large (LII) and random (RII) lines by the same method from the "LI" and "RI" lines, respectively. The third generation of the small line (SII) was lost by accident in 1981. Subsequent generations of selection lines (Fig. 1) for shell convexity were obtained in the same way as those selected for shell width as described above.

Shell length, hinge line length and shell width were recorded at 2 and 3 years, and dried shell weight was recorded at 3 years in each generation of every line. The number of selected parents which spawned and number of offspring measured for every generation of each line are shown in Fig. 1.

Spawning of the selected parents was induced by thermal stimulation of the oysters in trays, and the individuals which spawned were recorded. Fertilized eggs were washed and embryos were placed in 20-l tanks until the D-shaped larval stage. Larvae were transferred 24 h after fertilization to 60-l tanks and reared in filtered sea water at 27 $\pm$ 1 °C. Initial density of D-shaped larvae was about 1 /ml. Larvae were fed on algal diet (a mixture of *Pavlova lutheri* and *Chaetoceros calcitrans*) for 30 days at rates of 5×10^3 cells/l on days 0—5, 2×10^4 cells/l on days 11—20 and 3×10^4 cells/l on days 21—30. Every other day 30 l of sea water were replaced in each tank and water was completely changed on days 5, 10 and 15. At approximately 20 days, knitted plastic film was suspended in each culture to provide substrate for metamorphosis. Juveniles were reared to a size of 1—2 mm in the hatchery and then suspended in floating nets in Ago Bay. Animals as large as 5—10 mm were removed from films and reared in the nets with larger mesh. Shells were cleaned and placed in new nets at appropriate intervals.

RESULTS AND DISCUSSION

The changes in mean shell width (SW), shell height (SH), hinge line length (HL) and shell weight (LWT,RWT) for the three experimental lines of selection on shell width are shown in Fig. 2. The generation 2 mean shell width in the line selected for large width (LII) was significantly greater than that of generation 1 (LI) at 2 years, but not at 3 years. The decreased response of the large line at 3 years can be ascribed principally to adverse environmental conditions between the second and third years, because the response of the random line also markedly decreased at 3 years. The changes in other traits such as shell height, hinge line length and shell weight showed the same tendency as did shell width. Thus, high genetic correlations might be expected between these traits and shell width.

Shell convexity of the large line from generation 2 (LII) to generation 3 (LIII) significantly increased at 2 years of age, although it was insignificant at

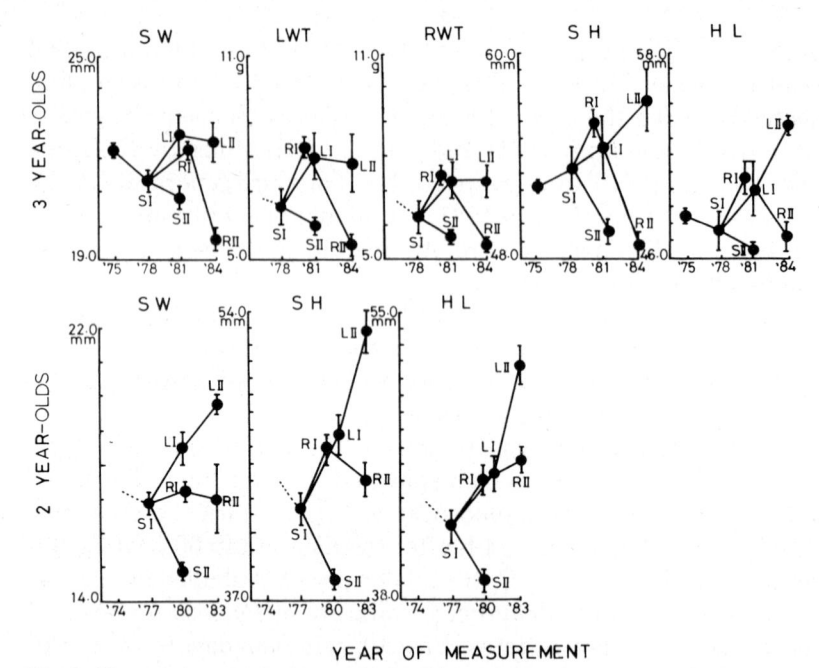

Fig. 2. Changes in means of several traits during the selection experiment for shell width (SW): LWT, weight of left valve of shell; RWT, weight of right valve; SH, shell height; HL, hinge line length. Vertical bars indicate 95% confidence interval. See Fig. 1 for explanation of line designations.

3 years (Fig. 3). Shell convexity for generation 2 of the small line (SII) significantly decreased at both 2 and 3 years. Changes in other traits in this experiment were unpredictable and showed no relation to the responses measured for shell convexity.

The deviation from the control of the large line shell width at 3 year intervals as a function of cumulative selection differential for the period 1978—1984 is shown in Fig. 4A. The realized heritability estimated by regression was 0.467 (Falconer, 1973), larger than that previously estimated on the basis of the response from generation 1 to generation 2 (Wada, 1984). These results mean that selective breeding for shell growth rate would be effective in pearl oyster hatcheries. Furthermore, individual selection might be recommended because of the high realized heritability for shell width.

The divergence between large and small lines shell convexity as a function of cumulative selection differential is presented in Fig. 4B. The realized heritability estimated from regression was 0.350. This value almost coincided with the result estimated from one generation (Wada, 1984). This realized heritability provides evidence that the shape of the pearl oyster shell may also be improved through selective breeding. In commercial pearl oyster hatcheries, seeds with larger shell widths are required. However, the mean size of the shell width was greater in the large shell width line than in the large shell convexity

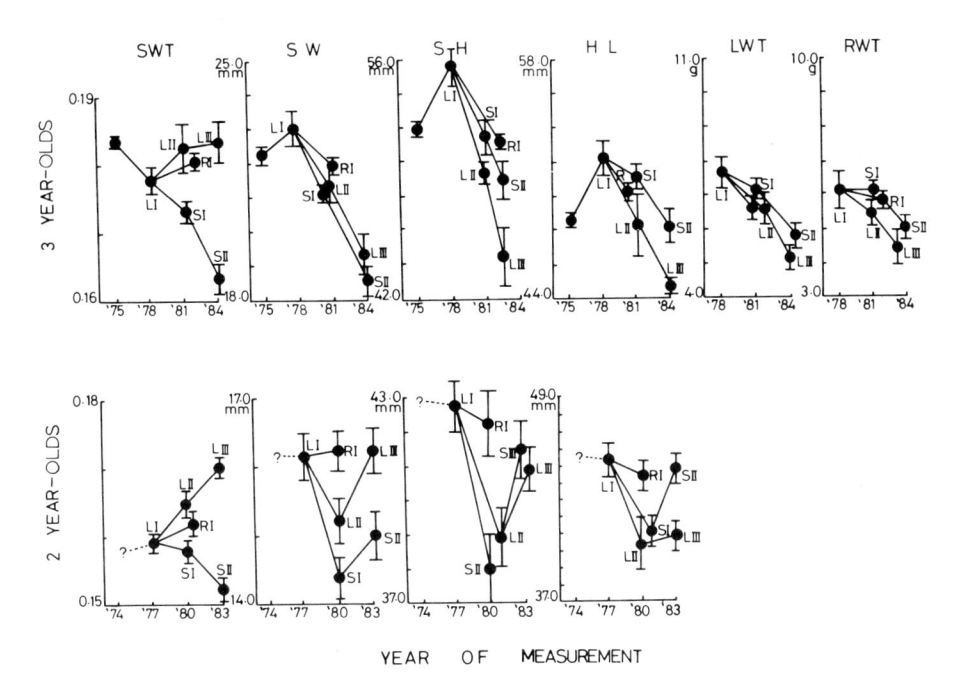

Fig. 3. Changes in means of several traits during the selection experiment on shell convexity (SWT). See Figs. 1 and 2 for explanation of abbreviations.

line at generation 3 (Figs. 2 and 3). This difference suggests that selective breeding for shell width may be more effective in production of lines with larger shell width than would be selection for shell convexity.

Rather small numbers of parents and offspring were measured each gener-

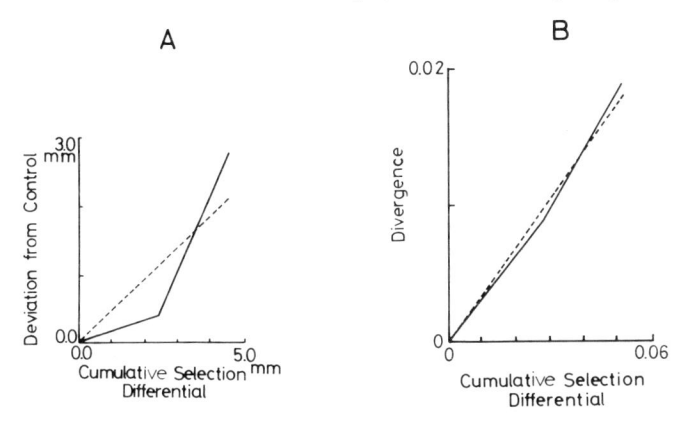

Fig. 4. Responses plotted against cumulative selection differential. A, Selection for shell width (response based on mean of large selected line as deviation from mean of control line). B, Selection for shell convexity (response based on divergence between means of large and small lines). Fitted regressions are shown as dotted lines.

ation in the present experiment. Between-generation variation in selection response may be accounted for by sampling error and genetic drift, besides the environmental variation. The small number of parents may have also caused some inbreeding depression. An attempt was made to estimate biochemical variation of the selected lines, and decreased variation has been observed in four loci of each line (Wada, 1984; K.T. Wada, unpublished data).

ACKNOWLEDGMENTS

I am much indebted to Messrs. Hiroya Maeda, Shigeya Yamamoto and Akira Komaru for doing the technical work of running these experiments. I am also grateful to Dr. Arlene Longwell, Milford Laboratory, NMFS, U.S.A. and Dr. Sidney Townsley, University of Hawaii, U.S.A. for critical reading of the manuscript.

REFERENCES

Beatie, J.H., Chew, K.K. and Hershberger, W.K., 1980. Differential survival of selected strains of Pacific oysters (*Crassostrea gigas*) during summer mortality. Proc. Natl. Shellfish. Assoc., 70: 184—189.
Chanley, P.E., 1961. Inheritance of shell marking and growth in the hard clam, *Venus mercenaria*. Proc. Natl. Shellfish. Assoc., 50: 163—169.
Falconer, D.S., 1973. Replicated selection for body weight in mice. Genet. Res., 22: 291—321.
Haskin, H.H. and Ford, S.E., 1978. Mortality patterns and disease resistance in Delaware Bay oysters. Proc. Natl. Shellfish. Assoc., 68:80 (Abstract).
Hershberger, W.K., Perdue, J.A. and Beatie, J.H., 1984. Genetic selection in Pacific oyster culture. Aquaculture, 39: 237—245.
Newkirk, G.F., 1980. Review of the genetics and the potential for selective breeding of commercially important bivalves. Aquaculture, 19: 209—228.
Newkirk, G.F. and Haley, L.E., 1983. Selection for growth rate in the European oyster, *Ostrea edulis*: response of second generation groups. Aquaculture, 33: 149—155.
Wada, K.T., 1984. Breeding study of the pearl oyster, *Pinctada fucata*. Bull. Natl. Res. Inst. Aquaculture, 6: 79—157 (in Japanese with English summary).

Mass Selection for Growth Rate in the Nile Tilapia (*Oreochromis niloticus*)[1]

GIDEON HULATA, GIORA W. WOHLFARTH and AMIR HALEVY

Agricultural Research Organisation, Fish and Aquaculture Research Station, Dor, Mobile Post Hof Hacarmel 30820 (Israel)

ABSTRACT

Hulata, G., Wohlfarth, G.W. and Halevy, A., 1986. Mass selection for growth rate in the Nile tilapia (*Oreochromis niloticus*). *Aquaculture*, 57: 177–184.

Two generations of mass selection for rapid growth were performed on juvenile and adult *Oreochromis niloticus*. A special procedure was developed for producing broods of equal-aged individuals. Response was tested after one and two cycles of selection by comparing the progenies of selected fish to those of two controls. Mean weight gains of the "random sample" control groups were 4% and 11% lower than those of the "original parent" control group in the first and second growth tests, respectively. Mean weight gain of the "original parent" control was equal to that of progeny of fish selected at adult age, but higher than weight gain of progenies of fish selected as juveniles. No improvement over the original base population was achieved by selecting for growth.

INTRODUCTION

Breeding methods based on quantitative genetics have contributed little to the improvement of tilapias. Most applied research in tilapia genetics has been aimed at producing monosex populations for control of reproduction (Wohlfarth and Hulata, 1983). Mass selection for growth in tilapias has given different results in different studies and with different species. Thien (1971) reported realized heritabilities of 0.10 and 0.16 for male and female *Tilapia mossambica*. Bondari et al. (1983) obtained a realized heritability of 0.23 ± 0.05 in *Tilapia aurea*. Tave and Smitherman's (1980) estimated heritability, by half-sib analysis, in *Tilapia nilotica* (Ivory Coast strain) was less than 0.10, while the realized heritability was -0.10 for rapid early growth in that strain (Teichert-Coddington and Smitherman, 1986).

A major concern in performing mass selection in tilapias is the age structure of the population. Groups of tilapias do not spawn synchronously but pairwise (Fryer and Iles, 1972), preceded by nest digging, choice of mate and courtship.

---

[1]Part of a study supported by a grant from the U.S.-Israel Binational Agricultural Research and Development Fund (BARD U.S.-164-79).

When a number of mature tilapias of both sexes are stocked into a spawning pond, their fry consist of groups of individuals of different ages. Such mixed populations are not suitable for selection for growth, since the larger individuals are likely to be progenies of earlier spawns, rather than the faster growers. In the common carp, for example, small differences in initial weight, resulting from a 1-day difference in age, cause large differences in growth (Hulata et al., 1976). Thus, a hatchery procedure must be developed to obtain equal-aged fry populations. The method used by Teichert-Coddington and Smitherman (1986), i.e., 'skimming" the swim-up fry from the surface water of spawning pools daily, was not used due to a lack of such facilities.

This study deals with the Ghana strain of *Oreochromis (Tilapia) niloticus*, commonly used in Israel for hybridization with *O. aureus*. Selection was carried out separately on juvenile and adult fish. An attempt was also made to select for late sexual maturity by culling individuals showing external signs of sexual maturity.

MATERIALS AND METHODS

History of broodstock

The Ghana strain of *Oreochromis niloticus* from the Dor station stock was used. This stock was founded in 1974 when around 50 fingerlings were introduced from Ghana (Mires, 1977). Offspring of these founder individuals were later transferred to the station and to several commercial fish farms.

Spawning procedures

One-year-old *O. niloticus* spawners were stocked in 700-l plastic containers, each equipped with an air-lift operated, internal gravel filter. Eight spawning families were established, each consisting of six to eight females and two males. The females were individually tagged using different coloured anchor tags (Floy-Tag) for identification. Brooding females were identified by daily inspection, and the eggs of each were removed 2—3 days after spawning for further incubation in a separate jar (Rothbard and Hulata, 1980; Hulata et al., 1983). This enabled determination of the age of each fry group. Pooling of fry from different spawns was restricted to spawns of the same or two consecutive days. The female and male parents of these spawns were kept for respawning in subsequent years as one of the controls in the selection experiment. The pooled groups of fry were each stocked into a 400-m^2 earthen pond.

Method of selection

Each group of fry was transferred to a separate earthen pond for nursing, 2 or 3 days after hatching was completed. The nursery ponds were drained after

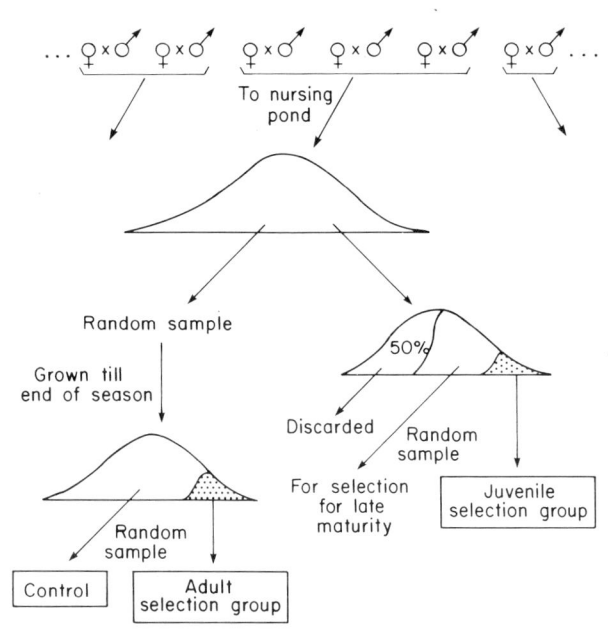

Fig. 1. Procedure of selection performed in each pond.

about 10 weeks and all fish removed. Four samples of fish were taken from each group: a random sample, a sample selected for late maturity, and two selected samples taken before and after sexual maturity. The selection procedures applied to each group of fry were as follows (Fig. 1).

(1) All fry in each pond were sexed and samples taken separately from females and males.

(2) A random sample was taken from each group, transferred to another pond and grown till the end of the season.

(3) From the remaining fish the smaller individuals (about half the number of fingerlings) were culled. The larger fish were then randomly divided into two samples.

(4) One of these samples was stocked into a new pond and retained for a separate experiment on selection for late maturity.

(5) The largest fish were selected from the other sample and fish from the different groups were pooled and grown in one pond. In the following year they served as parents for the "juvenile selected group".

(6) At the age of about 7 months the ponds stocked in step (2) were drained. Random samples were taken from each pond and pooled to serve as "random sample control". The larger individuals were selected from among the remaining fish and pooled to serve as parents for the "adult selected group". The second control was produced by respawning the original parents. For the second cycle of selection, samples of the two selection lines were again spawned

in containers in the hatchery, and the nursing and selection procedures repeated.

Method of growth testing

The growth tests included two selection and two control lines. These fish, on which no further selection was practised were mass spawned in earthen ponds. Samples of fry were seined out of the four spawning ponds, counted and transferred to nursery ponds, with the aim of nursing them to at least 10 g previous to stocking into the testing ponds. Growth testing was carried out in separate ponds, i.e., samples of each of the four test groups were stocked separately into three replicate ponds in 1982 (first generation of selection) and into four replicate ponds in 1983 (second generation of selection). Growth tests lasted about 4 months and response to selection was evaluated. All ponds were stocked with common carp, silver carp, grass carp and freshwater prawns, as well as the tilapia groups. Total fish stocking rate was 14 000/ha of which 9 000/ha were tilapia. The major nutrient input consisted of dry poultry manure applied 6 days a week at amounts increasing from 50 to 175 kg dry matter of manure per ha per day (Wohlfarth, 1978). In addition, relatively small amounts of 25% protein pellets were applied daily.

RESULTS

First cycle of selection — 1981

Sixteen full-sib groups of fry were collected and pooled into seven groups consisting of two or three individual spawns. The parents of these spawns consisted of 15 females and 11 males. Pre-maturation selection was carried out at 9—12 weeks after stocking. A total of 5113 males and 1913 females were recovered. By random sampling, 2059 males and 761 females were taken to form the control groups which were stocked into 14 ponds for further growth. From the remainder, 181 males (5.9%) and 59 females (5.1%) were selected. Selection distances were 8.9 and 7.4 g, respectively (Table 1). Adult selection was carried out at the age of about 7 months. A total of 958 males and 366 females were recovered. The largest 4.0% males and 16.0% females, averaging 89 and 46 g above their respective population means were selected. Selection intensity for females was lower than for males in order to obtain a sufficient number of individuals for generating the next generation.

Second cycle of selection — 1982

Unexpected difficulties in obtaining spawns were experienced. Six suitable spawns, obtained from fish selected as juveniles, were pooled into three groups.

TABLE 1

Details of first cycle of selection

Total No. of fish	Mean wt. (g)	Random sample		Selected		Selection distance (g)	Selection intensity (%)
		No. of fish	Mean wt. (g)	No. of fish	Mean wt. (g)		
Juvenile males							
5113	43.4	1029	43.8	181	52.3	8.9	5.9
Juvenile females							
1913	36.7	380	37.8	59	44.1	7.4	5.1
Adult males							
958	323	103	323	34	412	89	4.0
Adult females							
366	166	72	172	47	212	46	16.0

Only three suitable spawns were obtained from fish selected as adults, and each was grown separately.

The technique of selection was simplified. No attempt was made to select for late sexual maturity in the second cycle, since no variation for this trait was found in the first cycle.

Selection intensity of 10% was practised, and the selected fish averaged 11.4, 8.1, 27.3 and 25.1 g above population means, for males and females selected as juveniles and adults (Table 2).

Testing selection response

The results of the two growth tests are presented in Table 3. Large differences in survival occurred among the tested groups, particularly in the response test of the first cycle of selection. Unexplained deviations from the expected 1:1 sex ratio also occurred in several groups. In both cycles of selection there was a considerable surplus of males in the parent groups. In the first response test, on the other hand, the proportion of males was low in both progenies of selected groups.

Weight gain of the "random control" was lower than that of the "original parent control" by 4% and 11% in the first and second generations, respectively. The mean weight gain of the "adult-selected" group was very similar to that of the "original parent" control after both cycles of selection. The mean weight gain of the "juvenile-selected" group was lower than that of the "original parent" control after one (significant) and two (not significant) cycles of selection.

TABLE 2

Details of second cycle of selection

Total No. of fish	Mean wt. (g)	Selected		Selection distance (g)	Selection intensity (%)
		No. of fish	Mean wt. (g)		
Juvenile males					
2354	58.0	230	69.4	11.4	9.8
Juvenile females					
1423	54.6	142	62.7	8.1	10.0
Adult males					
1507	123.5	153	150.8	27.3	10.1
Adult females					
842	92.7	85	117.8	25.1	10.1

DISCUSSION

No real differences were found between the progenies of the selected fish and the controls after one or two cycles of selection. No improvement over the original base population was achieved by selecting for growth. This lack of response may be due to depletion of variation in the *O. niloticus* population on which selection was practised, as a result of inbreeding or drift. Similar results were reported by Tave and Smitherman (1980) and Teichert-Coddington and Smitherman (1986) for a different strain of *O. niloticus*.

Relative growth rates of the tested groups may have been affected by two sources of bias, differences between groups in proportion of males and in survival. Groups with, presumably random, higher proportion of males are expected to show a higher mean weight gain, due to the faster growth of males (Wohlfarth and Hulata, 1983). Similarly, groups suffering from higher mortalities, resulting in an effective decrease in density, are likely to grow faster than groups with lower mortalities. Yields are also sensitive to these two factors. We regard the results of the first response test as unreliable based on the contradictions in relative ranks of growth and yield between the tested groups. The conclusions of this investigation are therefore based on the second response test, carried out after two successive generations of selection.

The lower mean weight gain of the "random sample" control group, compared to the "original parent" control group, is most reasonably explained by the lower proportion of males in the "random sample" (40 vs. 51%). This difference in male percentage is unlikely to be a result of differential mortality

TABLE 3

Testing the response to one and two cycles of selection in *O. niloticus* (Ghana strain)

Group	First selection cycle							Second selection cycle						
	Initial wt. (g)	Weight gain (g)			% Males	Survival (%)	Yield (kg/ha)	Initial wt. (g)	Weight gain (g)			% Males	Survival (%)	Yield (kg/ha)
		Females	Males	Total					Females	Males	Total			
Selected adults	13.5	140.5	194.5	161.0	38	50	553	9.5	124.4	226.0	183.1	55	86	1396
Selected juveniles	24.5	135.9	165.5	143.9	27	92	982	6.5	119.2	209.3	171.0	56	78	1207
Original parents	11.1	132.9	185.2	159.0	51	64	737	13.5	122.9	230.6	177.0	51	93	1468
Random control	12.5	117.3	204.1	152.5	40	82	921	12.5	126.8	209.4	159.3	40	93	1330

since survival was high (82 and 93%). We do not have any explanation for this deviation.

The effect of further generations of inbreeding is another possible cause for the differences between the two control groups. The selected groups and the "random control" should have a similar inbreeding level, which was higher than in the original population. The "adult-selected" group did perform better than the "random control" in both tests. It seems as if selection was effective in counteracting inbreeding depression in growth rate, though it did not result in improvement over the original population.

Mass selection for rapid growth in *O. niloticus* is not a promising method for improvement, unless genetic variation is increased in the base population and measures are taken to avoid inbreeding, such as within-family selection. The non-synchronous spawning of tilapia makes mass selection more difficult to perform with tilapia than with some other fishes.

REFERENCES

Bondari, K., Dunham, R.A., Smitherman, R.O., Joyce, J.A. and Castillo, S., 1983. Response to bidirectional selection for body weight in blue tilapia. In: L. Fishelson and Z. Yaron (Compilers), Proceedings of the International Symposium on Tilapias in Aquaculture, Nazareth, Israel. Tel Aviv University Press, pp. 300—310.

Fryer, G. and Iles, T.D., 1972. The Cichlid Fishes of the Great Lakes of Africa — Their Biology and Evolution. Oliver and Boyd, Edinburgh, 641 pp.

Hulata, G., Moav, R. and Wohlfarth, G., 1976. The effects of maternal age, relative hatching time and density of stocking on growth rate of fry in the European and Chinese races of the common carp. J. Fish Biol., 9: 499—513.

Hulata, G., Wohlfarth, G. and Rothbard, S., 1983. Progeny-testing selection in tilapia brood stocks producing all-male hybrid progenies—preliminary results. Aquaculture, 33: 263—268.

Mires, D., 1977. Theoretical and practical aspects of the production of all-male tilapia hybrids. Bamidgeh, 29: 94—101.

Rothbard, S. and Hulata, G., 1980. Closed system incubator for cichlid eggs. Prog. Fish Cult., 42: 203—204.

Tave, D. and Smitherman, R.O., 1980. Predicted response to selection for early growth in *Tilapia nilotica*. Trans. Am. Fish. Soc., 109: 439—445.

Teichert-Coddington, D.R. and Smitherman, R.O., 1986. Bidirectional selection for growth in *Tilapia nilotica*. Trans. Am. Fish. Soc., in press.

Thien, C.M., 1971. Estimation of realized weight heritability in tilapia (*Tilapia mossambica* Peters). Genetika, 7(12): 53—59 (in Russian with English abstract, English translation in: Sov. Genet., 7: 1550—1554).

Wohlfarth, G., 1978. Utilization of manure in fish farming. In: C. Pastakia (Editor), Proc. Conference on Fish Farming and Wastes, University College, London, pp. 78—95.

Wohlfarth, G.W. and Hulata, G., 1983. Applied genetics of tilapias. ICLARM Studies and Reviews, 6, 26 pp. (second revised edition).

Factors Affecting Growth in Rainbow Trout (*Salmo gairdneri*) Stocks

LIISA SIITONEN

Agricultural Research Centre, Department of Animal Breeding, 31600 Jokioinen (Finland)

ABSTRACT

Siitonen, L., 1986. Factors affecting growth in rainbow trout (*Salmo gairdneri*) stocks. *Aquaculture*, 57: 185–191.

Factors affecting growth in rainbow trout were studied. Fish from separate broodstocks were reared under two environmental conditions till slaughter at the age of 2.5 years. There were statistically significant differences between stocks and between sexes for fish weight measured at different ages but there were also significant interactions. All stocks grew better in brackish than in fresh water and there were slight interactions between stocks and environments but it could not be shown that some of the stocks were clearly better adapted to one or the other of the two environments. Fish size and sexual maturity seemed to be correlated; the fish which were heaviest at 2 years of age initiated sexual maturation to spawn in their third year of life.

INTRODUCTION

The commercial production of rainbow trout (*Salmo gairdneri*) in Finland has increased rapidly, from 1000 t in 1970 to 7500 t in 1983, the latter produced in nearly 300 fish farms located at fresh or brackish water sites. In order to plan a scheme for centralized selective breeding, it was important to estimate variation in growth between and within stocks, relationships between fish size measured at different ages and the effect of crossing stocks.

This paper deals with differences in growth between stocks, between sexes and between environments and the relationship between sexual maturation and fish weight. In two experiments, the fish were reared until slaughter at the age of 2.5 years.

MATERIALS AND METHODS

In Experiment 1 (1980–1982), the fish were from 10 stocks. Five (I–V) of the stocks were distinct breeding populations from a fish culture station (LFC) belonging to the Finnish state and five (VI—X) were separate broodstocks from five commercial farms. The breeding populations of LFC were derived from distinct imports made between 1960 and 1970. Experiment 2 (1981–1983) was a partial replicate of Experiment 1. The stocks involved were I, II, IV, VI and VII (symbols as in Experiment 1). In both experiments, the eggs were taken at the source of the broodstock and hatched at LFC.

In period I (July–April/May) the stocks were reared separately in indoor tanks, two tanks per stock but in periods II and III the fish were within stocks randomly divided into all experimental groups used. In period II (May of first year to May of second year for Experiment 1 and May–Nov./Dec. for Experiment 2) they were reared intermingled in four earthen outdoor ponds. For period III (May–Nov./Dec. in Experiment 1, Dec.–Nov./Dec. in Experiment 2) each stock was divided into two groups, with one left in fresh water, and the other transferred to brackish water cages.

The fish were weighed individually at the end of each period. Although in Experiment 2 the measurements for period II were made as early as Nov./Dec. 1982, the results closely represent growth for period II because the fish do not grow during the severe winter period. The weight of period III was gutted weight. The largest fish were individually tagged and the rest marked by stock during weighing at the end of period I. The remaining fish were individually tagged at the end of period II.

At slaughter at the age of 2.5 years (period III), the fish were sexed and the stage of sexual maturity and weight were recorded for each fish. The maturity score was 0 if the fish was not assumed to spawn the following spring and 1 if it was assumed to spawn. The information on sex and stage of sexual maturity at slaughter was related to the weight records of periods I and II according to the tag numbers. In Experiment 1 all fish could be identified as females or males, but in Experiment 2 there were immature fish for which sex could not be identified.

The data were analyzed by analysis of variance following the procedure of Harvey (1966). All effects were assumed to be fixed. The significance between paired least squares constants was calculated by the Student–Newman–Keuls test (Sokal and Rohlf, 1969) at 5% risk level.

RESULTS AND DISCUSSION

At slaughter the number of the fish in the stocks varied from 274 to 369 in Experiment 1 and from 280 to 342 in Experiment 2. Mean weights and coefficients of variation are presented in Table 1. The clearly higher average gutted weight in Experiment 2 was for the great part due to the longer brackish water period in this experiment compared to Experiment 1. In Experiment 1 the fish were transferred into brackish water in June while in Experiment 2 they were transferred in May. The mean gutted weight in fresh water was 1152 g in

TABLE 1

Absolute mean weights (*W*) and coefficients of variation (CV) at the end of three periods for all fish

	Period I		Period II		Period III	
	W (g)	CV (%)	W (g)	CV (%)	W (g)	CV (%)
Exp. 1	46	30	420	36	1290	33
Exp. 2	33	33	343	38	1693	41

Experiment 1 and 1193 g in Experiment 2, in brackish water the weights were 1371 g and 1895 g in Experiments 1 and 2, respectively.

There were statistically significant ($P<0.05$) weight differences between stocks and between sexes in every period. Stock × sex interaction effects were also significant ($P<0.05$) for all weights except in period I for Experiment 2. The results of the analysis for gutted weight are presented in Table 2. The least squares constants for sex classes are presented in Table 3.

The superiority of the stocks ranked highest according to gutted weight was already evident at period I in Experiment 1 and at period II in Experiment 2. The stock (II) with the highest gutted weight in Experiment 2 was, however, significantly lighter than the heaviest stocks in Experiment 1 (Fig. 1). This could be the result of a genotype × experiment (environment) interaction or due to poor sampling of fish.

In every stock, the fish grew better in brackish than in fresh water. The two environments differed not only in the salinity of the water but also in fish density. The density was lower in the brackish water cages but the densities used were typical of fresh and brackish water farms in Finland. The absolute

TABLE 2

The analysis of variance of the factors affecting gutted weight at the end of period III

Source of variation	Exp. 1		Exp. 2	
	df	F	df	F
Stocks	9	81.75***	4	55.44***
Sexes	1	93.02***	2	91.31***
Stocks × sexes	9	5.99***	8	4.53***
Groups III[a]	5	58.70***	3	206.21***
Groups II[b]			3	10.26***
Residual	3189		1535	

[a] Ponds or cages in period III.
[b] Ponds in period II.

TABLE 3

Least squares constants for the effect of sex on fish weight (g)

	Period I		Period II		Period III	
	Exp. 1	Exp. 2	Exp. 1	Exp. 2	Exp. 1	Exp. 2
Sex unknown		−4.0	—	−91	—	−519
Females	−0.66	1.3	−25	28	−65	194
Males	0.66	2.7	25	63	65	326

growth (g) in brackish water was 928±360 and 1552±611 in Experiments 1 and 2, respectively. The corresponding figures for growth in fresh water were 740±215 and 851±292. Whether better growth was due to better appetite or better feed efficiency is not known. The fat% of the fish was higher in brackish than in fresh water (Sumari et al., 1984).

There were slight environment and stock interactions for growth but the results were not consistent for the stocks involved in both the experiments.

Fish size and sexual maturation

Ninety-six percent and 94% of males and 64% and 47% of females in Experiments 1 and 2, respectively, were scored as mature at slaughter at the age of 2.5 years. Seven percent of all fish in Experiment 2 were immature and sex not determined. Nine percent of the males in Experiment 1 were found to be mature

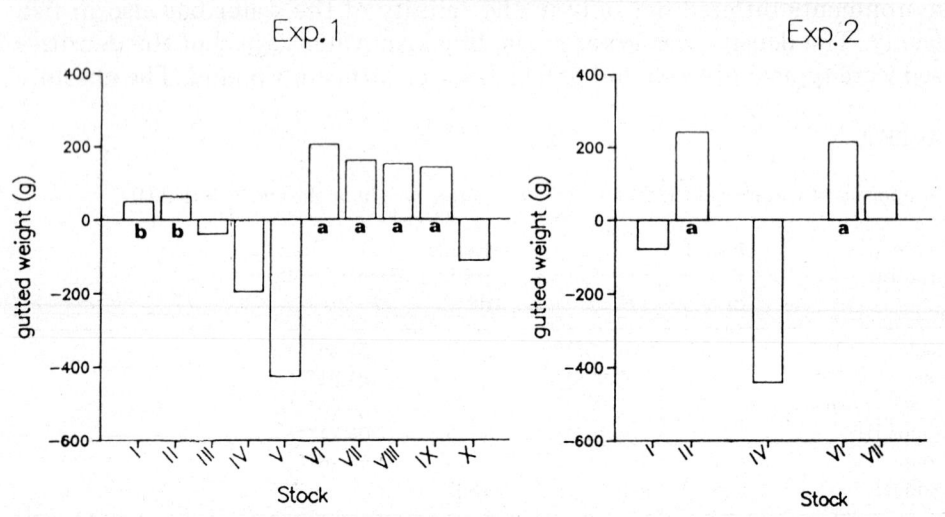

Fig. 1. Differences (LS-constants) between stocks in gutted weight. Differences marked with the same sign (a or b) were not significant at 5% level (signs within the experiments).

TABLE 4

The distribution of fish weight in Exp. 2 at the end of period II and absolute growth in period III for the fish recorded mature at slaughter at the end of period III (N = number of mature fish and % = percent mature of all fish in the class)

Weight (g)	Males		Females		Growth (g)	Males		Females	
	N	%	N	%		N	%	N	%
0–99	—	0	1	9	−199	3	50	1	11
100–199	22	46	11	12	200–499	10	50	8	23
200–299	124	82	57	29	500–799	62	81	27	31
300–399	208	93	111	58	800–1099	123	84	60	42
400–499	187	98	100	68	1100–1399	114	93	52	54
500–599	95	99	36	68	1400–1699	108	96	48	42
600–699	19	95	11	92	1700–1999	85	90	50	50
700–799	3	100	—	0	2000–2299	80	95	55	66
800–899	—	0	2	100	2300–2599	48	98	24	86
					2600–2899	18	100	3	100
					2900–	7	100	1	50

while weighing the fish at the end of period II based on the presence of milt. Therefore we probably underestimated the number of males mature in their second year of life.

The stage of sexual maturation and individual fish size seemed to be strongly correlated within both experiments. The fish which were heaviest at the end of period II matured during period III (Table 4). The proportion of mature fish was higher in males than in females in most size classes and it seemed that the size threshold to start maturation was lower for males than for females. Also during period III the maturing fish grew faster than the immature ones.

In Experiment 2 the rank correlation between the mean weights of the stocks and proportions of mature fish (Table 5) was high (0.9) and significant ($P<0.1$); in Experiment 1 the correlation was 0.5 and not significant. However, correlation in Experiment 1 clearly increased by grouping the fastest and slowest growing stocks.

In spite of the faster growth rate in brackish water, the proportion of mature fish was lower (63%) than it was in fresh water (71%). This could indicate genetic control of maturation independent of environmental factors, as discussed by Naevdal (1983), or that fish size at the end of period II was sufficient for the onset of maturation.

The present data suggested that if a fish reached a certain minimum weight by the end of period II, it matured during period III, but a lower weight did not always prevent maturation. For example, in stocks VII and X of Experiment 1, the proportion of mature fish and the proportion of males was about the

TABLE 5

Proportions (%) of sexually mature fish at the end of period III and mean weights (g) for all fish at the end of period II

Stock	I	II	III	IV	V	VI	VII	VIII	IX	X
Experiment 1										
Weight	448	442	406	397	251	499	492	475	465	381
Mature (%)[a]	91	55	65	57	59	86	73	73	81	75
Experiment 2										
Weight	318	347		210		360	299			
Mature (%)[a]	77	66		46		87	63			
Sex unknown (%)[b]	3	5		20		3	7			

[a]Unweighted mean for males and females.
[b]Proportion of immature fish of all fish in the stock for which sex could not be determined.

same but the mean weight of mature fish of stock X was no higher than the mean weight of immature fish in stock VII.

From the above results it is clear that there is considerable opportunity for improvement of rainbow trout broodstocks in Finland. Selection can be performed both between and within stocks. A possible need for separate breeding stocks for fresh and brackish water should be further evaluated and preferably on the basis of sib records.

The need to correct data for sex differences in ranking fish groups was pointed out by Gjerde (1981). The correction can be rather complicated if there are strong interactions between genotypes and sexes. The existence of this kind of interaction should also be considered when material is chosen for producing all-female populations. In the data of this paper, the gutted weight of the males, within experiments, was very similar for stocks II and VI (Table 6). However, the weight of the females of stock II was clearly lower than that of the males while the difference between the males and the females of stock VI was insignificant.

TABLE 6

Least squares means of gutted weight (g) for males and females in stocks II and VI

Stocks	Exp. 1		Exp. 2	
	II	VI	II	VI
Sex unknown	—	—	1254	1123
Females	1157	1443	1916	2009
Males	1431	1428	2106	2059

REFERENCES

Gjerde., B., 1981. Genetic variation in production rates of Atlantic salmon and rainbow trout. 32nd Annual Meeting of the European Association for Animal Production, Zagreb, Yugoslavia, Commission on Animal Genetics, IV-15.

Harvey, W.R., 1966. Least squares analysis of data with unequal subclass number. ARS 20-8, Agric. Res. Service. U.S. Department of Agriculture, Beltsville, MD, 157 pp.

Naevdal, G., 1983. Genetic factors in connection with age at maturation. In: N.P. Wilkins and E.M. Gosling (Editors), Genetics in Aquaculture. Aquaculture, 33: 97–106.

Sokal, R.R. and Rohlf, F.J., 1969. Biometry. W.H. Freeman and Company, San Francisco, CA, 776 pp.

Sumari, O., Siitonen, L. and Linder, D., 1984. Valtakunnallinen kirjolohen rodunjalostusohjelma. (Nationwide breeding scheme for rainbow trout.) Riista- ja Kalatalouden Tutkimuslaitoksen Monistettuja Julkaisuja, 30: 82 pp. (Series of mimeo publications by Fisheries Division of the Finnish Game and Fisheries Research Institute, Helsinki.)

Comparison of Strains, Crossbreeds and Hybrids of Channel Catfish for Vulnerability to Angling

REX A. DUNHAM, R. ONEAL SMITHERMAN, RANDELL K. GOODMAN and PHILIP KEMP*

Department of Fisheries and Allied Aquacultures, Alabama Agricultural Experiment Station, Auburn University, AL 36849 (U.S.A.)
*Present address: Kemp Fisheries, Dudley, NC 28333 (U.S.A.)

ABSTRACT

Dunham, R.A., Smitherman, R.O., Goodman, R.K. and Kemp, P., 1986. Comparison of strains, crossbreeds and hybrids of channel catfish for vulnerability to angling. *Aquaculture*, 57: 193–201.

Fee-fishing is a major outlet for commercially grown catfish in the United States. Success at these operations is partially dependent upon the vulnerability of catfish to angling. Genetic differences in catchability have been previously demonstrated. The interspecific hybrid produced by mating a channel catfish, *Ictalurus punctatus*, female with a blue catfish, *I. furcatus*, male was more catchable than its parent species. However, the effect of parental strain on heterosis was not evaluated. In the present experiment, several strains of channel catfish were compared for vulnerability to angling and strain differences were evident ($P<0.05$). The channel—blue hybrid was more catchable than parental strains of channel catfish ($P<0.05$). A crossbreed of the Marion strain and Kansas strain exhibited heterosis ($P<0.05$) for vulnerability to angling. This crossbred channel catfish was as catchable as the channel—blue hybrid. Results were similar in ponds of varying size, and in communal and separate evaluation. The skill of angler and concentration of dissolved oxygen caused genotype—environment interactions in tests which included the channel—blue hybrid.

INTRODUCTION

Sport-fishing is a multi-billion dollar industry in the United States and is a potential outlet for significant quantities of aquaculture products. McCoy and Crawford (1975) found that in 1975 up to 50% of the catfish harvested in the United States came from fee-fishing ponds. Despite industry growth and market development, producer sales of food-size catfish to live haulers and to fee and recreational operations was still 14 million pounds (7% of total production) in 1984 (Jensen, 1984; Knight and Ishee, 1985). Marketing through these recreational outlets increased 38% in Alabama during 1984 (Jensen, 1984). The fish-out market is sufficiently large that it supports a number of individuals who purchased fish exclusively for live-hauling to these operations (Jensen and Brown, 1985).

A large part of the success of a fee-fishing operation depends upon creating a rewarding experience for customers (Watt, 1974). Fishermen seek food from their catch but they also seek entertainment and desire a high quality of fishing (number of fish caught per man-hour) (Gray, 1970). The catch rate can be affected by environmental factors such as stocking rate (Prather, 1959, 1969), water quality, frequency of feeding and incidence of disease as well as genetic factors (Tave et al., 1981).

Tave et al. (1981) compared the channel catfish, *Ictalurus punctatus*, blue catfish, *I. furcatus*, and their reciprocal hybrids for hook-and-line vulnerability in a communal test. Communal evaluation is desirable since all genotypes are represented in each environment which reduces the component of environmental error. The hybrids were more catchable than the parent species and the channel female×blue male the more vulnerable of the two hybrids. However, Auburn was the parent strain of channel catfish in this evaluation. This strain is difficult to catch by seining (Dunham and Smitherman, 1984) and if catchability by angling and seining are correlated, the heterosis exhibited by the hybrid in Tave's experiment could have been greater than expected when other strains of channel catfish are utilized as the parent species.

The objectives of the present experiment were to compare strains, crossbreeds and hybrids of channel catfish for vulnerability to angling in ponds of varying size, and in communal and separate evaluation.

METHODS

The channel×blue hybrid (C×B) was compared to several strains of channel catfish, including Auburn (A) from Tave et al. (1981), and to the Marion×Kansas crossbreed (M×K) to determine if the hybrid was superior when compared to different channel catfish strains and crosses. M×K was compared to its parental types. Several strains were compared among themselves and Marion×Kansas. C×B, M×K and/or A were standards in all tests. Varying pond sizes and communal or separate evaluation were utilized. Data were collected from 1981 to 1984. Crosses compared, type of evaluation, fish density and pond size are summarized in Table 1 and individual experiments are described below.

Experiment 1

C×B were communally stocked with Kansas strain of channel catfish (K) or M×K channel catfish in duplicate 0.90-ha ponds. Stocking density was

TABLE 1

Crosses compared, type of evaluation, fish density and pond size in experiments to compare hook-and-line vulnerability of different genetic groups of catfish

Crosses compared[a]	Type of evaluation and fish density/ha		Size of pond (ha)
	Communal	Separate	
C×B, M×K, K	9880		0.90
		19 660	0.04
C×B, K		14 820	0.40
C×B, M×K	19 600		0.04
C×B, M	7400		0.04
	14 800		0.04
C×B, A, M	47 000		0.04
M×K, MS×KS, A, Commercial, FFES-1, LSU, MK-3, MSU, Tif[+]	13 000		0.04

[a]Channel×blue = C×B, Marion×Kansas = M×K, Marion Select×Kansas Select = MS×KS, M = Marion, A = Auburn.

9880 fish/ha. Each pond was subjected once to 4.5 man-hours/ha of fishing pressure on 15 October 1981.

C×B and K were also separately stocked at 14 820 fish/ha in two 0.4-ha earthen ponds. The pond containing C×B was fished four times from 24 April to 24 June 1982, totalling 560 man-hours/ha and the pond containing K was fished five times during the same period totalling 800 man-hours/ha.

Experiment 2

C×B, K or M×K were stocked in 0.04-ha ponds at 19 660 fish/ha. C×B and K were stocked separately in duplicate ponds, and M×K was stocked separately in one pond. A sixth pond was communally stocked with C×B and M×K. Each pond was fished once for 740 man-hours/ha on 6 March 1982. One replicate from each treatment had an additional 740 man-hours/ha of fishing pressure applied 1 month later. Angling pressure was applied five times, 6 March—31 May 1982, for a total of 2990 man-hours/ha of fishing in the communal pond.

Experiment 3

C×B hybrids were communally stocked with A and Marion (M) strains of channel catfish at 47 000 fish/ha in a 0.04-ha pond and 540 man-hours/ha of fishing pressure applied once. C×B and M were also stocked communally at

7400 ha and 14 800/ha in two 0.04-ha ponds, and 300 and 450 man-hours/ha fishing pressure applied, respectively. All fishing pressure was applied on 12 October 1984.

Experiment 4

M×K, M select×K select (MS×KS), A, commercial, FFES-1, LSU, MK-3, MSU, and Tif$^+$ channel catfish (Dunham and Smitherman, 1984) were communally stocked at 13 000/ha in a 0.04-ha pond. A total of 2170 man-hours/ha of angling were applied on four occasions from 15 October 1983 to 15 May 1984.

Fishermen angled in more than one pond in each experiment to allow fishermen of all skill levels to spend an equal amount of time at each treatment. No. 6 hooks and red worms were utilized in all experiments. At the conclusion of each experiment the pond was drained and the remaining fish inventoried.

Data analysis

Number of fish caught, percentage of population caught and percentage of population captured/man-hour of fishing were calculated for each experiment. Chi-square or t-tests were applied to evaluate genetic differences in each experiment.

RESULTS

Experiment 1

The catch rates between C×B and K in communally stocked 0.90-ha replicate ponds were consistent resulting in a ratio greater ($P<0.05$) than 1:1 (1.63 C×B:1 K). More ($P<0.01$) M×K were caught than hybrids (1.8:1). The first replicate pond had a low level of dissolved oxygen in the morning, the fish population had a low-grade parasite infection and the catch ratio was 1.1 C×B:1 M×K. In the second replicate, the pond had supersaturated dissolved oxygen concentration in the morning, the ratio was 4.0 M×K:1 hybrid.

C×B were more vulnerable to angling, 1.6:1, than K in separately stocked 0.4-ha ponds. Fourteen percent fewer K were caught despite 43% greater fishing pressure in that treatment.

Experiment 2

Catch rate was not significantly different ($P>0.05$) among M×K, C×B and K in separately stocked 0.04-ha ponds after 740 man-hours/ha of angling (1.4:1:1) and after 1480 man-hours (1.35:1.07:1). More ($P<0.01$) M×K were caught than hybrids (1.97:1) when the first 15% of the total population was removed by angling in the communally stocked pond. The experimental fishing was continued and when more than 15% of the total population was removed, the cumulative percentage of M×K and C×B caught was equal and at the conclusion of 2990 man-hours of angling, 81.3% of M×K and 78.7% of the hybrids had been captured.

After completion of the experiment in separate ponds, it was determined that a greater proportion of the fishing pressures exerted in the ponds containing C×B was by novice anglers compared to the other treatments. Experienced fishermen who angled in both M×K and K or C×B and K treatments caught more ($P<0.05$) fish in the M×K and C×B treatments (14 of 18 and 12 of 17 paired comparisons, respectively). No difference ($P>0.05$) was found between number of fish caught by experienced anglers in the M×K and C×B treatment, 14 fishermen catching more M×K and 12 catching more C×B. A sample of five to eight novice fishermen had equal difficulty in catching the three genetic groups.

Experiment 3

C×B were more ($P<0.01$) vulnerable to angling than A or M channel catfish in communally stocked 0.04-ha ponds. Catch rates were 2.4 C×B:1 A and 2.1 C×B:1 M.

Experiment 4

Greater ($P<0.05$) numbers of M×K and MS×KS were captured and fewer ($P<0.05$) FFES-1 and Tif$^+$ were captured than expected in the communally stocked 0.04-ha ponds during 25% removal of the total population (Table 2). After 50% of the population was harvested by angling, more MS×KS had been removed and fewer A, Commercial, FFES-1 and Tif$^+$ had been removed than for the mean of the total population. At the conlusion of the test, 75% removal of the total population, fewer A, Tif$^+$, and FFES-1 had been caught than expected.

DISCUSSION

Vulnerability of hybrids

The C×B was more vulnerable to angling than A, K and M strains of channel catfish, which confirms the results of Tave et al. (1981). Sunfish hybrids (Childers, 1967) and tilapia hybrids (Dunseth, 1977) are also more catchable than their parent species. The difference in catchability between the C×B and K was not as great as the relative difference between the C×B and the other

TABLE 2

Percentage of nine genetic groups of channel catfish caught by hook-and-line in a communal pond when cumulative catch was 25, 50 or 75% of the total population

Genotype[a]	Percentage of total population captured					
	25%		50%		75%	
	U[b]	A	U	A	U	A
MS×KS	42.1[x]	42.1[x]	70.2[y]	70.2[y]	84.2	84.2
M×K	32.2	35.4[x]	55.1	60.8	75.9	83.5
MK-3	15.9	23.3	39.7	58.1	57.1	83.7
MSU	30.0	32.7	43.3	47.3	70.0	76.4
A	15.7	29.6	21.6[y]	40.7	41.0[z]	77.8
LSU	18.3	21.2	43.3	50.0	60.0	69.2
COM	20.0	21.8	33.3[y]	36.4	58.3	63.6
TIF+	14.8[x]	28.6	18.5[y]	35.7	33.0[z]	64.3
FFES-1	15.0[x]	27.3	26.7[y]	48.4	28.3[z]	51.5[z]

[a]Marion Select × Kansas Select (MS×KS), Marion × Kansas (M×K), Auburn (A), Commercial (COM).
[b]U = Unadjusted for mortality, A = adjusted for mortality.
[x]Mean different from 25%; [y]mean different from 50%; [z]mean different from 75% at $P = 0.05$.

two strains. The amount of hybrid vigor exhibited was affected by the strain compared to the hybrid.

The relative vulnerability of the C×B compared to the A strain in this experiment was similar to that found by Tave et al. (1981). Auburn strain was difficult to catch by hook-and-line as well as by seine (Dunham and Smitherman, 1984); this may be related to the practice of retaining brood replacements from fish left in experimental fishing ponds at the conclusion of experiments at Auburn during the nineteen-sixties. M strain has high seinability (Dunham and Smitherman, 1984) but had low angling vulnerability in this study. Apparently, the two traits are not strongly correlated.

One variable which caused genotype—environment interactions in the present tests was skill of the angler. The hybrid was equal in catchability to the K strain when angling pressure was applied by novice fishermen.

Vulnerability of M×K crossbreeds

The M×K crossbreed expressed overdominance for vulnerability to angling. M×K was easier to capture by hook-and-line than either parent in direct and indirect comparisons. The M×K crossbreed was also equal to or greater than the C×B hybrid for catchability.

Genotype—environment interactions involving M×K and C×B were evident. Smitherman and Dunham (1985) have also found that these interactions are more likely to occur for the trait body weight in genetically distant catfishes than in closely related catfishes. In the current example, the M×K was more vigorous and catchable than the hybrid in ponds where high oxygen concentrations were found. In ponds where stress from low oxygen or diseases occurred the hybrid was equal in vulnerability to M×K. The C×B hybrid is more tolerant of low dissolved oxygen than channel catfish (Dunham et al., 1983) and it is not suprising that its rate of growth and hook-and-line vigor is less affected in environments with low oxygen than that of channel catfish.

Communal and separate evaluation

Catch rate of C×B:K in communally stocked 0.90-ha ponds and in separately stocked 0.40-ha ponds were equal, 1.63:1 and 1.60:1, respectively. Separately stocked 0.04-ha ponds yielded similar results after adjustment for the relative fishing pressure applied by experienced and novice fishermen.

Catch rate of M×K:C×B in communally stocked 0.90-ha ponds was 4:1 and 1:1.1 in high and low oxygen environments, respectively. On the initial day of fishing in 0.04-ha environments catch rate of M×K:C×B was 1.97:1 and 1.60:1 in high oxygen communal and separate environments, respectively. Results of tests in communal and separate environments were similar.

Strain comparisons

M×K crossbreeds were more catchable than seven strains of channel catfish in Experiment 4 (Table 2). Auburn, Commercial, FFES-1 and Tif[+] generally had lower than expected hook-and-line vulnerability. Mean catchabilities of LSU, MK-3 and MSU were closer to expected values. These three genotypes were of F_2, F_3 and F_2 generations, respectively, after crossing of two strains. If crossbreeding improved catchability in their corresponding F_1's as in M×K, epistasis could be an explanation for this higher vulnerability to angling compared to the remaining test groups. Tif[+] has a similar but more complex breeding history (Dunham and Smitherman, 1984) compared to LSU, MK-3 and MSU but was less catchable than expected.

Strain differences were moderated when percent of population captured was adjusted for mortality (Table 2). However, mass mortality was not observed during the several months the fish were held in the experimental pond. The unadjusted data may be more realistic since in a commercial fee-fishing operation fish not harvested by hook-and-line would be more likely to experience other forms of mortality resulting in economic loss.

Effect of total man-hours of fishing

The total man-hours of fishing could affect the results and interpretation of experiments on angling. Depending upon the frequency of restocking of fish or the frequency of harvest by seining of hook-wary fish, short-term and long-term catchability are of interest for commercial fee-fishing ponds.

In the comparison of the C×B and M×K, the M×K was generally more catchable than the C×B on the first day of fishing. When the comparison was extended over several repeat fishing periods the total percentages caught were equal. In contrast, C×B was more vulnerable than K both initially and for extended periods of time.

In the comparison of M×K with the various strains of channel catfish, M×K was more vulnerable intially; however, in several paired comparisons the magnitude of difference decreased with time (Table 2). For all genotypes, the correlation between initial numbers captured and total captured was moderate (0.49) for the adjusted data but high (0.87) for the unadjusted data. Generally, the initial harvest by angling was similar to the long-term harvest, however, a certain proportion of the M×K population appears exceptionally aggressive during the initial angling pressure. The M×K crossbreed was the most catchable genotype initially; the C×B hybrid equal to M×K over the long term and C×B was the most catchable in low oxygen environments.

REFERENCES

Childers, W.F., 1967. Hybridization of four species of sunfishes (Centrarchidae). Ill. Nat. Hist. Surv., Bull., 29(3):159—241.

Dunham, R.A. and Smitherman, R.O., 1984. Ancestry and breeding of catfish in the United States. Circular 273. Ala., Agric. Exp. Stn., Auburn Univ., AL, U.S.A., 100 pp.

Dunham, R.A., Smitherman, R.O. and Webber, C., 1983. Relative tolerance of channel × blue hybrid and channel catfish to low oxygen concentrations. Prog. Fish Cult., 45: 55—56.

Dunseth, D.R., 1977. Polyculture of Channel Catfish *Ictalurus punctatus*, Silver Carp *Hypophthalmichthys molitrix*, and Three All-Male Tilapias *Saratherodon* spp. Ph.D. Diss., Auburn Univ., AL, 50 pp.

Gray, D.L., 1970. Put and take catfish programs. Proc. Calif. Catfish Conf., Sacramento, CA, U.S.A., 20—20 Jan. 1970, pp. 81—83.

Jensen, J.W., 1984. Fisheries processing and marketing (Alabama). Ala. Coop. Ext. Serv. NARS-Accomplishment Rep. FY 1984, AL 21A, Auburn, AL, 2 pp.

Jensen, J.W. and Brown, S., 1985. Live haulers. Ala. Coop. Ext. Serv. Nat. Resource Ser., 12 June 1985, Auburn, AL, U.S.A., 2 pp.

Knight, D. and Ishee, H.S., 1985. Mississippi catfish growers survey 1 January 1985. USDA, Mississippi Crop and Livestock Reporting Serv., 25 Jan. 1985, Washington, D.C., 1 pp.

McCoy, E.W. and Crawford, K.W., 1975. Alabama catfish producers and processors 1974, a directory. Dept. Agric. Economics and Rural Sociology, Auburn Univ., AL, 25 pp.

Prather, E.E., 1959. The use of channel catfish as sport fish. Proc. Southeast. Assoc. Game Fish Comm., 13: 331—335.

Prather, E.E., 1969. Fishing success for channel catfish and white catfish in ponds with daily feeding. Proc. Southeast Assoc. Game Fish Comm., 23: 480—490.

Smitherman, R.O. and Dunham, R.A., 1985. Genetics and breedings. In: C.S. Tucker (Editor), Channel Catfish Culture, Elsevier Scientific Publ. Co., Amsterdam, pp. 283—322.

Tave, D., McGinty, A.S., Chappell, J.A. and Smitherman, R.O., 1981. Relative harvestability by angling of blue catfish, channel catfish and their reciprocal hybrids. North Am. J. Fish Manage., 1: 73—76.

Watt, C.E., 1974. Location and design of a commercial fish-out operation. Proc. Fish Farming Conf. and Annu. Conv. Catfish Farmers of TX, 3—4 January 1974. Texas Agric. Ext. Serv., Dept. of Wildl. Fish. Sci., Texas A&M Univ., College Station, TX, U.S.A., pp. 74—79.

Interspecific Hybridization of the Species of *Macrocystis* in California

RAYMOND J. LEWIS, MICHAEL NEUSHUL[1] and BRUCE W.W. HARGER

Neushul Mariculture Inc., 475 Kellogg Way, Goleta, CA 93117 (U.S.A.)
[1]*Department of Biological Sciences, University of California, Santa Barbara, CA 93160 (U.S.A.)*

ABSTRACT

Lewis, R.J., Neushul, M. and Harger, B.W.W., 1986. Interspecific hybridization of the species of *Macrocystis* in California. *Aquaculture*, 57: 203–210.

Three species of *Macrocystis* occur in California: *M. pyrifera*, *M. angustifolia* and *M. integrifolia*. Gametophytes isolated from them were grown in the laboratory. Male and female gametophytes were crossed in the laboratory in all possible combinations. Normal F_1 sporophytes were obtained from all crosses and eventually grown in the sea near Santa Barbara. After 9 months, plants from seven of nine lines reproduced and viable F_1 gametophytes were obtained. Growth measurements of the F_1 sporophytes showed significantly different growth rates in different genetic groups, with the highest growth rates obtained in plants with *M. angustifolia* or *M. pyrifera* as parents. *M. integrifolia* self-crossed individuals showed the lowest growth. This suggests that the observed differences in growth rate between genetic groups was due to genetic adaptations to the conditions of the locality of growth of the parental plants.

INTRODUCTION

Several attempts have been made recently to produce hybrids of marine macroalgae. Crosses have been made between different populations or forms within a species (Nakahara and Yamada, 1974; Lüning, 1975; Polanshek and West, 1975; Rueness and Rueness, 1975; Mueller, 1979; Clarke and Womersley, 1981), between species within a genus (Rueness, 1973; Lüning et al., 1978; West et al., 1978; Bolton et al., 1983) and between genera (Bolwell et al., 1977; Sanbonsuga and Neushul, 1978, 1979; Neushul, 1982). These studies have been useful in helping to delineate species and suggest taxonomic relationships.

Hybridization within the brown algal order Laminariales, which contains several species used as crop plants, has been of particular interest (Saito, 1972; Chapman, 1974; Lüning, 1975; Lüning et al., 1978; Sanbonsuga and Neushul, 1978; Bolton et al., 1983). The life histories of these kelps, with dioecious microscopic haploid gametophytes that can be grown and crossed in the laboratory, make it possible to combine specific gametophytes to produce specific hybrid macroscopic sporophytes (Sanbonsuga and Neushul, 1979). An evaluation of hybrids should involve the demonstration of inheritance of specific traits, but should also demonstrate whether the hybrids are able to reproduce. However, the cultivation of the diploid sporophyte progeny to maturity is much more difficult because of the large size attained by these kelps. This is particularly difficult with the giant kelp, *Macrocystis*. In this study, this was accomplished by starting genetically defined plants in the laboratory and then outplanting them onto farming structures moored in the sea.

In California, three species of *Macrocystis* are recognized (Neushul, 1971): *M. pyrifera* (L.) C. Agardh, *M. angustifolia* Bory and *M. integrifolia* Bory. These species are distinguished by characteristics of the holdfast (Neushul, 1971). *M. pyrifera* holdfasts have haptera arising from the base of the primary stipe below the basal dichotomy. *M. angustifolia* holdfasts have terete or slightly flattened rhizomatous branches (the basal dichotomy) with haptera arising from all sides. *M. integrifolia* holdfasts have rhizomatous branches that are strongly flattened with haptera only produced at the edges. *M. pyrifera* and *M. angustifolia* in California are considered to be conspecific by Nicholson (1976) while Brostoff (1977) recommended that *M. angustifolia* be recognized as a form of *M. pyrifera*. The ability or inability of these species to hybridize and characteristics of the resultant hybrids would provide important information on the genetic basis for these characteristics. This paper reports on the attempted hybridization of the three species of *Macrocystis*.

MATERIALS AND METHODS

Sporophylls were collected from a sporophyte of *M. pyrifera* from Santa Catalina Island on 6 May 1979, from *M. angustifolia* at Goleta Point near the University of California at Santa Barbara on 1 May 1979 and from *M. integrifolia* at Cannery Row, Monterey Bay on 23 September 1979. Spore release and gametophyte isolation were achieved using the methods described by Sanbonsuga and Neushul (1979). Several female and male gametophytes were successfully isolated from each of the three species. These were grown to produce clumps of branched filaments that could be fragmented for use in hybridization studies.

Female and male gametophytes were prepared for crossing and selfing by placing a gametophyte clump into 500 ml sterile Provasoli's Enriched Seawater (McLachlan, 1973) with Iodine (Tatewaki, 1966; PESI) in a sterilized blender and blending for 10 s. The resulting slurry was poured through a 43-μm screen and the gametophyte fragments retained on the screen were rinsed off in 50 ml of PESI. All possible crosses were made between one female and one male gametophyte isolate of each of the three species by inoculating a 5-ml volume of the suspension of gametophyte fragments into 1000 ml erlenmeyer flasks containing 800 ml PESI. This yielded nine female × male combinations, identified as Mp × Mp, Mp × Ma, Mp × Mi, Ma × Mp, Ma × Ma, Ma × Mi, Mi × Mp, Mi × Ma, and Mi × Mi, where Mp = *M. pyrifera*, Ma = *M. angustifolia* and Mi = *M. integrifolia*. These plants were grown at 15°C under cool-white fluorescent lights. After 14 days they were transferred to 3-l bottles in PESI supplied with bubbled air. After 40–50 days some of the individual sporophytes that had been produced were attached to glass rods with plastic clips, or by inserting their holdfasts into split tubing, and transplanted to a water table supplied with flowing seawater as described by Charters and Neushul (1979). Here, the genetically defined plants continued to grow until they were large enough to be planted in the sea.

Plants that were initiated on 28 May 1980 were outplanted in the sea in July and August on PVC pipe farm structures that were anchored to the sea floor at 6–8 m depth. The plants were attached to short split sections of PVC pipe with a short segment of rope attached to them, by inserting the holdfasts in the rope strands. These split PVC collars could be attached to the farm structures and easily removed for periodic weighings. Sporophytes were grown at two sites, Goleta Point and Ellwood Pier, until October 1980. Mi × Mi, Ma × Ma, Ma × Mp, Mi × Mp, Mi × Ma, and Ma × Mi were grown at Goleta Points, and Mp × Ma, Ma × Mp, Mi × Mp, Mi × Ma, and Ma × Mi were grown at Ellwood Pier. After October 1980 all sporophytes were grown at Ellwood Pier until they were lost in catastrophic winter storms in January 1983.

Growth rates were calculated from periodic weighings according to the formula

$$\text{Growth rate (\%/day)} = 100 * [(e^{(\ln W_2 - \ln W_1)/t}) - 1]$$

where W_1 = weight at time 1, W_2 = weight at time 2, and t = time in days.

The growth data were analyzed using analysis of variance (ANOVA) and analysis of covariance (ANCOVA). Homogeneity of variances for these analyses was determined using Bartlett's test. Only plants that were grown in the same place during the same period and of the same developmental stage were compared in these analyses. Statistically significant differences revealed by ANOVA or ANCOVA were determined using the Student—Newman—Keuls procedure (Sokal and Rohlf, 1969).

RESULTS

Sporophytes were obtained from all nine crosses of gametophytes attempted. The sporophytes from all species combinations grew normally and were outplanted to the sea where they grew to reproductive maturity. Gametophytic progeny of these F_1 sporophytes were obtained 9 months after initiation of the crosses from five out of six interspecific hybrids produced and two of three intraspecific crosses. The dates of isolation and numbers of isolates obtained

TABLE 1

Numbers of female and male gametophytes isolated from genetically defined sporophytes and the dates of isolation

	Date	Female	Male
Mp × Mp	2 Jan. 1981		2
Ma × Ma	4 March 1981	10	12
Mp × Ma	21 Dec. 1980	20	13
Ma × Mp	31 Dec. 1980	10	10
Ma × Mi	23 Dec. 1980	10	10
Mi × Mp	5 March 1981	9	8
Mi × Ma	5 March 1981	8	12

are listed in Table 1. None were obtained for Mp × Mi and Mi × Mi since few plants from these lines survived and none of these survived long enough to produce spores.

Survival of sporophytes

Growth and survival of outplanted sporophytes were periodically measured from the time they were outplanted in July and August 1980 until February 1981, after which the plants had grown too large for the collars and were attached to gravel bags on the sea floor. The survival of these plants is summarized in Table 2. Survival was high for all genetic combinations except Mi × Mi from the time they were outplanted until December 1980. There were several winter storms from December 1980 to February 1981 in which many plants were lost. Larger plants were lost more frequently than smaller ones (Harger and Neushul, 1983, fig. 2).

Growth rates

The sporophytes grown at Ellwood Pier from October to December showed good growth during this period. However, they varied widely with respect to initial plant size, which could significantly affect the subsequent growth rate.

TABLE 2

Survival of sporophytes of genetically defined *Macrocystis* plants from August 1980 to February 1981

	Mp/Mp	Mi/Mi	Ma/Ma	Mp/Ma	Ma/Mp	Mi/Mp	Mp/Mi	Mi/Ma	Ma/Mi
Aug. 1980	11	10	14	30	66	67	2	49	36
Dec. 1980	9	2	12	27	59	53	2	39	33
Aug. 1980–Dec. 1980	82%	20%	86%	90%	89%	79%	100%	80%	92%
Feb. 1981	7	1	6	14	20	24	1	28	24
Dec. 1980–Feb. 1981	78%	50%	50%	52%	34%	45%	50%	72%	73%

TABLE 3

Pairwise comparisons of mean exponential growth rates (%/day) for sporophytes of six different genetic makeups from October to December 1980 [means with the same superscript were not significantly different ($P < 0.05$)]

Genetic cross	Sample size	Growth rate
Ma × Ma	11	3.774^a
Ma × Mp	42	4.055^a
Mp × Ma	21	2.866^{bc}
Ma × Mi	43	2.834^c
Mi × Ma	47	2.688^c
Mi × Mp	47	3.157^b

A regression of growth rate on initial plant size showed that growth rate significantly decreased for plants with larger initial weight in all genetic groups. This relationship was essentially linear between initial weights of 50–600 g. In order to be able to determine whether differences in growth rate could be attributed to genetic makeup independent of initial plant size, an analysis of covariance (ANCOVA) was performed on these data to compare regression lines of growth rate on initial plant size. The data on six of the crosses were found to be suitable for this analysis. The ANCOVA showed that the slopes of the regression lines were all similar ($F = 1.837$, df = 5, 199), indicating that the effect of initial plant size on growth rate was the same for all groups tested. This allowed the testing of homogeneity of intercepts of the regression lines, which showed that the intercepts were dissimilar ($F = 26.27$, df = 5, 204). A multiple range test, using the Student—Newman—Keuls procedure, was performed on the growth rates obtained from the regression lines at the mean initial plant size (225 g) to determine which groups were significantly different from the others. The results of this test are presented in Table 3.

The data on the three crosses not included above were not suitable to be included in the ANCOVA. Including Mp × Mp in the ANCOVA yielded a rejection of the null hypothesis, according to Bartlett's test, that the variances were homogeneous. The growth rate of these plants standardized to an initial plant size of 225 g was 3.717 ($n = 9$). This appears to be not significantly different from the growth rate of Ma × Mp and Ma × Ma. For Mp × Mi, the mean growth rate was 3.129 ($n = 2$), but because of the small sample size, this rate is not significantly different from any of the other rates. Plants of Mi × Mi had a mean growth rate of 2.120 ($n = 2$), which is lower than any of the above rates, but because of the small sample size could not be adequately compared to the other groups.

DISCUSSION

The results of the attempted hybridizations among the three species of *Macrocystis* show that these species are fully interfertile, with the resultant F_1

progeny able to produce viable spores. Viable spores were not obtained from two crosses, Mi × Mi and Mp × Mi, since no plants survived to reproductive maturity. The *M. integrifolia* self-cross (Mi × Mi) was presumably fertile but perhaps unable to grow well in the Santa Barbara area. The Mp × Mi hybrid was also probably fertile since the reciprocal cross, Mi × Mp, was successfully raised to maturity.

The fact that all three species were fully interfertile raises questions regarding the distinction of these three entities as different species. However, further hybridization attempts using isolates of *Macrocystis* from other parts of its distribution from Alaska to Mexico need to be made before such a decision can be rendered. In addition, hybridization studies, using isolates of *Macrocystis* from various regions of the Southern Hemisphere where it is found and these Northern Hemisphere isolates, would be useful to determine whether these are also interfertile and able to produce viable progeny.

Growth rate of young sporophytes raised in this study was found to be different in groups of different genetic composition. Plants from the Ma × Mp and Ma × Ma crosses had much higher growth rates than those of other genetic backgrounds. The *M. angustifolia* gametophyte isolates used in these investigations were isolated from an adult sporophyte collected at Goleta Point. As a consequence, sporophytes derived from these gametophytes may be better suited for growth in the Santa Barbara vicinity than sporophytes derived from gametophytes from other localities. The data also suggest that Mp × Mp plants grew at a similar fast rate. Interestingly, Mp × Ma plants grew at a significantly lower rate than Ma × Mp plants. This suggests that the male and female genetic contributions are different. Similar differences between reciprocal crosses in kelps have been noted in intergeneric hybridization (Sanbonsuga and Neushul, 1978; Neushul, 1982). This could possibly be due to sex chromosomes or cytoplasmic inheritance.

Mi × Mi plants grew least well of all crosses, with plants showing weight loss during the summer growth period. This may be due to the large difference between the environment from which the parental sporophytes were collected and the environment in which the F_1 sporophytes were grown. Water temperature may have been the most important environmental factor here, being higher than where *M. integrifolia* naturally occurs. In addition, the light levels at the depth of outplanting may have been significantly lower than where it naturally occurs. Lüning et al. (1978) similarly found that *Laminaria* progeny from Helgoland or hybrids with Helgoland isolates grew better at Helgoland than *Laminaria* from other places in the North Atlantic Ocean.

Since genetic background significantly affects growth rate, this can be used as a basis for breeding strains of kelp for fast growth as a marine crop plant. Results from an experimental farm of transplanted wild plants, used to determine the biomass yield of *Macrocystis* in the Santa Barbara vicinity, showed that certain plants consistently exhibited higher yields than others (Neushul and Harger, 1985). This can probably be attributed to genetic constitution. However, the development of improved strains of *Macrocystis*, and indeed all crop plants, is dependent on the conditions at the locality at which the plants are cultivated.

ACKNOWLEDGEMENTS

Thanks are due to Mr. S. Fain for providing the technical oversight for this work. Valuable discussions on plant genetics were held with Drs. R. Snow, S. Jain, T.C. Fang and Y. Sanbonsuga. This research was supported under Gas Research Institute contract #5083-226-0802.

REFERENCES

Bolton, J.J., Germann, I. and Lüning, K., 1983. Hybridization between Atlantic and Pacific representatives of the Simplices section of *Laminaria* (Phaeophyta). Phycologia, 22: 133–140.

Bolwell, G.P. Callow, J.A., Callow, M.E. and Evans, L.V., 1977. Cross-fertilization in fucoid seaweeds. Nature, 268: 626–627.

Brostoff, W.N., 1977. A Taxonomic Revision of the Giant Kelp *Macrocystis* in Southern California Based on Morphometric and Transplant Studies. M.S. thesis, San Diego State University, 75 pp.

Chapman, A.R.O., 1974. The genetic basis of morphological differentiation in some *Laminaria* populations. Mar. Biol., 24: 85–91.

Charters, A.C. and Neushul, M., 1979. A hydrodynamically defined culture system for benthic seaweeds. Aquat. Bot., 6: 67–78.

Clarke, S.M. and Womersley, H.B.S., 1981. Cross-fertilization and hybrid development of forms of the brown alga *Hormosira banksii* (Turner) Decaisne. Aust. J. Bot., 29: 497–505.

Harger, B.W.W. and Neushul, M., 1983. Test-farming of the giant kelp, *Macrocystis*, as a marine biomass producer. J. World Maricult. Soc., 14: 392–403.

Lüning, K., 1975. Kreuzungsexperimente an *Laminaria saccharina* von Helgoland und von der Isle of Man. Helgol. Wiss. Meeresunters., 27: 108–114.

Lüning, K., Chapman, A.R.O. and Mann, K.H., 1978. Crossing experiments in the non-digitate complex of *Laminaria* from both sides of the Atlantic. Phycologia, 17: 293–298.

McLachlan, J., 1973. Growth media — marine. In: J.R. Stein (Editor), Handbook of Phycological Methods, Vol. I. Culture Methods and Growth Measurements. Cambridge University Press, New York, NY, pp. 25–51.

Mueller, D.G., 1979. Genetic affinity of *Ectocarpus siliculosus* (Dillw.) Lyngb. from the Mediterranean, North Atlantic and Australia. Phycologia, 18: 312–318.

Nakahara, N. and Yamada, I., 1974. Crossing experiments between four local forms of *Agarum cribosum* Bory (Phaeophyta) from Hokkaido, northern Japan. J. Fac. Sci., Hokkaido Univ., Ser. V, 10: 49–54.

Neushul, M., 1971. The species of *Macrocystis* with particular reference to those of North and South America. In: W.J. North (Editor), The Biology of Giant Kelp Beds (*Macrocystis*) in California. Nova Hedwigia, 32: 211–222.

Neushul, M., 1982. Morphology, structure, systematics and evolution of the giant kelp *Macrocystis*. In: C.K. Tseng (Editor), Proceedings of the U.S.—China Phycological Symposium. Science Press, Beijing, People's Republic of China, pp. 1–27.

Neushul, M. and Harger, B.W.W., 1985. Studies of biomass yield from a near-shore macroalgal test farm. J. Sol. Energy Sci. Eng., 107: 93-96.

Nicholson, N., 1976. Order Laminariales. In: I.A. Abbott and G.J. Hollenberg (Editors), Marine Algae of California. Stanford University Press, Stanford, CA, pp. 28-57.

Polanshek, A.R. and West, J.A., 1975. Culture and hybridization studies on *Petrocelis* (Rhodophyta) from Alaska and California. J. Phycol., 11: 434-439.

Rueness, J., 1973. Speciation in *Polysiphonia* (Rhodophyceae, Ceramiales) in view of hybridization experiments: *P. hemisphaerica* and *P. boldii*. Phycologia, 12: 107-109.

Rueness, J. and Rueness, M., 1975. Genetic control of morphogenesis in two varieties of *Antithamnion plumula* (Rhodophyceae, Ceramiales). Phycologia, 14: 81-85.

Saito, Y., 1972. On the effects of environmental factors on morphological characteristics of *Undaria pinnatifida* and the breeding of hybrids in the genus *Undaria*. In: I.A. Abbott and M. Kurogi (Editors), Contributions to the Systematics of Benthic Marine Algae of the North Pacific. Japanese Society of Phycology, Kobe, Japan, pp. 117-132.

Sanbonsuga, Y. and Neushul, M., 1978. Hybridization of *Macrocystis* (Phaeophyta) with other float-bearing kelps. J. Phycol., 14: 214-224.

Sanbonsuga, Y. and Neushul, M., 1979. In: E. Gantt (Editor), Handbook of Phycological Methods. Vol. III. Developmental and Cytological Methods. Cambridge University Press, New York, NY, pp. 69-76.

Sokal, R.R. and Rohlf, F.J., 1969. Biometry. W.H. Freeman and Co., San Francisco, CA, 776 pp.

Tatewaki, M., 1966. Formation of crustaceous sporophyte with unilocular sporangia in *Scytosiphon lomentaria*. Phycologia, 6: 26-36.

West, J.A., Polanshek, A.R. and Shevlin, D.E., 1978. Field and culture studies on *Gigartina agardhii* (Rhodophyta). J. Phycol., 14: 416-426.

Genetic Differentiation among Seasonally Distinct Spawning Populations of Chum Salmon, *Oncorhynchus keta*

ROSS F. TALLMAN

Institute of Animal Resource Ecology, University of British Columbia, Vancouver, B.C. V6T 1W5 (Canada) and
Pacific Biological Station, Nanaimo, B.C. V9R 5K6 (Canada)
Mailing address: Department of Fisheries and Oceans, Government of Canada, Pacific Biological Station, Nanaimo, B.C. V9R 5K6 (Canada)

ABSTRACT

Tallman, R.F., 1986. Genetic differentiation among seasonally distinct spawning populations of chum salmon, *Oncorhynchus keta. Aquaculture*: 57: 211–217.

Fry from three Vancouver Island chum salmon populations were reared under identical conditions from egg onward to determine if genetic divergence occurs among populations that spawn in different seasons. Comparisons of genetic traits, such as incubation rate and external morphology, revealed significant differences between the autumn spawning stock and the two winter spawning stocks. At constant 6°C, simulated autumn spawning regime and simulated winter spawning regime incubation rates of winter stock embryos were more rapid than those of the autumn spawning population. However, genotype—environment interaction occurred at constant 10°C such that autumn stock embryos incubated more rapidly. Discriminant analysis of 10 morphometric features of the progeny separated groups corresponding to the three stocks. Environmental factors were important as the key characteristics of separation varied with temperature regime. Both genotype and genotype—environment interactions contribute to divergence among seasonally distinct spawning populations.

INTRODUCTION

Resource-use patterns of marine biota are in transition from a hunting—gathering economy to a farming economy (Towle, 1983). Generation of domesticated or semi-domesticated stock from wild organisms is an important step in this process. The success of the transition will thus depend on the genetic variability of many traits of wild organisms. Elucidation of stock distinctness, genetic flexibility and adaptations among wild populations will provide a starting point for efforts in domestication and ultimately, in the genetic engineering of aquacultural animals (Colwell, 1982).

Anadromous salmonid aquaculture programs exercise the greatest control over the freshwater phase of production. According to Gjedrem (1983), aquaculture problems can partly be overcome by choosing stock suitable for a particular system. Investigations of wild stock genetic variability in traits such as incubation rate and external morphology of the post-emergent fry will be both beneficial and tractable.

Tallman and Healey (1986) hypothesized that genetic divergence in incubation rate has occurred among seasonal ecotypes of chum salmon, *Oncorhynchus keta*. Stabilizing selection for synchronized downstream fry migration among populations in the spring has resulted in more rapid embryonic incubation in the winter spawning populations compared to the autumn spawning stocks. In contrast, morphological characteristics of the fry did not appear to follow this pattern and might have been related to the variations in spawning environment. A critical test of this hypothesis is to compare the performance of these ecotypes under identical thermal conditions in the laboratory.

Few comparisons of the external morphology of emergent fry have been made among seasonal races. Surprisingly little information exists regarding adaptation of incubation rate. This study investigates the influence of genotype on incubation rate and fry morphological variation among seasonally distinct spawning populations of chum salmon.

MATERIALS AND METHODS

The wild parent stock "Autumn Bush" (AB) spawned mainly in October while "Winter Bush" (WB) and "Walker" (W) spawned in late November tomid-December (Tallman and Healey, 1986).

Sample populations from each stock were reared under identical conditions during 1983—1984 to determine the influence of genotype on incubation rate to hatch and emergence. To determine the influence of thermal regime and genotype—environment interaction, sub-samples from each sample population were reared under four different temperature regimes. The protocol was as follows: five males and five females from each population were mated to produce 25 families per population. These were pooled and then distributed among four different temperature regimes: constant 6°C; constant 10°C; 'early spawning' regime which simulated a natural fall—winter—spring progression in temperature and a 'late spawning' regime which simulated a natural winter—spring temperature pattern. A cross-mated population consisting of the progeny from mating between five late Bush males and five Walker females was also tested. Each population—temperature regime treatment was replicated to estimate tank-to-tank variation. Observations were made on the experimental groups each week until hatch or emergence was imminent, then observations were made daily to record the number of individuals hatched or emerged each day.

To determine the contribution of maternal and additive genetic variance to time to hatch and emergence within the populations, samples from the 25 families produced per population were also reared individually at 8°C during 1983—1984. The sire, dam and sire plus dam heritabilities were estimated using the intraclass correlation (Becker, 1975).

Maternal effects were also measured by comparing egg weights of preserved water-hardened eggs from different females by analysis of covariance, using the female length as a covariate.

To compare the influence of genotype, environment and genotype—environment interaction on post-emergent juvenile morphology, temporally stratified samples of 50 emergent fry from each replicate were preserved in 5% buffered formaldehyde.

Total length (TL), standard length (STL), head length (HL), snout length (SNL), pectoral fin (PFL), eye diameter (ED), head depth (HD), body depth (BD) and wet weight (WT) were measured as described by Hubbs and Lagler (1958) plus the number of parr marks (PM) on each fish. A parr mark was defined as any discrete vertical bar of dark pigment exceeding 40% of the body depth at that point. Thus, acceptable parr mark size varied with location on the fish.

Two-way analysis of variance with interaction was performed to estimate the effect of population and temperature regime on the time to hatch and emerge.

Heritability estimates and their standard errors were calculated by the factorial method described by Becker (1975). The lower confidence limits of the heritability estimates were generated via a Monte Carlo simulation technique similar to that of Rodda et al. (1977) assuming that the error entered when the data for hatch and emergence time were gathered. The error in the estimates of hatch time and emergence time of each family was assumed to be normally distributed. Two hundred heritability estimates were computed for hatch and emergence time of each sub-population from simulated random samples of 100 offspring per family. The tenth lowest simulated heritability estimate was taken as the lower confidence limit.

Stepwise discriminant analysis was used to determine the most important traits for separating the populations as well as estimating the degree of separation. The effect of incubation environment on the separation among the populations was compared by discriminant analysis of the sample populations stratified by temperature regime. Variables were added into the discriminant function until the multiple correlation coefficient, R^2, of each of the remaining variables with those already entered was greater than 0.40 (Dixon, 1981).

RESULTS

The general result from the laboratory rearing experiments was that late stocks incubated more rapidly. Two-way ANOVA showed significant differ-

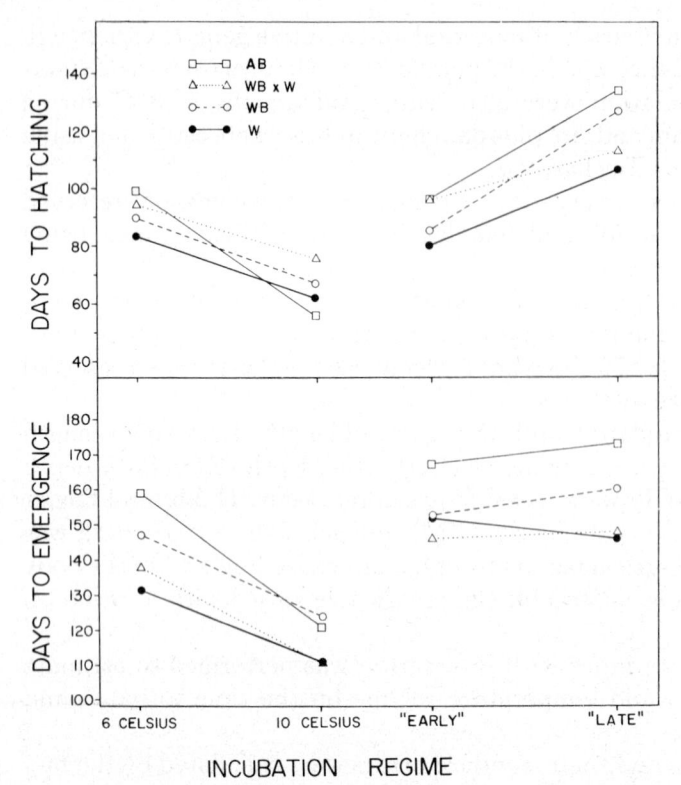

Fig. 1. Comparisons of the mean number of days to hatch and emergence among the experimental populations. Each population incubated under four temperature regimes. AB = Autumn Spawning Bush Creek Stock; WB = Winter Spawning Bush Creek Stock; W = Winter Spawning Walker Creek Stock. Temperature regimes: constant 6°C; constant 10°C; early = fall—winter—spring progression; late = winter—spring progression.

ences among populations for time to hatch and time to emergence ($P < 0.0001$, Fig. 1). However, significant genotype—environment interaction occurred at 10°C such that the WB progeny incubated more slowly than the AB progeny. This interaction was especially pronounced for time to hatch. In 1983—1984 deaths were under 5% in all the within-population matings. Mortalities were from 60 to 99% in the crossed population.

Families in which mortality exceeded 10% were not used to estimate heritability for time to hatch and time to emergence. To avoid missing cells in the analysis, rows and columns were dropped as necessary to maintain a factorial design. The sire by dam ($S \times D$) design for time to hatch of AB was reduced to a 4×4, WB to 4×3 and W to a 3×4. The $S \times D$ design for the time to emerge was 4×4 for AB; 2×5 for WB; 3×4 for W.

Heritability based on combined sire and dam components for time to hatch was 0.27 for AB, 0.35 for WB and 0.50 for W. Heritability for time to emergence

TABLE 1

F. Values testing the equality of means for each pair of populations using variables included in the discriminant functions [i.e., F(AB vs. WB) at 6°C = 89.23]

Temp. regime	AB vs. WB	AB vs. W	WB vs. W
6°C	89.23**	105.09**	20.77**
10°C	146.47**	146.47**	134.46**
Early	72.78**	229.60**	96.46**
Late	74.67**	74.52**	16.70**
Pooled	132.79**	185.77**	28.93**

(Degrees of freedom: pooled = 8, 590; single temperature = 7, 141).
**$P < 0.0001$.

exceeded that for time to hatch in all populations (0.50 — AB, 0.40 — WB, 0.54 — W). In all cases, except time to emergence of AB, h_S^2 was greater than h_D^2. The lower bounds of the 95% confidence limits estimated by simulations were greater than 0 in all cases.

Analysis of covariance for egg weight differences among stock revealed that WB had significantly smaller eggs than W or AB. There was no difference between the size of the W and AB adjusted egg weights.

Stepwise discriminant analysis of the populations with all temperature treatments pooled revealed that AB progeny were morphologically distinct from those of WB and W, although all paired population comparisons were significant (Table 1). Symptomatic of the relatively greater morphological overlap between WB and W was the much higher misclassification among these two groups by the discriminant functions (Table 2). The discriminators, in order of importance, were: ED, SNL, PM, PFL, HL, HD, and STL.

When the discriminant functions were calculated using fry reared under a single temperature the percentage misclassifications decreased (Table 2). In general, a marked discontinuity occurred between progeny of the early spawning populations and those of the late spawning populations (Table 1). The important discriminators changed with the temperature of incubation. The discriminators, in order of importance, were: ED, SNL, HD, BD, HL, PM, WT at 6°C; ED, PM, HL, PFL, SNL, BD, HD at 10°C; ED, SNL, HL, WT, HD, PFL, BD under the "early spawning" regime; and HL, PM, PFL, STL, ED, HD, TL under the "late spawning" regime.

DISCUSSION

The results show clearly that the late spawning stocks incubate more rapidly to both hatch and emergence. I believe that the relatively inefficient incubation of the WB and W progeny compared to AB at 10°C indicates adaptation to

TABLE 2

Jackknifed classification using discriminant functions

Temp. regime	Sample origin	Classified as:		
		AB	WB	W
6°C	AB	46	2	2
6°C	WB	0	44	6
6°C	W	0	2	48
10°C	AB	50	0	0
10°C	WB	0	36	14
10°C	W	0	6	44
Early	AB	50	0	0
Early	WB	2	48	0
Early	W	0	0	50
Late	AB	48	0	2
Late	WB	2	44	4
Late	W	0	2	48
Pooled	AB	178	20	2
Pooled	WB	16	138	46
Pooled	W	2	44	154

the season of reproduction. The late spawning populations would never encounter 10°C in their early incubation while AB progeny are likely to experience high temperatures in some years. Differences in the kinetic efficiencies and stabilities of enzyme morphs is the probable mechanism responsible (Hochachka and Somero, 1984).

The relatively high heritabilities of time to hatch and emergence suggest that incubation rate may be modified rather rapidly by selective breeding programs or by ecological forces in the wild. It confirms the hypothesis that chum salmon populations are genetically adapted to their season of reproduction with respect to incubation rate. I suspect that the higher heritability of time to emergence was due to a common environmental effect. Godin (1982) proposed that fry within a redd should behaviorally synchronize their timing of emergence in order to 'swamp' predators. Fry that emerge earlier or later than their fellows will have a higher probability of being eaten. To improve their chances of survival rapidly developing alevins may delay their emergence until others are prepared to leave while slower developing alevins move out of the gravel prematurely. The variance in the time to emergence would be reduced within families. This would drive heritability estimates up.

Early and late spawners produce distinctly different emergent fry. Egg size differences cannot account for this variation since AB and W had similarly sized eggs while their progeny had different external morphology. Therefore, the differences observed are probably a result of genetic divergence.

I conclude that both incubation rate and the resulting fry morphology are dependent on the season of reproduction of the parent population. The high heritability of incubation rate is encouraging for the selective breeding prospects for this species. For example, a rapid incubating broodstock could be established using a winter spawning wild stock as seed.

ACKNOWLEDGEMENTS

M.C. Healey, T.G. Northcote and B. Riddell provided helpful advice and criticism of this study. M.C. Healey kindly supplied laboratory space and helped collect experimental stock. I was supported for this study by a Natural Sciences and Engineering Research Council of Canada Post-graduate Scholarship.

REFERENCES

Becker, W., 1975. A Manual of Quantitative Genetics, 3rd edition. Washington State University, Pullman, WA, 170 pp.
Colwell, R.R., 1982. Potential genetic engineering for aquaculture. ICES Council Meeting, ICES, Copenhagen, 10 pp.
Dixon, W.J., 1981. Biomedical Computer Programs, P-Series. University of California Press, Berkeley, CA, 880 pp.
Gjedrem, T., 1983. Genetic variation in quantitative traits and selective breeding in fish and shellfish. Aquaculture, 33: 51—72.
Godin, J.-G., 1982. Migrations of salmonid fishes during early life history phases: daily and annual timing. In: E.L. Brannon and E.O. Salo (Editors), Proc. Salmon and Trout Migratory Behaviour Symposium, 3-5 June 1981. School of Fisheries, University of Washington, Seattle, WA, pp. 22-50.
Hochachka, P. and Somero, G., 1984. Biochemical adaptation. Princeton University Press, 537 pp.
Hubbs, C.L. and Lagler, K.F., 1958. Fishes of the Great Lakes. University of Michigan Press, Ann Arbor, MI, 213 pp.
Rodda, D., Schaffer, L., Mullen, K. and Friars, G., 1977. Measuring the precision of genetic parameters by simulation technique. Theor. Appl. Genet., 51: 35—40.
Tallman, R.F. and Healey, M.C., 1986. Stabilizing selection of phenotype among seasonal ecotypes of chum salmon (*Oncorhynchus keta*). Can. J. Fish. Aquat. Sci. (in review).
Towle, J.C., 1983. The Pacific salmon and the process of domestication. Geogr. Rev., 73: 287—300.

Variability of Embryo Development Rate, Fry Growth, and Disease Susceptibility in Hatchery Stocks of Chum Salmon

WILLIAM W. SMOKER[1]

Department of Fisheries and Wildlife, Oregon State University, Corvallis, OR (U.S.A.)

[1]Present address: School of Fisheries and Science, University of Alaska-Juneau, 11120 Glacier Hwy, Juneau, AK 99801 (U.S.A.)

ABSTRACT

Smoker, W.W., 1986. Variability of embryo development rate, fry growth, and disease susceptibility in hatchery stocks of chum salmon. *Aquaculture*, 57: 219–226.

Three hatchery stocks of chum salmon (*Oncorhynchus keta*) contributed sperm to a factorial breeding experiment. Matings with four females from the Netarts stock produced 60 sibling groups which were uniquely marked and combined as fry.

There was an effect of paternal hatchery stock and of dam on time between spawning and hatching, an effect of sires in one paternal stock and of dam on fry lengths after 1 month of rearing, and an effect of sires within paternal stocks on susceptibility to vibriosis.

INTRODUCTION

Little is known about genetic variation of quantitative characters within and between populations of chum salmon (*Oncorhynchus keta*), a species important to aquaculture production on the northern Pacific Rim (Kobayashi, 1980; Konovalov, 1980; FRED, 1985). Chum salmon are cultured as embryos and either immediately released into the sea or cultured for several weeks before release, a practice that has been responsible for a doubling of ocean survival to over 2% in Japan (Kobayashi, 1980). Traits such as age at maturity, size at age, annual timing of spawning migration, and larval development rate are known to vary over stocks (Koski, 1975), but genetic contribution to that variability has not been precisely described. Such a description could be made from analysis of phenotypic variation in members of different stocks reared in common environments. This study explored the variability of several traits in embryos and fry smaller than 2 g, fish which were the product of the combination of gametes from three hatchery stocks and which were cultured simultaneously in a common environment.

Fig. 1. Oregon and Washington locations of Washington Dept. Fisheries chum salmon hatcheries at Hoodsport and Nemah and the Oregon State University Aquaculture Station at Netarts.

MATERIALS AND METHODS

Three populations of chum salmon in Oregon and Washington (Fig. 1) contributed gametes to a factorial breeding experiment at the Oregon State University (OSU) Aquaculture Experiment Station at Netarts Bay, Oregon. Parents for the experiment were randomly chosen from among mature fish at each hatchery on the day of the experiment. Gametes from five males at each hatchery and from four females at Netarts were combined by dividing each female's eggs evenly into 20 lots of about 150 eggs, fertilizing each lot with milt from a different male. Milt from Netarts males fertilized two lots per female each, milt from Hoodsport and Nemah populations fertilized one lot per female each. The 80 lots of eggs were randomly assigned to an array of 80 replicate incubators fabricated from 10-1 plastic dishpans. Forty incubators held 20 purebred Netarts sibling groups, each group in two replicate incubators. Twenty incubators held the 20 crossbred Netarts by Nemah sibling groups. Twenty incubators held 20 crossbred Netarts by Hoodsport sibling groups. The eggs were incubated on a 6-mm mesh screen above a 2-cm layer of gravel, one egg per aperture in the screen. Hatched embryos dropped through the screen into the gravel where they completed their development into fry. A common source

(Whiskey Creek) provided each incubator with 0.3 l/min of water. The number of eggs in each incubator was reduced to 100 live individuals soon after fertilization.

Development rate

Development rate for each embryo was measured as the number of days between fertilization and hatching. During hatching, about 60–80 days post-fertilization, hatched eggs in each incubator were counted on each day.

The average number of days to hatching in each of the 80 incubators was analyzed. A hatching egg on a incubator screen tended mechanically to stimulate its immediate neighbors to hatch, so that hatching proceeded in patches of 2–10 eggs. Because of this, number of days to hatching was not independent for different eggs in an incubator. Variance ratio tests of significance assume that errors are independent. Therefore, the average number of days to hatch was assumed to be unbiased and was analyzed.

Growth

Growth was measured as standard length (nearest mm) of fry after feeding for 28 days (one tank) and 33 days (replicate tank). Initial size was not considered because it is determined by egg size and egg size varied little within females.

Twenty-five fry from each incubator were pooled and fed together in replicate 200-l tanks. Each lot of fry received a unique mark made with a blunt probe chilled in liquid nitrogen (0, 1, or 2 dots above the lateral line, anterior or posterior to the dorsal fin, on the left or right side). They were fed a ration of Oregon Moist Pellet (Crawford and Law, 1972) according to the manufacturer's schedule.

Analyses of mean length are reported here because, after the month of feeding, there were unequal numbers of fish in each lot. Behavioral interactions between members of the paternal-stock groups might have contributed to any effect of paternal stock in this design but tank effects would have influenced all groups equally.

Disease susceptibility

The susceptibility of sibling groups to the pathogenic marine bacterium *Vibrio anguillarum* (Ranson et al., 1984) was determined in a controlled challenge test. Twenty fry from each of 36 sibling groups (offspring of the four dams but of only three sires from each of the three populations) were marked by nine combinations (one mark per sire) of fin excisions. Limitations on the number of available excisions restricted the number of sires which could be tested. The offspring of each of the four dams were cultured separately in four tanks similar to those described above. This confounding of effects of dams and of tanks, imposed by limitations on the number of available challenge tanks, allowed tests only of the effects of stock and of sires within stocks. The fish, grown in freshwater for 2 months to a mean size of 1.75 g, were transferred to the OSU Fish Disease Laboratory at Corvallis, housed in 20-l challenge tanks, and exposed in freshwater to 500 000 cells of V. anguillarum (LS-1-74 isolate, Dept. of Microbiology, OSU) per ml for 15 min in static water. Dead fish, collected from the tanks twice daily until mortality ceased (17 days), and surviving fish were examined; kidney tissue was aseptically streaked on brain–heart infusion agar medium and incubated at room temperature. Bacterial colonies growing on the medium were tested by rapid slide agglutination against rabbit anti-LS174 serum.

Analyses of survival, transformed by ArcSin $[(r+1/4)/(n+1/2)]^{1/2}$ (K.E. Rowe, personal communication, 1980) where r is the number of survivors and n the initial number (20) in each group, are reported here.

Analyses

Observations of these traits were analyzed by three models.

Model A

$$Y_{ijk} = \mu + P_i + S_{j(i)} + D_k + (PD)_{ik} + (SD)_{j(i)k} + e_{ijk}$$

where Y_{ijk} is the mean of the trait measured in the kth dam (D) mated to the jth sire (S) in the ith paternal population (P); $i=1,...,3$; $j=1,...,5$; $k=1,...,4$ and e_{ijk} is the independent normally distributed error.

Model B

$$Y_{jkl} = \mu + S_j + D_k + SD_{jk} + e_{jkl}$$

where Y_{jkl} is the mean of the trait measured in the lth replicate of the kth dam (D) mated to the jth sire (S); $j=1,...,5$; $k=1,...,4$; $l=1,2$

Model C

$$Y_{jk} = \mu + S_j + D_k + (SD)_{jk} + e_{jk}$$

where Y_{jk} is the mean of the trait measured in the mating of the kth dam (D) to the jth sire (S); $j=1,...,5$; $k=1,...,4$.

Model A was applied twice for each trait in parallel analyses, with data from one or the other of the two replicates of Netarts Sires; each analysis encompassed the three stocks and 60 sibling groups. Because the two analyses used common data from the unreplicated crossbred sibling groups, the analyses were not independent. In a test of the significance of a factor, two variance ratios, one in each of the parallel analyses, were computed. The smaller variance ratio was chosen for the test.

Model B used only the data from Netarts sires and included both replicates of each group. Model C was used twice, once with data from the 20 Nemah sires and once with data from the 20 Hoodsport sires. The expectations of mean squares and analyses of variance are in Smoker (1981).

All analyses were computed by the program BMDP2V (Sampson, 1975). F-tests were based on the expectations of mean squares for a random effects model. (Fixed effects models might also have been appropriate; but these tests of significance of effects were more conservative.) In the tests for effects of paternal population and of sires in paternal populations, approximate F-statistics (Snedecor and Cochran, 1967) were computed.

Two females had small eggs (0.2322 g SE=0.0044 and 0.2283 g SE=0.0020), and two females had large eggs (0.3017 g SE=0.0025 and 0.3185 g SE=0.0028), providing a comparison of traits in offspring of large- and small-egg dams. Two of the paternal populations were on the western slope of the coastal mountains (Netarts and Nemah) and the third was on the eastern slope in Puget Sound (Hoodsport), providing a comparison of traits in offspring of coastal paternal populations with those of the Puget Sound population which experiences a different hydrologic regime.

RESULTS AND DISCUSSION

Development rate

There was a significant effect of sires within paternal populations on development rate (Table 1). This indicated significant additive genetic variability in the three stocks. Variability was expressed significantly in crossbred Nemah and Hoodsport sired groups, suggesting that additive genetic variability of development rate might be enhanced in a hatchery stock by the use of gametes from a different stock. Evidence of dominance or epistasis was lacking, i.e., there was no evidence of effects of interaction between dams and sires within paternal populations.

The effect of dam and the effect of egg size on development rate indicated a maternal effect on development rate and the importance of egg size as a trait in selection programs.

The development rate difference between coastal and Puget Sound sires may be related to local temperature regimes. The average incubation temperature at Hoodsport (6.7°C) is colder than a both Netarts and Nemah (7.8°C) (Smoker, 1981). Rate of salmon embryo development is generally slower at cooler temperatures, but after having been spawned in November fry from all stocks enter the marine environment in March and April in both Puget Sound

TABLE 1

Development rate, growth, and disease susceptibility in chum salmon and significance of effects (probability of observed F-ratio)

Average	Significance of effects						
	Paternal stock	Coast vs. Sound	Sires	Dams	Egg size	Interaction	
Development rate							
69.306 days (all sires)	0.103	0.039	0.002	<0	<0	P×D:	0.956
69.509 days	(Netarts sires)		0.502	0.003		S×D:	0.192
69.548 days	(Nemah sires)		0.003	0.002		S×D:	Insign.
68.862 days	(Hoodsport sires)		0.016	0.009		S×D:	Insign.
Growth							
46.18 mm (all sires)	0.077		0.129	0.084	0.025	P×D	0.779
						S×D	0.446
46.34 mm	(Netarts sires)		<0	0.099		S×D:	0.410
45.66 mm	(Nemah sires)		0.530	0.036		S×D:	0.077
46.43 mm	(Hoodsport sires)		0.185	0.036		S×D:	0.100
Susceptibility to vibriosis							
0.25 (all sires)	0.226		0.037	0.144		P×D:	0.184
0.24	(Netarts sires)		0.127	0.021			
0.19	(Nemah sires)		0.379	0.589			
0.29	(Hoodsport sires)		0.340	0.689			

and coastal bays (Koski, 1975; Poon, 1977). The action of genes from the putatively cold-adapted Puget Sound population may have allowed embryos to develop faster at the Netarts temperature.

Growth

There was no evidence of effect of paternal population on growth (Table 1). There was evidence of an effect of sires in the purebred offspring in Netarts sires but not in the crossbred fish, suggesting that additive genetic variability may contribute to variability of growth in the Netarts stock. Heritability of growth in the Netarts population estimated from these data is 0.9 ± 0.5.

Offspring of larger-egg females were larger after 1 month's growth, which reinforces arguments that maternal effects of egg size should be recognized in management of chum salmon broodstocks.

Disease susceptibility

A quarter of the fry survived the challenge by *V. anguillarum* (Table 1). The bacterium was isolated from each of the dead fish and from none of the survivors. A significant effect of sires within stocks suggested the presence of additive genetic variability, but there is no evidence of effects of paternal stock. The reduced experimental power of only three sires per paternal stock may explain the lack of significance. Gjedrem and Aulstad (1974) found similar evidence of additive genetic variability for susceptibility to vibriosis in Atlantic salmon (*S. salar*), and McIntyre and Amend (1978) found that a population of sockeye salmon (*O. nerka*) has significant genetic variability for susceptibility to a viral disease.

ACKNOWLEDGEMENTS

Supported by Oregon Sea Grant College grant number 04-6-158-44094,R/Aq-6. I thank D.R. Ransom and W.E. Groberg for supervising microbiological procedures, J. Trimble and R. Schwab for access to their broodstocks, and J.E. Lannan for advice. From a Ph.D dissertation at Oregon State University.

REFERENCES

Crawford, D.L. and Law, D.K., 1972. Mineral composition of Oregon pellet production formulations. Prog. Fish Cult., 34: 126-130.
FRED (Fisheries Rehabilitation and Enhancement Division), 1985. FRED 1984 Annual Report to the Alaska State Legislature, J.A. Hansen (Editor), Alaska Department of Fish and Game, Juneau, 75 pp.
Gjedrem, T. and Aulstad, D., 1974. Selection experiments with salmon. I. Differences in resistance to vibrio disease of salmon parr (*Salmo salar*). Aquaculture, 3: 51-59.
Kobayashi, T., 1980. Salmon propagation in Japan. In: J.E. Thorpe (Editor), Salmon Ranching. Academic Press, New York, NY, pp. 91-108.
Konovalov, S.M., 1980. U.S.S.R.: Salmon ranching in the Pacific. In: J.E. Thorpe (Editor), Salmon Ranching. Academic Press, New York, NY, pp. 63-90.
Koski, K.V., 1975. The Survival and Fitness of Two Stocks of Chum Salmon (*Oncorhynchus keta*) From Egg Deposition to Emergence in a Controlled-stream Environment at Big Beef Creek. Doctoral Dissertation, University of Washington, Seattle, WA, 213 pp.
McIntyre, J.D. and Amend, P., 1978. Heritability of tolerance for infections hematopoietic necrosis in sockeye salmon (*Oncorhynchus nerka*). Trans. Am. Fish. Soc., 107: 305-308.
Poon, D., 1977. Quality of Salmon Fry from Gravel Incubators. Doctoral Dissertation, Oregon State University, Corvallis, OR, 253 pp.
Ransom, D.P., Lannan, C.N., Rohovec, J.S. and Fryer, J.L., 1984. Comparison of histopathology caused by *Vibrio anguillarum* and *Vibrio ordali* in three species of Pacific salmon. J.Fish Dis., 7: 107-115.
Sampson, P., 1975. BMDP2V: Analysis of variance and covariance including repeated measures. In: J.W. Dixon (Editor), BMDP Biomedical Computer Programs. University of California Press, Los Angeles, CA, pp. 711-734.

Smoker, W.W., 1981. Quantitative Genetics of Chum Salmon, *Oncorhynchus keta*. Doctoral Dissertation, Oregon State University, Corvallis, OR, 170 pp.

Snedecor, G.W. and Cochran, W.G., 1967. Statistical Methods, 6th edition. Iowa State University Press, Ames, IA, 593 pp.

Genetic Studies Connected with Artificial Propagation of Cod (*Gadus morhua* L.)

KNUT E. JØRSTAD

Department of Aquaculture, Institute of Marine Research, Bergen (Norway)

ABSTRACT

Jørstad, K.E., 1986. Genetic studies connected with artificial propagation of cod (*Gadus morhua* L.). *Aquaculture*, 57: 227–238.

In connection with the artificial production of cod fry in Austevoll, Norway, genetic analyses were carried out using polymorphic proteins detectable in yolksac larvae. The eggs used for propagation were collected from natural spawning in a large pen. Yolksac larvae hatched from egg samples collected at different times during the spawning period demonstrated significant genetic heterogeneity among different groups of offspring. Several groups of yolksac larvae were released into seminatural environments and survived based on natural food available in these ponds. Genetic studies of the fry produced in one pond showed, however, that they differed genetically from the broodstock. The results emphasize the importance of propagating multiple egg samples during the maximum spawning period in the artificial propagation of cod, which aims to enhance cod stocks while maintaining the genetic characteristics of the stocks. In addition, large-scale production of fry should be based on pond systems which are similar to natural environments.

INTRODUCTION

Artificial rearing of cod aimed at enhancement of natural stocks has a long tradition in Norway. Such work, started about a hundred years ago, included artificial rearing and release of large quantities of yolksac larvae into coastal and fjord environments. The significance of this work as well as the extent of success is controversial (Solemdal et al., 1984). Recently, however, large seminatural enclosures (ponds), which supply sufficient amounts of natural food, have been successfully used to achieve a high survival rate of artificially reared yolksac larvae (Kvenseth and Øiestad, 1984).

The success of producing large quantities of cod fry opens several possibilities such as cod farming and release of cod fry to enhance natural stocks. As discussed at several meetings (Hynes et al., 1981; Ryman, 1981), artificial production often causes unintended genetic changes in the offspring. Changes due to genetic drift are often connected to a low number of parental fish used.

In addition, different selective forces in the artificial environment compared to the natural one are likely to be present. Genetic changes can reduce the possibility of a successful enhancement program. In addition, the release of large quantities of artificially reared fish may cause several unwanted effects on the natural stocks (Hynes et al., 1981; FAO, 1982).

During the last few years several studies have clearly demonstrated genetic changes in hatchery stocks of cutthroat trout (Allendorf and Phelps, 1980), brown trout (Ryman and Ståhl, 1980), Atlantic salmon (Cross and King, 1983) and black sea bream (Tanigushi et al., 1983). Such observations are especially relevant to the artificial propagation of cod because of the very high fecundity of this species. As suggested by Nævdal and Jørstad (1984), it is important to study the genetic characteristics of the parental fish, the offspring produced and the fry which are released into natural environments. Several polymorphic enzymes in cod are expressed in yolksac larvae (Jørstad et al., 1980); genetic studies can therefore be carried out at early stages.

MATERIAL AND METHODS

The artificial propagation program for cod was carried out at the Aquaculture Station Austevoll in Western Norway.

The cod which were used as broodstock consisted of wild fish caught in the coastal region of Austevoll. In 1983 twenty females and 45 males were used, compared to 65 females and 66 males in the 1984 experiment. Each parental fish was individually tagged and a small piece of white muscle for electrophoretic analyses was taken by biopsy. Blood samples were carefully taken from blood vessels in the gills by means of a sterile needle.

In February the broodstock was transferred to a large plastic spawning pen in which the fish spawned naturally. Fertilized eggs were collected daily from the spawning pen and incubated in the hatchery. Details concerning the spawning pen, egg collection system and egg incubators are described by Huse and Jensen (1983). The eggs were hatched after 20 days and samples of yolksac larvae were transferred to Bergen and analysed by electrophoresis at the Institute of Marine Research following the methods described by Jørstad et al. (1980).

Several groups of 5-day-old yolksac larvae were released into enclosures or ponds where they survived on natural food (Kvenseth and Øiestad, 1984). Only one pond, Hyltropollen, was used in 1983. In the following year a freshwater pond, Svartatjønn (Øiestad et al., 1984), was also used. Both ponds produced significant numbers of cod larvae and fry. Samples were collected in June and September.

All samples of white muscle, including those taken by biopsy, were analysed by using horizontal starch-gel electrophoresis followed by selective staining of the enzymes (Harris and Hopkinson, 1976). Haemoglobin typing was carried out by means of agar electrophoresis according to the method described by Sick (1961). Polymorphic enzymes in cod have been described elsewhere (Cross and Payne, 1978; Mork et al., 1982).

Comparisons of the genotypic distributions in the different samples with Hardy–Weinberg expectations were carried out using the test described by Christiansen et al. (1976) or the G-test (Sokal and Rohlf, 1969). The G-test was also used when testing for sample group homogeneity and in pairwise tests between broodstock and different samples of offspring. All the tests were based on the genotypic distributions in the different samples.

RESULTS

Offspring genetic variation during the spawning season

In the 1983 spawning season only three different groups of yolksac larvae were analysed for the *LDH-3* locus. The samples consisted of larvae hatched from eggs spawned in the middle part of March, and the genotypic distributions were very close to that of the broodstock used.

In 1984 a more detailed study was carried out. Collection of egg samples started in late February and continued throughout the spawning season, a period of nearly 2 months. During this time 17 samples, about 2400 individual yolksac larvae, were analysed. The genotypic distributions and allele frequencies are shown in Table 1. The G-test demonstrated significant heterogeneity among samples ($G=163$; df$=32$; $P<0.001$). The frequency of *LDH-3(100)* varied from 0.446 to 0.812. All samples which differed from the broodstock (pairwise test, Table 1) were spawned in April, late in the spawning season.

Only one sample (March 17, Table 1) differed significantly ($G=3.77$; df$=1$; $P=0.05$) from Hardy–Weinberg expectations. The overall distribution of genotypes, however, showed a significant deficiency of heterozygotes at the *LDH-3* locus ($G=6.83$; df$=1$; $P=0.009$).

Some of the samples of yolksac larvae could also be reliably scored for variation in the *PGM* locus. As seen in Table 2, three different alleles were present in the broodstock, two of them in low frequency. All three alleles were found in only one of the samples analysed, and in two samples only a single allele was observed. When the overall distribution of genotypes was compared to the broodstock (pairwise test), a significant difference, mainly due to a deficiency of heterozygotes, was detected ($G=9.86$; df$=2$; $P=0.007$). The data for the samples of offspring clearly demonstrated that alleles present at low frequencies are more likely to be lost when a limited number of parental fish are used as broodstock.

Fig. 1 compares the spawning intensity and the frequency of the most common allele for *LDH-3* and *PGM* during the spawning season. Clearly, the significant variation is connected with the late period with relatively low spawning

TABLE 1

LDH-3 variation in samples of yolksac larvae from different times during the spawning season in 1984

Spawning date	N	Genotype distributions			Allele frequencies		Test against broodstock	
		70/70	70/100	100/100	70	100	G	P
27 Feb.	135	8	63	64	0.293	0.707	2.01	0.37
29 Feb.	143	17	56	70	0.315	0.685	1.06	0.59
9 March	139	21	55	63	0.349	0.651	1.51	0.47
14 March	135	16	71	48	0.381	0.619	1.83	0.40
17 March	125	22	48	55	0.368	0.632	2.79	0.25
19 March	128	15	57	56	0.340	0.660	0.06	0.97
20 March	138	24	65	49	0.409	0.591	3.19	0.20
23 March	138	13	65	60	0.330	0.670	0.15	0.93
27 March	135	14	49	72	0.285	0.715	2.51	0.29
30 March	142	13	62	67	0.310	0.690	0.37	0.83
3 April	139	22	65	52	0.392	0.608	1.94	0.38
6 April	213	69	98	46	0.554	0.446	29.0	<0.001
11 April	127	22	59	46	0.406	0.594	2.83	0.24
13 April	125	7	33	85	0.188	0.812	14.8	0.001
16 April	143	17	75	51	0.381	0.619	1.8	0.401
17 April	142	13	63	66	0.313	0.687	0.28	0.87
20 April	137	10	44	83	0.234	0.766	7.3	0.03
Total larvae	2384	323	1028	1033	0.351	0.649		
Broodstock	121	13	55	53	0.335	0.665		

TABLE 2

PGM variation during the spawning season in 1984

Spawning date	N	Genotype distributions			Allele frequencies		
		30/100	100/100	100/150	30	100	150
9 March	144	0	141	3		0.990	0.010
23 March	144	2	142	0	0.007	0.993	
27 March	143	0	140	3		0.990	0.010
30 March	143	0	143	0		1.000	
3 April	144	0	143	1		0.997	0.003
6 April	236	1	234	1	0.002	0.996	0.002
11 April	144	2	142	0	0.007	0.993	
13 April	144	1	143	0	0.003	0.997	
16 April	144	0	143	1		0.997	0.003
17 April	144	0	144	0		1.000	
Total larvae	1530	6	1515	9	0.002	0.995	0.003
Broodstock	121	4	115	2	0.017	0.975	0.008

activity in the pen. No variation was observed in the maximum spawning period in the first part of March. As seen from Fig. 1c, all the samples analysed have a low frequency of heterozygotes at the PGM locus compared to the broodstock (broken line).

Comparisons between broodstock and cod fry produced in ponds

Several million 5-day-old yolksac larvae were released in Hyltropollen in 1983 (Kvenseth and Øiestad, 1984). High survival rates due to natural food in the pond permitted sampling of cod larvae and fry at different stages. The results of genetic analyses of fry collected from the pond in June and September are summarized in Table 3. G-tests for sample group homogeneity based on the genotype distribution at each locus revealed no significant values at any of the loci. Pairwise tests between the samples also showed no heterogeneity.

Two different types of ponds (Hyltropollen and Svartatjønn) were used in the 1984 season. Several groups of yolksac larvae hatched from eggs spawned during the maximum spawning period (see Fig. 1) were released in Hyltropollen. Samples of fry were taken from the pond in June and August and analysed for three different enzymes. The results are shown in Table 4. No significant differences were detected in either tests of sample group homogeneity or in pairwise comparisons with the broodstock.

The yolksac larvae which were released in Svartatjønn were hatched from eggs mainly spawned in the late spawning period (April). Table 5 shows the

TABLE 3

Comparisons between broodstock and fry produced in Hyltropollen in 1983

Sample	Month	N	LDH-3		PGI-1			PGM		GPD			HbI	
			70	100	30	100	150	30	100	90	100	120	1	2
Broodstock	—	74	0.304	0.696	0.007	0.721	0.271	0.014	0.986	0.014	0.953	0.034	0.561	0.439
Yolksac larvae	April	480	0.303	0.697										
Fry	June	81	0.426	0.574	0.012	0.685	0.302	0.013	0.987	0	0.981	0.019	—	—
Fry	Sept.	191	0.327	0.673	0.051	0.716	0.233	0.008	0.992	0.005	0.948	0.047	0.558	0.442
Homogeneity test based on genotype distributions			$G = 11.9$ $df = 6$ $P = 0.064$		$G = 12.27$ $df = 6$ $P = 0.056$			$G = 0.44$ $df = 2$ $P = 0.802$		$G = 3.04$ $df = 2$ $P = 0.219$			$G = 0.11$ $df = 2$ $P = 0.946$	

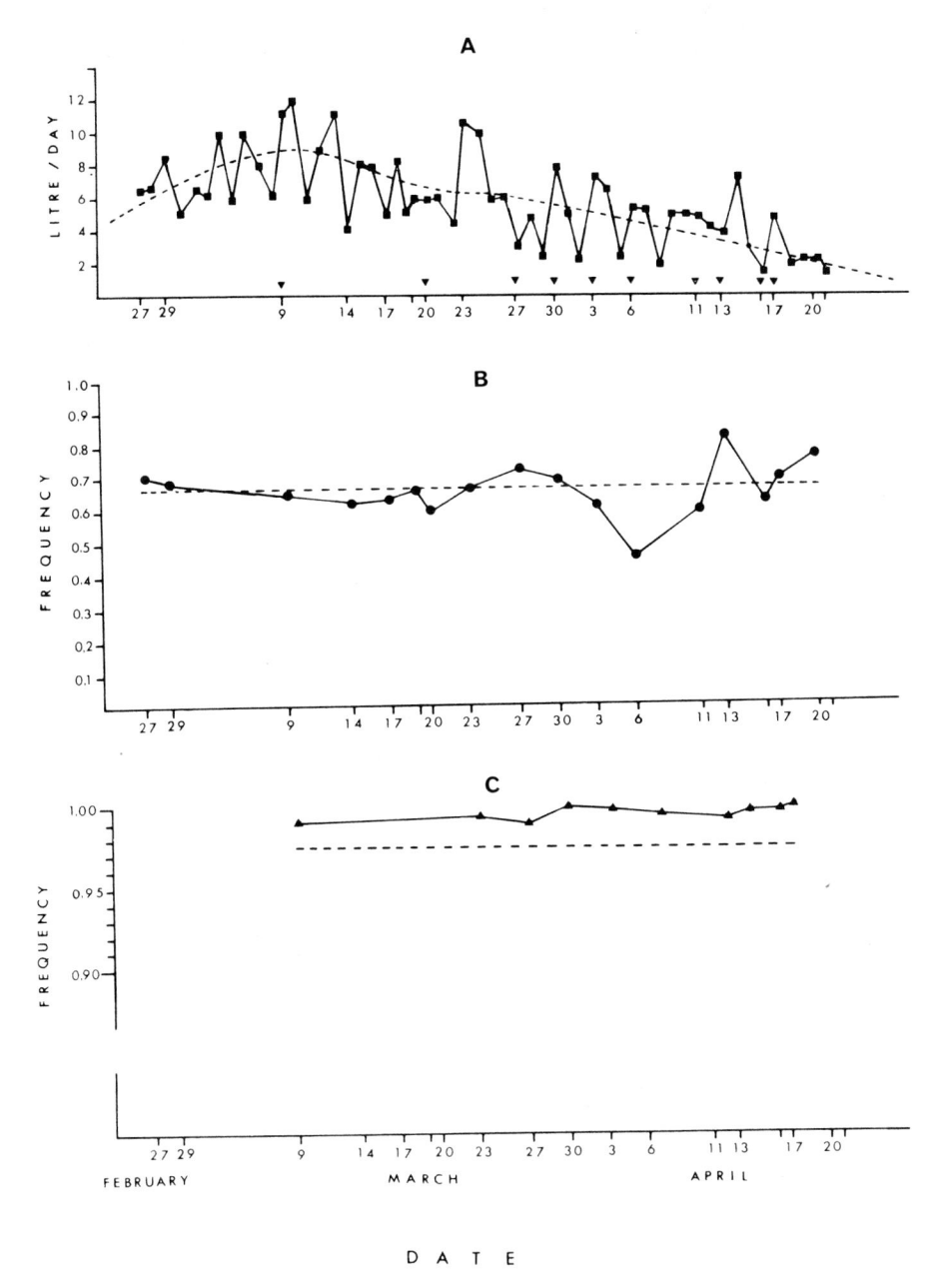

Fig. 1. Genetic variation in yolksac larvae hatched from eggs spawned at different times in the spawning season. A, Spawning intensity in litres of eggs per day; B, Frequency of LDH-$3(100)$; C, Frequency of $PGM(100)$. The broken lines in B and C represent the frequency estimated for the broodstock used.

TABLE 4

Comparisons between broodstock and fry produced in Hyltropollen in 1984

Sample	Month	N	LDH-3		PGI-1			PGM		
			70	100	30	100	150	30	100	150
Broodstock	April	121	0.335	0.665	0.041	0.674	0.258	0.017	0.975	0.008
Yolksac larvae		112	0.357	0.643						
Fry	June	144	0.274	0.726	0.014	0.651	0.335	0.017	0.969	0.014
Fry	August	65	0.377	0.603	0.031	0.708	0.262	0.015	0.969	0.015
Homogeneity test based on genotype distribution			$G = 6.84$ $df = 6$ $P = 0.336$		$G = 11.8$ $df = 8$ $P = 0.160$			$G = 0.54$ $df = 4$ $P = 0.969$		

TABLE 5

Comparisons between broodstock and fry produced in Svartatjønn in 1984

Sample	Month	N	LDH-3		PGI-1			PGM		
			70	100	30	100	150	30	100	150
Broodstock		121	0.335	0.665	0.041	0.674	0.258	0.017	0.975	0.008
Yolksac larvae	April	101	0.351	0.649						
Fry	June	144	0.142	0.858	0.021	0.535	0.444	0.066	0.931	0.003
Fry	August	144	0.271	0.729	0.008	0.609	0.383	0.028	0.965	0.007
Homogeneity test based on genotype distribution			$G = 40.14$ df = 6 $P < 0.001$		$G = 21.66$ df = 8 $P = 0.006$			$G = 10.55$ df = 4 $P = 0.032$		

allele frequencies observed in fry samples from this pond. Testing for within sample group homogeneity revealed significant variation for all enzymes analysed. Pairwise comparisons with the broodstock demonstrated that the heterogeneity was mainly due to the sample of fry taken from the pond in June. At this time the pond was emptied and the fry produced were transferred to net pens where they were kept feeding until August. The June sample of fry, which differed from the broodstock at two loci, is believed to be representative of the fry produced in Svartatjønn.

DISCUSSION

Allendorf and Phelps (1980) discussed the genetic changes observed in hatchery stocks of cutthroat trout in relation to a small number of parental fish used. They also emphasized that changes in enzyme loci possibly reflect changes in the total genome and, if so, polymorphic loci could be used as indicators of genetic processes.

Observed genetic changes in trout (Ryman and Ståhl, 1980) and Atlantic salmon (Cross and King, 1983) are most likely explained by genetic drift due to few parental fish. Sophisticated breeding schemes with sufficient numbers of parental fish are necessary to prevent inbreeding and genetic drift in salmonids.

In contrast to the artificial crossing commonly used for salmonids, propagation of black sea bream (Tanigushi et al., 1983) was accomplished by natural spawning of about 100 fish in a tank. Based on analyses of a large number of enzyme loci, the authors concluded that only a fraction of the fish present in the spawning tank have actually contributed to the next generation. The work on cod described here suggests a similar explanation for the genetic variation observed during the spawning season.

When large variation in spawning intensity from day to day exists (Fig. 1), it seems obvious that only 1-day samples of eggs from the spawning pen do not necessarily represent the genetic characteristics of the broodstock. As mentioned earlier, the differences are most clearly expressed in the offspring from April when nearly 50% of the samples analysed differed significantly from the broodstock. The most likely explanation is that only a few parental fish are actually involved in spawning at this time.

No genetic changes were detected in the experiments where cod fry were produced in Hyltropollen. This pond is a natural marine environment which has been closed off from the sea by dams and was treated with rotenone to remove predators before releasing the yolksac larvae (Kvenseth and Øiestad, 1984). This natural environment may be important for the successful survival and rapid growth of cod fry in the pond. In both years, the different groups of yolksac larvae which were released in the pond were from eggs mainly spawned in March, in the maximum spawning period. The Hyltropollen offspring seemed to reflect the genetic characteristics of the broodstock.

In comparison, the Svartatjønn pond is a freshwater pond located near the sea. The freshwater was replaced by saltwater and the environmental conditions are at the moment difficult to define accurately. The genetic changes observed for the fry produced in this pond are possibly the result of different selection pressure due to unstable environmental conditions. In addition the yolksac larvae released into this pond were offspring from the late spawning period, which also may be of importance considering the observed genetic changes.

Based mainly on culture practice and experience from freshwater species, Hynes et al. (1981) proposed a number of recommendations for the management of hatchery stocks. Although very few data on artificial propagation of marine fish species exist, the guidelines are relevant for new species which are going to be under culture management. In cod, as described here, several potential genetic problems were investigated. The results clearly suggest some critical points which have to be considered for the artificial production of cod. First, multiple egg samples from different times during the spawning period should be propagated. Offspring only from the last period of spawning do not reflect characteristics of the broodstock and should not be used in programs for enhancement of natural stock. Second, from the genetic point of view, the most successful pond investigated here was Hyltropollen, which has conditions similar to the natural marine environment.

ACKNOWLEDGEMENTS

The studies concerning spawning intensity and genetic variation during the spawning period were carried out in cooperation with Ingvar Huse. The samples of cod fry from the different ponds were kindly supplied by Victor Øiestad and Per Gunnar Kvenseth. I am indepted to the technical assistance of Ole Ingar Paulsen, the typing work of Anette Lyssand and linguistic corrections of Eva Marie Wagner.

REFERENCES

Allendorf, F.W. and Phelps, S.R., 1980. Loss of genetic variation in a hatchery stock of cutthroat trout. Trans. Am. Fish. Soc., 109: 537–543.
Christiansen, F.B., Frydenberg, O., Hjorth, J.P. and Simonsen, V., 1976. Genetics of *Zoarces* populations. IX, Geographic variation at three phosphoglucomutase loci. Hereditas, 83: 245–256.
Cross, T.F. and King, I., 1983. Genetic effects of hatchery rearing in Atlantic salmon. Aquaculture, 33: 33–40.
Cross, T.F. and Payne, R.H., 1978. Geographic variation in Atlantic cod, *Gadus morhua* L., off

eastern North America: a biochemical systematics approach. J. Fish. Res. Board Can., 35: 117-123.
FAO, 1982. Conservation of genetic resources in fish: problems and recommendations. Report of the Expert Consultation on Genetic Resources in Fish, Rome, 9-13 June 1980. FAO Fish. Tech. Pap., 217: 1-43.
Harris, J. and Hopkinson, D.A., 1976. Handbook of Enzyme Electrophoresis in Human Genetics. North-Holland, Amsterdam.
Huse, I. and Jensen, P., 1983. A simple and inexpensive spawning and egg collection system for fish with pelagic eggs. Aquacult. Eng., 2: 165-171.
Hynes, J.D., Brown, E.B., Helle, J.H., Ryman, N. and Webster, D.A., 1981. Guidelines for the culture of fish stocks for resource management. Can. J. Fish. Aquat. Sci., 38: 1867-1876.
Jørstad, K.E., Solberg, T. and Tilseth, S., 1980. Enzyme polymorphism expressed in newly hatched cod larvae and genetic analyses of larvae exposed to hydrocarbons. Counc. Meet. Int. Counc. Explor. Sea, (F:22): 1-16.
Kvenseth, P.G. and Øiestad, V., 1984. Large-scale rearing of cod fry on the natural food production in an enclosed pond. In: E. Dahl, D.S. Danielssen, E. Moksness and P. Solemdal (Editors), The Propagation of Cod, *Gadus morhua* L. Flødevigen Rapportser., 1: 645-655.
Mork, J., Reüterwoll, C., Ryman, N. and Ståhl, G., 1982. Genetic variation in Atlantic cod (*Gadus morhua* L.): a quantitative estimate from a Norwegian coastal poulation. Hereditas, 96: 55-61.
Nævdal, G. and Jørstad, K.E., 1984. Importance of genetic variation in the propagation of cod. In: E. Dahl, D.S. Danielssen, E. Moksness and P. Solemdal (Editors), The Propagation of Cod, *Gadus morhua* L. Flødevigen Rapportser., 1: 733-743.
Øiestad, V., Kvenseth, P.G. and Pedersen, T., 1984. Mass production of cod fry (*Gadus morhua* L.) in a large basin in western Norway — a new approach. Counc. Meet. Int. Counc. Explor. Sea, (F:16): 6 pp.
Ryman, N. (Editor), 1981. Fish gene pools. Preservation of genetic resources in relation to wild fish stocks. Ecol. Bull., 34: 1-111.
Ryman, N. and Ståhl, G., 1980. Genetic changes in hatchery stocks of brown trout (*Salmo trutta*). Can. J. Fish. Aquat. Sci., 37: 82-87.
Sick, K., 1961. Haemoglobin polymorphism in fishes. Nature, 192: 894-896.
Sokal, R.R. and Rohlf, F.J., 1969. Biometry. W.H. Freeman and Co., San Fransisco, CA, 776 pp.
Solemdal, P., Dahl, E., Danielssen, D.S. and Moksness, E., 1984. The cod hatchery in Flødevigen, background and realities. Flødevigen Rapportser., 1: 17-45.
Tanigushi, N., Sumantadinata, K. and Iyama, S., 1983. Genetic change in the first and second generation of hatchery stock of black seabream. Aquaculture, 35: 309-320.

Bottleneck Effects and the Depression of Genetic Variability in Hatchery Stocks of *Penaeus japonicus* (Crustacea, Decapoda)

V. SBORDONI[1], E. DE MATTHAEIS[2], M. COBOLLI SBORDONI[2], G. LA ROSA[2] and M. MATTOCCIA[2]

[1]*Department of Biology, 2nd University of Rome "Tor Vergata" (Italy)*
[2]*Department of Animal and Human Biology, 1st University of Rome "La Sapienza" (Italy)*

ABSTRACT

Sbordoni, V., De Matthaeis, E., Cobolli Sbordoni, M., La Rosa, G. and Mattoccia, M., 1986. Bottleneck effects and the depression of genetic variability in hatchery stocks of *Penaeus japonicus* (Crustacea, Decapoda). *Aquaculture*, 57: 239–251.

Aquaculture of *Penaeus japonicus* is developing in Italy at a production level. Genetic analysis of the founder stock and five subsequent hatchery generations revealed a constant reduction in levels of allozyme polymorphism. Average heterozygosity decreased from 0.102 to 0.039. The magnitude of the reduction in heterozygosity was much higher than expected from the numbers of breeders placed into spawning tanks at each reproductive cycle. We estimated, under the assumption of neutrality, that the effective number of parents contributing to each broodstock might have been as low as four, although the number of shrimp pairs held in spawning tanks varied from 50 to 300 after an initial bottleneck occurred in the first generation. This discrepancy may be explained as the combined effect of some common farming practices and it points out the importance of a careful check of the number of spawners actually contributing to each reproductive cycle.

INTRODUCTION

The value of electrophoretic allozyme markers as a powerful tool for monitoring the genetic structure of hatchery populations in aquaculture has been stressed by several authors (Utter et al., 1974; Wilkins, 1975; Moav et al., 1976; Hedgecock, 1977). We used this technique to investigate the genetic structure of a series of hatchery stocks of the Penaeid shrimp *Penaeus japonicus* Bate. Samples from various localities and generations, all descendants from the same broodstock introduced into Italy from Japan, were studied.

Penaeus japonicus is native to a vast area in the Indo-West Pacific region, and in the past 20 years it has been actively spreading in the Eastern Mediterranean coasts through the Suez Canal. This species is widely used in mariculture in Japan (Shigueno, 1975) and it is also considered a very promising shrimp for Italian aquaculture (Lumare, 1981). *P. japonicus* was first introduced into Italy in 1979 and acclimatized in the Lesina lagoon (Southeast coast of Italy) (Lumare and Palmegiano, 1980). This founder stock was the F_1 generation of a population sample collected in Southern Japanese coastal waters. Each subsequent reproductive cycle started from laboratory produced larvae which were released and allowed to grow in Lesina and in other natural lagoons. In 1985 the Italian population reached the F_7 generation. The inferred effective population size of the parental generation was reasonably high. By contrast, a severe bottleneck ($N=4$) occurred between F_1 and F_2. F_1 electrophoretic data for 31 loci were reported previously (De Matthaeis et al., 1983).

The aim of this paper is to study the genetic structure at a common set of loci of different stocks from the F_1 to the F_6 generations. The data show a notable progressive reduction in levels of genetic variability. These results are discussed with respect to production planning.

MATERIALS AND METHODS

P. japonicus *cultivation in Italy*

The introduction and successive production strategies of *P. japonicus* in Italy were managed by Lumare and co-workers in the Istituto per lo Sfruttamento Biologico delle Lagune at Lesina, National Research Council (CNR). Italian broodstock started in March 1979 from 10 000 hatchery-produced larvae representing the F_1 generation from naturally spawned parents collected in the wild (presumably in Kyushu coastal waters, South Japan). This initial stock was subjected to various acclimatization trials both in the Lesina lagoon and in the laboratory. In early autumn approximately 1000 adult shrimps were captured and subjected to different techniques for inducing spawning, including eyestalk ablation (Lumare, 1982). In the spring of 1980, as a result of these first attempts, 2000 eggs were gathered for reproductive purposes, presumably derived from only two to three fertilized females (Lumare and Palmegiano, 1980; F. Lumare, personal communication, 1985). These F_2 individuals were released in the Lesina lagoon as 22-day post-larvae (P22) in controlled conditions (Lumare, 1984). In the following years improvements of culture techniques have resulted in an appreciable production of shrimps, and release experiments of this kind have been extended to other coastal lagoons (Lumare, 1985).

The number of breeding parents has been increased yearly up to 300 pairs; however, the mean hatching rate decreased regularly from 1980 (F_2) to 1985 (F_7) (Table 1). These seeding programs were carried out both in semi-extensive farming conditions in brackish water lagoons and in controlled extensive breeding schedules in ponds with yields of around 300 kg/ha (Lumare, 1985).

TABLE 1

Data concerning Lesina broodstock of *Penaeus japonicus* at various generations

Year	Generation	No. of breeding pairs in previous generation	No. of hatchery-produced larvae (thousands)	Mean hatching rate
1980	F_2	2	2000	0.50
1981	F_3	50	80 000	0.40
1982	F_4	100	500 000	0.32
1983	F_5	300	750 000	0.33
1984	F_6	300	1 250 000	0.20
1985	F_7	300	1 500 000	0.10

In 1983 a new stock of *P. japonicus* post-larvae was introduced from Japan (Seto Inland Sea) and cultivated in net enclosures in the Lesina lagoon. However, due to acclimatization problems this stock experienced a drastic reduction in number (F. Lumare, personal communication, 1985).

Study samples

The following *P. japonicus* samples have been studied electrophoretically:
JLE1 ($N=20$), JLE4 ($N=30$), JLE5 ($N=39$), JLE6 ($N=30$), representing F_1, F_4, F_5 and F_6 generations, respectively, cultivated in the Lesina lagoon (Foggia Southeastern Italy);
JLV6 ($N=28$), F_6 cultivated in net fence at Lesina;
JVA6 ($N=30$), F_6 from Varano lagoon (Foggia);
JFO5 ($N=36$) and JFO6 ($N=30$), F_5 and F_6 from Valle Fosse lagoon (Venice, Northeastern Italy);
JSP6 ($N=30$), F_6 from Valle Sparesera (Venice);
JCH6 ($N=30$), F_6 from Valle Chiara (Grado, Northeastern Italy);
JNGS ($N=40$), F_1 (1983) from a wild caught sample, Seto Inland Sea, Japan.

Electrophoresis

Electrophoresis was carried out using 12% starch-gels from Connaught Laboratories. The proteins studied, the tissues analyzed, the routine electrophoretic systems used, the staining techniques, and the abbreviations used for the corresponding genetic loci are detailed in Table 2. The electrophoretic systems are designated as: A, discontinuous tris citrate pH 8.6 (Poulik, 1957); B, tris versene borate pH 9.1 (as buffer B of Ayala et al., 1973); B1, as buffer B but

TABLE 2

A summary of all proteins investigated, enzyme codes, procedures used and loci detected in *Penaeus japonicus* samples (see text for buffer system)

Enzyme	E.C.no.	Tissue	Buffer system	Staining procedure	Loci scored Monomorphic	Polymorphic
Acid phosphatase	3.1.3.2	m	A	Tracey et al. (1975)	Acph*	–
Adenosine deaminase	3.5.4.4	c	E	Ward and Beardmore (1977)	Ada-1,Ada-2,Ada-3	–
Aldeyde oxidase	1.2.3.1	c	B	Ayala et al. (1974)	Ao-1*	Ao-2*
Aldolase	4.1.2.13	c	B1	Ayala et al. (1972)	Aldo-1	Aldo-2
Alkaline phosphatase	3.1.3.1	c	A	Ayala et al. (1972)	Aph*	–
Carbonic anhydrase	4.2.1.1	c	G	Brewer and Sing (1970)	–	Ca-1*,Ca-2,Ca-3
Esterase	3.1.1.2	c	A	Ayala et al. (1972)	Est-1*,Est-3,Est-4*	Est-2*
Fructokinase	2.7.1.4	c	H	Brewer and Sing (1970)	Fk	–
Glucose-6-phosphate dehydrogenase	1.1.1.49	m	B1	Ayala et al. (1974)	G6pd-1*,G6pd-2*	–
Glutamic-oxalacetic transaminase	2.6.1.1	m	B1	Ayala et al. (1975)	Got*	–
Glyceraldehyde-3-phosphate dehydr.	1.2.1.12	m	B1	Ayala et al. (1974)	Gly-3-pdh	–
Isocitric dehydrogenase	1.1.1.42	m	F	Brewer and Sing (1970)	Idh	–
Lactate dehydrogenase	1.1.1.27	m	A	Selander et al. (1971)	Ldh-2*	Ldh-1*
Leucine amine peptidase	3.4.11.1	m	A	Ayala et al. (1972)	–	Lap*
Malate dehydrogenase	1.1.1.37	m	A	Ayala et al. (1972)	–	Mdh-1*,Mdh-3
Malic enzyme	1.1.1.40	m	F	Ayala et al. (1972)	Me*	–
Mannose phosphate isomerase	5.3.1.8	c	F	Harris et al. (1977)	–	Mpi
Phosphohexose isomerase	5.3.1.9	m	C	Brewer and Sing (1970)	–	Phi*
Phosphoglucomutase	2.7.5.1	m	D	Brewer and Sing (1970)	Pgm-1*	Pgm-2*
Triosephosphate isomerase	5.3.1.1	c	B1	Ayala et al. (1972)	Tpi	–
Xanthine dehydrogenase	1.2.1.37	c	A	Selander et al. (1971)	Xdh*	–
Muscle protein	–	m	A,C	Ayala et al. (1973)	Pt-1,Pt-2,Pt-3, Pt-7,Pt-8,Pt-9	–

c, Cephalothorax; m, abdominal muscle.
*Common set of loci considered in this paper. A locus is considered polymorphic when the frequency of the common allele is ≤0.99.

with NADP+ added; C, continuous tris citrate pH 8.4 (modified from buffer C of Ayala et al., 1972); D, tris maleate pH 7.4 (Brewer and Sing, 1970); E, continuous tris citrate pH 8.0 (Ward and Beardmore, 1977); F, continuous tris citrate pH 7.1 (modified from buffer C of Ayala et al., 1973); G, borate pH 8.0 (Brewer and Sing, 1970); H, tris versene borate pH 8.0 (Brewer and Sing, 1970).

RESULTS

Twenty-eight inferred loci have been consistently scored but only the 20 loci shared by all the samples are considered here (Table 2). Eight loci were found to be polymorphic in at least one sample (Table 3): Ao-2, Ca-1, Est-2, Lap, Ldh-1, Mdh-1, Pgm-2, Phi. Table 4 presents variability estimates for each of the different stocks. A progressive reduction of genetic variability levels through time is evident. Loci which appeared polymorphic in the F_1 generation (i.e., Lap, Mdh-1, Pgm-2 and Ao-2), became fixed for the more frequent allele in subsequent generations.

Wright's fixation index values were calculated for each generation to test the possible effects of inbreeding. Average F values for polymorphic loci were 0.011, 0.024 and 0.161, respectively, for the F_1, F_4 and F_5 generation samples from Lesina. These figures suggest increasingly higher levels of inbreeding up to the F_5 generation. However, the average value calculated for the F_6 Lesina sample was much lower than that of F_5 ($F=0.028$), suggesting that genetic admixture resulting from the new Japanese stock (JNGS) counterbalanced that inbreeding depression.

Actually the genetic structure of the JNGS stock was consistently different from that of the F_1 introduced into Italy. JNGS was characterized by different allele frequencies at most polymorphic loci and by a lower heterozygosity level. Moreover, two alleles, Phi^{98} and $Ldh-1^{102}$, were peculiar to this new stock. Such differences could be due to the different geographic origin of the new stock. Therefore a detailed study of the geographic variation of this species throughout its range would be of interest.

Different geographic samples (hatcheries) of the F_6 stock display a significant heterogeneity of the allele frequencies (Table 3). Variations among hatcheries in the allele frequencies at the Ca-1 and Est-2 loci are partly responsible for such heterogeneity together with the alleles introduced by the new stock, Phi^{98} and $Ldh-1^{102}$. Their frequency varies from 0.033 to 0.182 for Phi^{98} and from 0.023 to 0.058 for $Ldh-1^{102}$. F_{st} values (Nei, 1977) were also calculated to estimate the extent of genetic divergence among hatcheries. The F_{st} values ranged from 0.015 (Ldh-1 locus) to 0.066 (Est-2 locus) with an average of 0.043. These values are of the same order of magnitude of F_{st} values reported for different hatchery F_2 stocks of black seabream (Taniguchi et al., 1983). Comparative estimates available for natural populations of black seabream

TABLE 3

Allele frequencies and heterozygosity at polymorphic loci in study samples of *P. japonicus*

Locus		JLE1	JLE4	JLE5	JFO5	JLE6	JLV6	JVA6	JFO6	JCH6	JSP6	JNGS
Ao-2	N	26	60	66	66	54	28	58	48	48	60	66
	H_o	0.077	-	-	-	-	-	-	-	-	-	-
	H_a	0.073	-	-	-	-	-	-	-	-	-	-
	100	0.962	1	1	1	1	1	1	1	1	1	1
	98	0.038	-	-	-	-	-	-	-	-	-	-
	SE	0.037										
Ca-1	N	16	22	54	52	38	20	30	24	28	4	68
	H_o	0.625	0.636	0.333	0.346	0.105	0.100	0.133	0.333	0.214	0.500	0.412
	H_a	0.594	0.541	0.475	0.410	0.098	0.255	0.124	0.278	0.375	0.375	0.485
	107	0.500	0.455	0.388	0.288	0.053	0.150	0.067	0.167	0.250	0.250	0.412
	102	0.375	0.500	0.612	0.712	0.947	0.850	0.933	0.833	0.750	0.750	0.588
	100	0.125	0.045	-	-	-	-	-	-	-	-	-
	SE	0.125	0.106	0.066	0.062	0.036	0.080	0.045	0.076	0.082	0.216	0.060
Est-2	N	28	46	54	48	30	24	34	38	24	22	48
	H_o	0.429	0.478	0.222	0.417	0.267	0.500	0.412	0.421	0.250	0.091	0.458
	H_a	0.426	0.500	0.483	0.469	0.444	0.500	0.389	0.499	0.496	0.235	0.499
	104	0.036	-	-	-	-	-	-	-	-	-	-
	102	0.714	0.500	0.592	0.625	0.667	0.500	0.735	0.526	0.542	0.864	0.521
	100	0.250	0.500	0.407	0.375	0.333	0.500	0.265	0.474	0.458	0.136	0.479
	SE	0.085	0.074	0.067	0.069	0.086	0.102	0.076	0.081	0.102	0.073	0.072
Lap	N	34	34	24	28	22	12	16	20	14	18	22
	H_o	0.294	-	-	-	-	-	-	-	-	-	-
	H_a	0.472	-	-	-	-	-	-	-	-	-	-
	100	0.618	1	1	1	1	1	1	1	1	1	1
	98	0.382	-	-	-	-	-	-	-	-	-	-
	SE	0.083										
Ldh-1	N	34	30	66	66	56	22	52	44	44	44	80
	H_o	-	0.067	-	-	0.071	-	0.115	0.045	0.045	-	0.050
	H_a	-	0.064	-	-	0.069	-	0.109	0.044	0.044	-	0.049
	102	-	-	-	-	0.036	-	0.058	0.023	0.023	-	0.025
	100	1	0.967	1	1	0.964	1	0.942	0.977	0.977	1	0.975
	98	-	0.033	-	-	-	-	-	-	-	-	-
	SE		0.033			0.025		0.032	0.022	0.022		0.017
Mdh-1	N	34	54	66	72	54	28	58	48	48	60	60
	Hp_o	0.412	-	-	-	-	-	-	-	-	-	-
	H_a	0.390	-	-	-	-	-	-	-	-	-	-
	102	0.735	1	1	1	1	1	1	1	1	1	1
	100	0.265	-	-	-	-	-	-	-	-	-	-
	SE	0.076										
Pgm-2	N	34	60	78	72	58	52	60	58	60	60	80
	H_o	0.118	-	-	-	-	-	-	-	-	-	-
	H_a	0.111	-	-	-	-	-	-	-	-	-	-
	101	0.059	-	-	-	-	-	-	-	-	-	-
	100	0.941	1	1	1	1	1	1	1	1	1	1
	SE	0.040										
Phi	N	22	60	42	36	34	30	36	34	36	22	38
	H_o	0.091	0.200	0.095	0.278	0.176	0.066	0.278	0.176	0.222	0.454	0.368
	H_a	0.086	0.320	0.172	0.239	0.164	0.064	0.239	0.258	0.248	0.368	0.387
	98	-	-	-	-	0.059	0.033	0.139	0.118	0.055	0.182	0.079
	96	0.955	0.800	0.905	0.861	0.912	0.967	0.861	0.853	0.861	0.773	0.763
	94	0.045	0.200	0.095	0.139	0.029	-	-	0.029	0.083	0.045	0.158
	SE	0.044	0.052	0.045	0.058	0.048	0.033	0.058	0.060	0.057	0.089	0.068

N = Number of genes sampled; H_o = observed heterozygosity; H_a = expected heterozygosity at Hardy—Weinberg equilibrium; SE = standard error for allele frequencies.

TABLE 4

Average estimates of genetic variability at 20 enzyme loci in hatchery samples of P. japonicus (average estimates for each generation are also indicated)

	F_1	F_4	F_5		F_6						
	JLE1	JLE4	JLE5	JFO5	JLE6	JLV6	JVA6	JFO6	JCH6	JSP6	JNGS
H_o	0.102	0.069	0.033	0.052	0.031	0.033	0.047	0.048	0.037	0.052	0.064
H_a	0.105	0.071	0.057	0.056	0.039	0.041	0.043	0.053	0.058	0.050	0.072
P	0.350	0.200	0.150	0.150	0.200	0.150	0.200	0.200	0.200	0.150	0.200
A	1.40	1.25	1.20	1.15	1.25	1.15	1.20	1.25	1.25	1.20	1.30
$\bar{H}_o$	0.102	0.069	0.042				0.039				0.064
$\bar{H}_a$	0.105	0.071	0.059				0.049				0.072

H_o = Observed heterozygosity; H_a = expected heterozygosity at Hardy–Weinberg equilibrium; P = proportion of polymorphic loci; A = mean number of alleles per locus.

revealed consistent genetic divergence among hatchery stocks. Our F_{st} estimates may also be considered unexpectedly high since all hatchery samples derived from the same broodstock of P. japonicus.

DISCUSSION

The most relevant result emerging from this study is the consistent, progressive reduction of heterozygosity levels when samples from the F_1 through the F_6 generations are compared. The initial reduction in gene variability is a very predictable phenomenon since the F_1 generation experienced a severe bottleneck; the number of actual spawners which produced the F_2 generation being very low ($N=4$). The effects of this bottleneck on the heterozygosity levels can be studied using the relationship

$$H_t = H_o \prod_{i=1}^{t} (1 - 1/2N_i) \text{ (see Crow and Kimura, 1970)},$$

where H_t is the heterozygosity at generation t, N is the effective population size and H_o is the initial heterozygosity. If, as in this case, the number of generations at study is relatively low, the sensitivity of this relationship is highly increased, since mutation can be considered a negligible factor. The above equation predicts, under neutral conditions, that average heterozygosity decreases by a fraction $1/2N$ each generation. In Fig. 1, experimental heterozygosity values (H_a) are compared with the ones predicted by this model, on the basis of the different number of spawners reported for each generation (Table 1). Expected heterozygosity values are relatively constant from the F_2 to the F_6 generation, after an initial decrease due to the bottleneck. In contrast, the experimental values reveal a progressive reduction in the heterozygosity estimates. The best fit between observed and expected values is obtained only

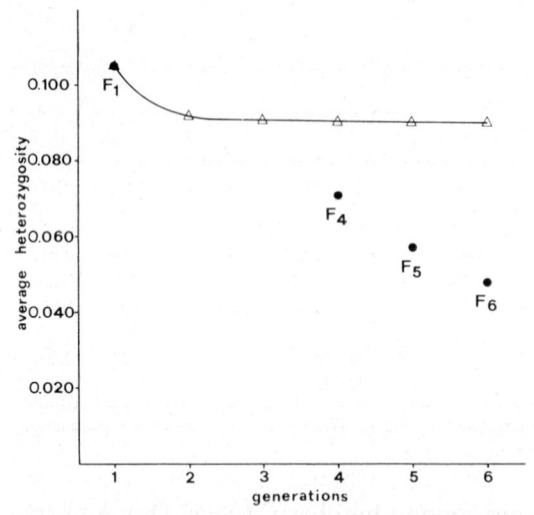

Fig. 1. A comparison of experimental (solid circles) and theoretical values (triangles) of average heterozygosity from F_1 to F_6 in hatchery stocks of *Penaeus japonicus*. Theoretical heterozygosities are those predicted under neutrality assumptions, from the number of breeders utilized in each generation (see Table 1 and text).

by assuming $N=2.5$ and $N=3.2$ for the F_4 and F_5 generations, respectively. These values are both much lower than the actual number of spawners ($N \simeq 600$). In Fig. 2, experimental values of heterozygosity for F_1, F_4, F_5 and F_6 generations are compared with four curves representing hypothetical pop-

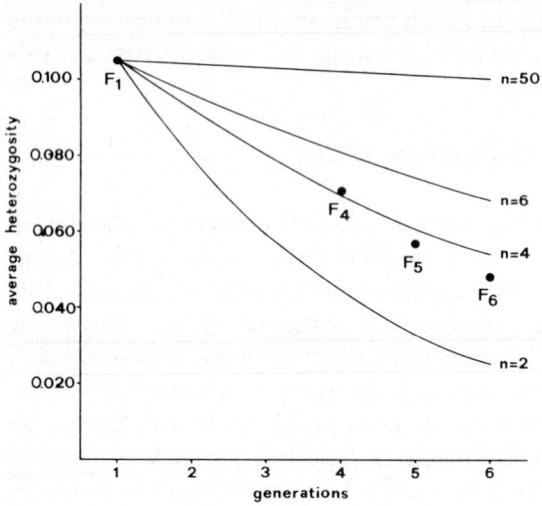

Fig. 2. Experimental values of average heterozygosity for different generations of *P. japonicus* (solid circles) compared with theoretical curves calculated for different effective breeding numbers, under neutrality conditions. For each plot a constant number of breeders is assumed.

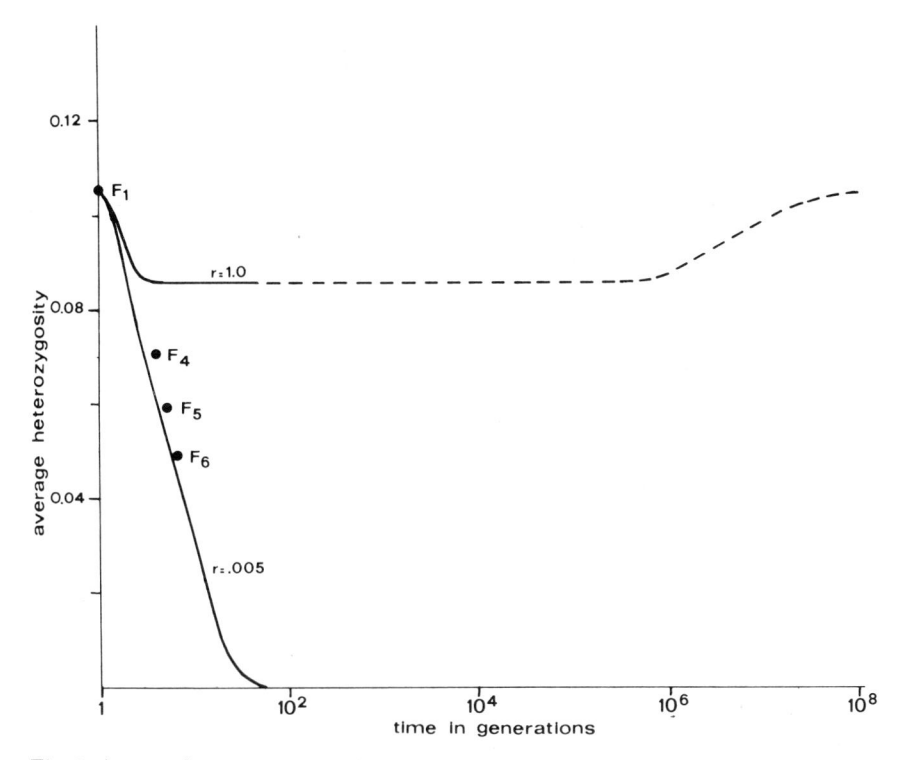

Fig. 3. Average heterozygosity values of P. japonicus hatchery generations (solid circles) compared with theoretical curves describing changes in heterozygosity after a bottleneck (Nei et al., 1975). A logistic population growth model is assumed, with two different r values. In both cases a bottleneck population size of $N=4$ is assumed at generation F_1; assumed average mutation rate and equilibrium population size are $\mu = 10^{-6}$ and $K = 50\,000$, respectively.

ulations with constant numbers of breeders at each generation ($N=2, 4, 16, 50$, respectively). The curve that best fits the observed values is the one based on a constant number of 4. This result suggests that even though the number of females utilized for reproduction oscillated from 50 to 300, a continuous bottleneck is at work and each generation genetically might have been produced by only two to four spawners.

The effects of a bottleneck on the average heterozygosity of a population were also studied theoretically by Nei et al. (1975) and Chakraborty and Nei (1977). Their models consider an hypothetical natural population which, after a reduction in effective number, grows logistically up to the carrying capacity of the environment. Despite some obvious differences in assumptions between natural and cultivated populations we found a good agreement between empirical and predicted values of heterozygosity of P. japonicus by considering a growth rate of 0.005 (Fig. 3). Such a growth rate is, of course, unrealistically low for a prawn but it stands as the necessary assumption to account for the observed decrease in heterozygosity just as the reduced effective breeding size

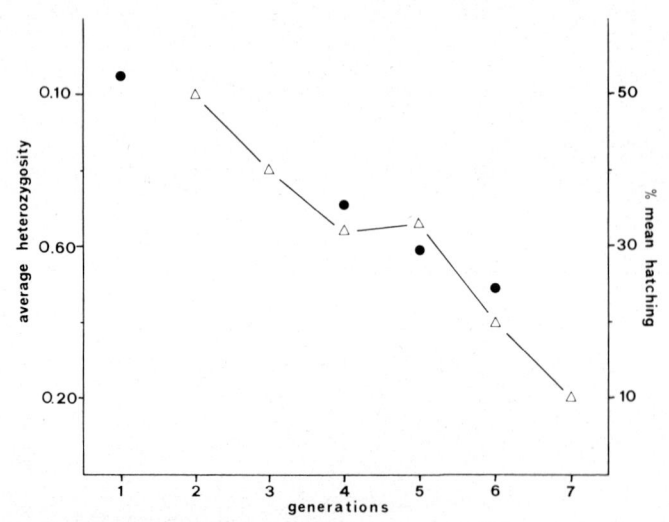

Fig. 4. Estimates of average heterozygosity (solid circles) plotted against mean hatching rates (triangles) for several hatchery generations of P. japonicus.

was in the previous model (see Fig. 2). Both these curves predict, if present farming conditions are maintained, that heterozygosity will decrease to 0 in a few tens of generations. Increase of homozygosity and associated inbreeding depression might well be causally related to the reduced hatching rate observed from F_1 to F_6 in the Italian broodstock of P. japonicus (Fig. 4).

Several observations on P. japonicus farming procedures can help explain the disproportion revealed between the number of prawns placed into a spawning tank and the estimated effective number of spawners. According to Lumare (1985) only the 47.5% of the 298 females isolated for reproduction (1984 data) were ready to spawn and, among these, 84.4% spawned an average of 35 690 eggs at a time. During the whole reproduction cycle, which can be repeated up to four times, each female spawned 1.4 times. The effective number of pairs contributing genetically to the next generation is then estimated to be 41.8 rather than 298.

In addition, two other important sources of bias may have been responsible for the reduced breeding size. First, even though several monthly lots of postlarvae were obtained in each reproduction cycle, only the first one was utilized for seeding into the Lesina lagoon, while the others were released in other hatcheries. Since the various broodstocks were taken only from Lesina harvests, this procedure may have caused severe indirect selection for genotypes associated with early spawning and hatching (see Doyle et al., 1983). Second, selection for size has been performed on spawners by selecting mature females exceeding 40 g in weight. The effects of this selection overlap to some extent the effects of selection for early spawning, but the two traits need not be correlated. If this interpretation is correct, our data suggest that the combined effect of selection for size and selection for early spawning and hatching may have reduced the number of effective parents by roughly 90% (see also Newkirk, 1978).

These last considerations allow the uncoupling of the reduction in variability at structural loci from possible selective presures that could have been acting directly upon these loci and/or hitch-hiking loci. In principle, selection effects (e.g., differential mortality of genotypes) could be detected from heterogeneity of allele frequencies among sites characterized by differing biotic and abiotic parameters. Such heterogeneity did exist among different F_6 samples of *P. japonicus*. However, since different monthly lots of post-larvae were released in different hatcheries, sampling error rather than differential mortality might account for such heterogeneity.

In conclusion, disregarding possible minor selective effects on enzyme loci, we maintain that random genetic drift (bottleneck effect) and inbreeding caused the consistent and undesirable reduction of genetic variability revealed in hatchery stocks of *Penaeus japonicus*.

To prevent a predictable future collapse of the hatchery stock due to genetic load, some appropriate methods should be considered to increase and preserve genetic variability. Introduction of new stocks from natural populations now seems advisable. Starting from a new stock, selection for survivorship should be performed before any other feature (see Doyle and Hunte, 1980). In addition, particular care should be paid to increase the effective number of parents in hatchery stocks; e.g., collection of eggs should be repeated over different days or months and/or breeders should be harvested from different hatcheries.

Results from this study demonstrate the utility of allozyme genetic markers in monitoring the genetic structure of cultivated stocks in aquaculture.

ACKNOWLEDGEMENTS

The authors are grateful to Dr. F. Lumare for loan of *P. japonicus* samples and for continuing assistance to this research. Drs. G. Palmegiano, G. Carchini, P. Fratini and Scovacricchi helped in collecting samples. Drs. G. Allegrucci, D. Cesaroni, P. Ferrazza, P. Fratini and G. Scozzafava gave substantial help in laboratory processing of shrimps. Thanks are also due to Drs. M. Turelli and A. Caccone for useful suggestions on the manuscript.

This research was supported by grants to V. Sbordoni from CNR, IPRA project, subproject 1 (paper no. 618).

REFERENCES

Ayala, F.J., Powell, J.R., Tracey, M.L., Mourao, C.A. and Perez-Salas, S., 1972. Enzyme variability in the *Drosophila willistoni* group. IV. Genetic variation in natural populations of *Drosophila willistoni*. Genetics, 70: 113–139.

Ayala, F.J., Hedgecock, D., Zumwalt, G.S. and Valentine, J.W., 1973. Genetic variation in *Tridacna maxima*, an ecological analog of some unsuccessful evolutionary lineages. Evolution, 27: 177–191.

Ayala, F.J., Tracey, M.L., Barr, L.G. and Ehrenfeld, J.G., 1974. Genetic and reproductive differentiation of the subspecies *Drosophila equinaxialis caribbensis*. Evolution, 28: 24–41.

Ayala, F.J., Valentine, J.W. and Zumwalt, G.S., 1975. An electrophoretic study of the Antartic zooplankter *Euphasia superba*. Limnol. Oceanogr., 20: 635–640.

Brewer, G.J. and Sing, C.F., 1970. An Introduction to Isozyme Techniques. Academic Press, New York, NY, 186 pp.

Chakraborty, R. and Nei, M., 1977. Bottleneck effects on average heterozygosity and genetic distance with the stepwise mutation model. Evolution, 31: 347–356.

Crow, J.F. and Kimura, M., 1970. An Introduction to Population Genetics Theory. Harper and Row, New York, NY, 591 pp.

De Matthaeis, E., Allegrucci, G., Caccone, A., Cesaroni, D., Cobolli Sbordoni, M. and Sbordoni, V., 1983. Genetic differentiation between *Penaeus kerathurus* and *P. japonicus* (Crustacea, Decapoda). Mar. Ecol. Prog. Ser., 12: 191–197.

Doyle, R.W. and Hunte, W., 1980. The importance of selecting for survivorship in genetic yield improvement programs in crustacean aquaculture. Proc. World Maricult. Soc., 11: 529–534.

Doyle, R.W., Singholka, S. and New, M.B., 1983. "Indirect selection" for genetic change: a quantitative analysis illustrated with *Macrobrachium rosenbergii*. Aquaculture, 30: 237–247.

Harris, H., Hopkins, D.A. and Edward, Y.H., 1977. Polymorphism and subunit structure of enzyme contribution to the neutralist—selectionist controversy. Proc. Natl. Acad. Sci. U.S.A., 74: 698–701.

Hedgecock, D., 1977. Biochemical genetic markers for broodstock identification in aquaculture. Proc. World Maricult. Soc., 8: 523–531.

Lumare, F., 1981. Artificial reproduction of *Penaeus japonicus* Bate as a basis for the mass production of eggs and larvae. J. World Maricult. Soc., 12: 335–344.

Lumare, F., 1982. Comparative study of induced reproduction in *Penaeus japonicus* Bate using unilateral eyestalk ablation and environmental conditioning. Boll. Doc. Ist. Sfrutt. Biol. Lagune CNR, Lesina, 2: 1–28.

Lumare, F., 1984. Stocking trials of *Penaeus japonicus* Bate (Decapoda, Natantia) post-larvae in Lesina lagoon (Southeast coast of Italy). FAO Stud. Rev. GFCM, 61 (2): 593–606.

Lumare, F., 1985. Marine shrimp culture in the world and present state and trends of kuruma prawn culture in Italy. III. Colloquium Crustacea Decapoda Mediterranea, Barcelona, 25–29 March 1985 (preprint).

Lumare, F. and Palmegiano, G.B., 1980. Acclimatazione di *Penaeus japonicus* Bate nella Laguna di Lesina (Italia sudorientale). Riv. Ital. Piscic. Ittiop., 15: 53–58.

Moav, R., Brody, T., Wohlfarth, G. and Hulata, G., 1976. Application of electrophoretic genetic markers to fish breeding. 1. Advantages and methods. Aquaculture, 9: 217–228.

Nei, M., 1977. *F*-statistics and analysis of gene diversity in subdivided populations. Ann. Hum. Genet., London, 41: 225–233.

Nei, M., Maruyama, T. and Chakraborty, R., 1975. The bottleneck effect and genetic variability in populations. Evolution, 29: 1–10.

Newkirk, G.F., 1978. A discussion of possible sources of inbreeding in hatchery stock and associated problems. Proc. World Maricult. Soc., 9: 93–100.

Poulik, M.D., 1957. Starch-gel electrophoresis in a discontinuous system of buffer. Nature, 180: 1477–1479.

Selander, R.K., Smith, M.H., Yang, S.Y., Johnson, W.E. and Gentry, J.B., 1971. Biochemical polymorphism and systematics in the genus *Peromyscus*. I. Variation in the old-field mouse *Peromyscus polionotus*. Studies in Genetics, VI. Univ. Texas Publ., 7103: 49–90.

Shigueno, K., 1975. Shrimp Culture in Japan. Assoc. Int. Tech. Prom., Tokyo, 153 pp.

Taniguchi, N., Sumantadinata, K. and Iyama, S., 1983. Genetic change in the first and second generations of hatchery stock of black seabream. Aquaculture, 35: 309-320.

Tracey, M.L., Nelson, K., Hedgecock, D., Shleser, R.A. and Pressick, M.L., 1975. Biochemical genetics of lobsters (*Homarus*): genetic variation and the structure of American lobster populations. J. Fish. Res. Board Can., 32: 2091-2101.

Utter, F.M., Hodgins, H.O. and Allendorf, F.W., 1974. Biochemical genetic studies of fishes: potentialities and limitations. In: D.C. Malins and J.R. Sargent (Editors), Biochemical and Biophysical Perspectives in Marine Biology, 1. Academic Press, New York, NY, pp. 213-238.

Ward, R.D. and Beardmore, J.A., 1977. Protein variation in the plaice, *Pleuronectes platessa*. Genet. Res., 30: 45-62.

Wilkins, N.P., 1975. Genic variability in marine Bivalvia: implications and applications in molluscan mariculture. In: G. Persoone and E. Jaspers (Editors), Mariculture. Proceedings of the 10th European Symposium on Marine Biology, Vol. 1. Universa Press, Wetteren, Belgium, pp. 549-563.

The Enzyme Gene Variation of Ten Finnish Rainbow Trout Strains and the Relation Between Growth Rate and Mean Heterozygosity

MARJA-LIISA KOLJONEN

Finnish Game and Fisheries Research Institute, Fisheries Division, P.O. Box 193, SF-00131 Helsinki 13 (Finland)

ABSTRACT

Koljonen, M.-L., 1986. The enzyme gene variation of ten Finnish rainbow trout strains and the relation between growth rate and mean heterozygosity. *Aquaculture*, 57: 253–260.

The electrophoretic survey of 10 rainbow trout (*Salmo gairdneri* Richardson) strains reared in Finland was made in connection with a simultaneous analysis of the growth rate of the strains. Five of the strains were from a State Fish Culture Station and the other five from commercial fish farms. Nine enzyme systems and 20 loci were examined. Eight loci were polymorphic. The genetic variation found was at about the same level as in earlier rainbow trout studies. Thus, Finnish rainbow trout strains have probably not lost much genetic variation since they were imported from the U.S.A. Genetically, the most distinct strain was the autumn spawning one. The genetic distances between the commercial strains were smaller than between the strains of the State Fish Culture Station. The genetic variation and the growth rate were greater in commercial strains than in the State strains, which have been kept more isolated from each other. In the nine spring spawning strains the correlation between the mean heterozygosity (H) and average growth rate in sea water was 0.62, between H and growth rate in fresh water 0.59, and between H and the average weight 0.54. When only eight stocks of the same origin were included, the corresponding correlations were $+0.83$ ($P < 0.05$), $+0.70$ ($P < 0.1$) and $+0.76$ ($P < 0.05$), and the correlation between H and the variation in weight was -0.66 ($P < 0.1$).

INTRODUCTION

The usefulness of electrophoretic studies for breeding purposes depends on (1) the correlation between the two measures of genetic variation: the heterozygosity of enzyme genes and the amount of genetic variation in quantitative characters, and (2) the relation between the genetic differentiation at the biochemical loci and that at the loci controlling the quantitative characters (Ryman, 1983), or (3) the relation between the mean heterozygosity and some fitness components or quantitative characters, such as growth rate.

Although the phenomenon of heterosis is well known and clearly accepted, the advantage of heterozygosity of enzyme loci in natural populations is not self-evident (Lewontin, 1974; Nei, 1983). If we suppose that in general the heterozygosity is an advantage and that there is a relation between the heterozygosity of enzyme genes and the level of variation in the whole genome, it should be possible to observe a relation between fitness and mean heterozygosity.

The aim of this study was to measure allelic differences and levels of heterozygosity in 10 Finnish rainbow trout strains to examine (1) the relation between heterozygosity and variation in length and weight, (2) the correspondence between genetic distances and the known histories of the strains, and (3) the correlation between the growth rate and heterozygosity and the possible correlation of individual alleles with the growth rate.

MATERIAL AND METHODS

The material consisted of 370 rainbow trout representing 10 strains reared in Finland. Each sample (27–42 fishes) was the progeny of 15 males and 15 females. Five of the strains were from the Laukaa Fish Culture Research Station (LFC) of the Finnish Game and Fisheries Research Institute and five from commercial fish farms. The strains from LFC were: (1) Donaldson, a strain imported from the University of Washington, U.S.A. in 1969, (2) Kamloops, a strain originating from the Kamloops River in Canada, but imported into Finland from Denmark; the present stock is a cross between the Finnish and Danish Kamloops strains, (3) Danish, a strain of mixed rainbow trout material imported into Finland from Denmark, (4) A13, a strain imported from the University of Washington, U.S.A., where it was selected for autumn spawning, and (5) American, a cross between the Donaldson strain and steelhead strains. The commercial strains were (6) Saimaanlohi, (7) Siikataimen, (8) Arvokala, (9) Hatsina, and (10) Saarioinen. The origin of the commercial strains is unknown, but they were probably developed from the same genetic material as the Donaldson and Danish strains.

The strains were reared separately for 1 year and then marked individually with Carling tags. After marking they were reared together, each pond containing fish from each strain. After 2 years each strain was divided into two groups, one being kept in fresh water as before and the other in sea water. Weight and length data were recorded at 2 years of age and the growth rate data at 2.5 years of age (Sumari et al., 1984; Siitonen, 1986).

Liver and muscle tissue samples were analysed for nine enzymes, known from earlier studies to be the most variable ones (Allendorf and Utter, 1973; Busack et al., 1979; Guyomard, 1981). The enzymes and the 20 loci screened for them were: alcohol dehydrogenase (ADH-1), α-glycerophosphate dehydrogenase (AGP-1), isocitrate dehydrogenase (IDH-3,4), lactate dehydrogenase (LDH-1,2,3,4), malate dehydrogenase (MDH-1,2,3,4), malic enzyme (ME-

1,2,3), phosphoglucoisomerase (PGI-1,2,3), phosphoglucomutase (PGM-1) and superoxide dismutase (SOD-1). The number of loci and the allele symbols were according to Guyomard (1981). In this study, new variation was found in the liver malate enzyme (ME-3), in which a slower allele 40 was observed. The buffer systems used were: Ridgway et al. (1970) for AGP, LDH, PGI, PGM; Clayton and Tretiak (1972) for ME, MDH; and Cross and Payne (1977) for ADH, IDH, SOD. The staining procedures were mainly according to Allendorf et al. (1977) and Shaw and Prasad (1970).

RESULTS

Genetic variation and genetic distances

Eight of the 20 loci examined were polymorphic (Table 1). Since loci IDH-3,4 and MDH-3,4 are duplicated, the allele frequencies for the individual loci could not be calculated directly. In order to solve this problem, the assumption was made that the same allele was not present at both the duplicated loci. The alleles were then assigned to the loci by using the genotype information as follows. When an individual was found with genotype 69, 69, 38, 38 in IDH-3,4, it was assumed that alleles 69 and 38 belonged to different loci of the IDH-3,4 duplicate. Further, for the genotype 69, 69, 100, 130, it was assumed that the allele 130 belonged to the same locus as allele 38. The same method was used for MDH-3,4. The numbering of the loci is of course arbitrary.

The deviation from the Hardy-Weinberg distribution was not statistically significant in any of the polymorphic loci. The greatest strain differences in allele frequency were at the ME-3 and AGP-1 loci of strain A13, the only autumn spawning strain, which deviated clearly from the other strains. Of the spring spawning strains, the most similar were three commercial strains, Hatsina, Saimaanlohi and Siikataimen, and those differing most from the other strains were the Danish and the Saarioinen strains. The three most similar strains did not differ from each other statistically. The average genetic distance (Nei, 1978) (based on the eight polymorphic loci) was 0.0513 for the State strains and 0.0165 for the commercial strains. If we exclude the autumn spawning strain the average genetic distance between the State strains was 0.0315.

The genetic diversity varied rather widely between the strains (Table 2). The mean heterozygosity (Nei, 1978) for the 20 loci varied between 6.8% and 12.8%. The diversity was greatest in the autumn spawning A13 strain and smallest in the American strain. On the average there was slightly more variation in the commercial strains (10.6%) than in the State strains (9.8%). The diversity varied less among the commercial strains than among the State strains.

Variation in length and weight, and mean heterozygosity

In the case of the eight genetically similar strains (excluding A13 and Danish strains), the correlation between the mean heterozygosity and the variation in

TABLE 1

Allele frequencies and sample size (n) for 10 Finnish rainbow trout strains
Strains 1–5 are from the Finnish State Fish Culture Station and 6–10 from commercial fish farms. The names of the strains are 1, Donaldson; 2, Kamloops; 3, Danish; 4, A13; 5, American; 6, Saimaanlohi; 7, Siikataimen; 8, Arvokala; 9, Hatsina; 10, Saarioinen

	Number of strain									
	1	2	3	4	5	6	7	8	9	10
n	38	39	38	40	33	41	36	27	42	37
AGP-1										
100	1.000	1.000	1.000	0.763	0.985	0.963	0.972	0.889	0.976	0.986
125	0.000	0.000	0.000	0.237	0.015	0.037	0.028	0.111	0.024	0.014
IDH-3										
100	0.947	0.846	0.632	0.500	0.727	0.878	0.806	0.889	0.893	0.905
69	0.039	0.154	0.368	0.500	0.121	0.122	0.194	0.111	0.107	0.081
63	0.013	0.000	0.000	0.000	0.152	0.000	0.000	0.000	0.000	0.014
IDH-4										
100	0.579	0.487	0.579	0.963	0.621	0.512	0.597	0.315	0.607	0.784
38	0.329	0.487	0.355	0.013	0.212	0.439	0.333	0.685	0.333	0.135
130	0.092	0.026	0.066	0.025	0.167	0.049	0.069	0.000	0.060	0.081
MDH-3										
100	0.855	0.987	0.711	0.663	0.924	0.756	0.903	0.759	0.881	0.662
79	0.145	0.013	0.289	0.338	0.076	0.244	0.097	0.241	0.119	0.338
MDH-4										
100	0.829	1.000	1.000	0.913	1.000	0.780	0.806	0.870	0.821	0.703
69	0.079	0.000	0.000	0.088	0.000	0.220	0.153	0.130	0.143	0.297
96	0.092	0.000	0.000	0.000	0.000	0.000	0.042	0.000	0.036	0.000
ME-3										
100	1.000	1.000	1.000	0.755	0.985	1.000	1.000	0.963	1.000	1.000
40	0.000	0.000	0.000	0.225	0.015	0.000	0.000	0.037	0.000	0.000
PGM-1										
100	1.000	0.692	0.632	0.913	0.970	0.854	0.792	0.833	0.810	0.905
70	0.000	0.308	0.368	0.088	0.030	0.146	0.208	0.167	0.190	0.095
SOD-1										
100	0.618	0.641	0.618	0.588	0.939	0.768	0.569	0.759	0.774	0.824
161	0.382	0.333	0.237	0.413	0.061	0.207	0.333	0.148	0.226	0.149
39	0.000	0.026	0.145	0.000	0.000	0.024	0.097	0.093	0.000	0.027

TABLE 2

Observed (H_o) and expected (H_e) mean heterozygosities (Nei, 1978) and their standard errors (SE), and mean number of alleles per locus (N_a) calculated for 20 loci in rainbow trout strains in Finland

		H_o	SE	H_e	SE	N_a
1	Donaldson	0.072	0.032	0.084	0.038	1.4
2	Kamloops	0.101	0.048	0.087	0.040	1.4
3	Danish	0.125	0.052	0.122	0.049	1.4
4	A13	0.140	0.048	0.128	0.043	1.5
5	American	0.070	0.035	0.068	0.034	1.5
6	Saimanlohi	0.104	0.036	0.109	0.039	1.5
7	Siikataimen	0.119	0.046	0.116	0.043	1.5
8	Arvokala	0.111	0.035	0.110	0.035	1.5
9	Hatsina	0.106	0.040	0.097	0.036	1.5
10	Saarioinen	0.088	0.032	0.096	0.036	1.5

length (measured as coefficient of variation) was -0.445, which is not significant. The correlation between mean heterozygosity and the variation in weight was -0.656 ($P<0.1$), and the level of heterozygosity explained 43% of the total variation (Fig. 1A).

Growth rate and heterozygosity

Comparison of the individual length and weight with individual level of heterozygosity within the strains failed to reveal any statistically significant correlation. The observed correlations varied from -0.300 to 0.305, being positive in five strains and negative in five strains.

A prerequisite for studying the effect of mean heterozygosity is that the strains are relatively closely related. Correlations between measures of growth and mean heterozygosity (H) were calculated excluding the autumn spawning strain (A13), and also the Danish strain. The correlations with H based on the nine strains, excluding A13, were: length, 0.553; weight, 0.540; growth rate in fresh water, 0.587; and growth rate in sea water, 0.662. Although the values were fairly high, none was significant. Correlations with H based on eight strains of similar origin were: length, 0.729; weight, 0.760; growth rate in fresh water, 0.701; and growth rate in sea water, 0.829 (Fig. 1B). All the correlations, except the one with growth in fresh water, were significant at the 5% level, and coefficients of determination ranged from 49.2% to 68.7%. There was also a difference between the sexes in the correlation of heterozygosities with length; the correlation was 0.708 for males and 0.415 for females. The correlation analysis was also conducted locus by locus within strains but the results did not show any clear trend.

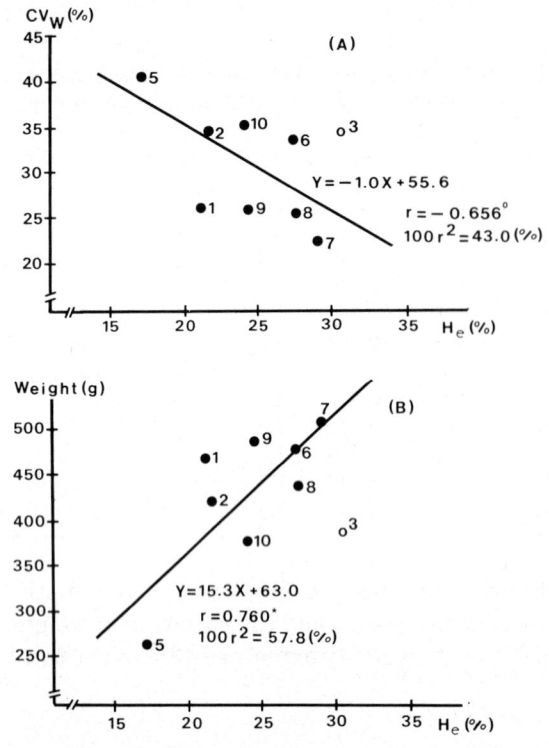

Fig. 1. Relationship between (A) mean expected heterozygosity (H_e; Nei, 1978) and the variation in weight (coefficient of variation CV_W), and (B) mean expected heterozygosity and mean weight in spring spawning Finnish rainbow trout strains. The regression lines and correlation coefficients have been calculated without the Danish strain (see the text). The strains were: 1, Donaldson; 2, Kamloops; 3, Danish; 5, American; 6, Saimaanlohi; 7, Siikataimen; 8, Arvokala; 9, Hatsina; and 10, Saarioinen.

DISCUSSION

Genetic distances and the total amount of variation

The observed genetic distances agreed well with the breeding histories of the strains. The most strongly differentiated autumn spawning strain has been subjected to intensive selective breeding and thus its genetic structure has changed. Two alleles, ME-3 (40) and AGP-1 (125) were more frequent in this strain than in spring spawning strains.

The five strains at Laukaa Fish Culture Station have intentionally been kept separate and intentional selection has been avoided. The breeding practice of the commercial farmers has been different: the strains have been mixed and some selection for improved growth rate has been practiced. The State strains had lower genetic diversity and a slower growth rate compared to the commercial strains. For the commercial strains it is difficult to say which is the cause and which is the effect; was it the selection for improved growth rate that has resulted in greater heterozygosity, or was the mixing of strains responsible for higher overall diversity and thus also the growth rate?

The genetic diversity of the rainbow trout strains was quite similar to estimates of earlier studies (Allendorf and Utter, 1973; Utter et al., 1973; Allendorf and Utter, 1979; Busack et al., 1979; Guyomard, 1981). In these studies the loci examined were not exactly the same and there were differences in the interpretation also. Thus, Finnish strains have probably not lost much of their genetic diversity since they were imported from the U.S.A. and Denmark.

Mean heterozygosity and the variation in length and weight

The negative correlation found between mean heterozygosity and the variation in length and weight was contrary to expectation. Soulé (1979) and Soulé et al. (1973) found a negative correlation between heterozygosity and scale morphology in lizards but only at intra-individual level. They concluded that heterozygosity within an individual balances development and diminishes morphological variation, but heterozygosity among individuals should increase diversity. Contrary to this, it seems that heterozygosity balances development among individuals as well, at least as regards to growth parameters. Variation in growth rate of fish probably depends more on developmental stability and overall viability than on specific growth rate genes. Zouros et al. (1980) also found a negative correlation between heterozygosity and the variation of the growth rate in the American oyster (*Crassostrea virginica*) and similar results were reported by Mitton (1978) for morphological characters of killifishes (*Fundulus heteroclitus*).

Connection between growth rate and heterozygosity

The lack of significant correlations between the degree of heterozygosity and growth rate at the individual level, but the presence of a significant positive correlation at the population level was probably a result of heterozygosity differences between individuals being small compared with differences between populations. It has been shown that the variance of number of heterozygous loci per individual in a random mating population is very low (Sved et al., 1967). In addition, the number of polymorphic loci studied was quite small and the number of fish sampled in each strain was not very large.

Two theories can explain the positive correlation between the mean heterozygosity and the growth rate: the enzyme loci studied are closely linked to loci affecting the growth rate, or enzyme heterozygosity reflects overall genetic heterozygosity. It is not very likely that the few loci examined in this study are linked to a significant number of loci determining growth rate. The more plausible explanation is that the loci studied reflect overall genetic heterozygosity. Studied strains cannot be regarded as natural populations, and it is possible that inbreeding has decreased both the growth rate and the mean heterozygosity. It was not possible to confirm a clear effect of any individual locus on growth rate; the IDH-3 locus was the only one which showed a slight trend in this direction.

REFERENCES

Allendorf, F.W. and Utter, F.M., 1973. Gene duplication within the family Salmonidae: disomic inheritance of two loci reported to be tetrasomic in rainbow trout. Genetics, 74: 647–654.
Allendorf, F.W. and Utter, F.M., 1979. Population genetics. In: W.S. Hoar, D.J. Randall and J.R. Brett (Editors), Fish Physiology, Vol. 8. Academic Press, New York, NY, pp. 407–454.
Allendorf, F.W., Mitchell, N., Ryman, N. and Ståhl, G., 1977. Isozyme loci in brown trout (*Salmo trutta* L.): detection and interpretation from population data. Hereditas, 86: 179–190.
Busack, C.A., Halliburton, R. and Gall. G.A.E., 1979. Electrophoretic variation and differentiation in four strains of domesticated rainbow trout (*Salmo gairdneri*). Can. J. Genet. Cytol., 21: 81–94.
Clayton, J.W. and Tretiak, D.N., 1972. Amine–citrate buffers for pH control in starch gel electrophoresis. J. Fish. Res. Board Can., 29: 1169–1172.
Cross, T.F. and Payne, R.H., 1977. NADP-isocitrate dehydrogenase polymorphism in the Atlantic salmon *Salmo salar*. J. Fish Biol., 11: 493–496.
Guyomard, R., 1981. Electrophoretic variation in four French populations of domesticated rainbow trout (*Salmo gairdneri*). Can. J. Genet. Cytol., 23: 33–47.
Lewontin, R.C., 1974. The Genetic Basis of Evolutionary Change. Columbia University Press, New York, NY, 346 pp.
Mitton, J.B., 1978. Relationship between heterozygosity for enzyme loci and variation of morphological characters in natural populations. Nature, 273: 661–662.
Nei, M., 1978. Estimation of average heterozygosity and genetic distance from a small number of individuals. Genetics, 89: 583–590.
Nei, M., 1983. Genetic polymorphism and the role of mutation in evolution. In: M. Nei and R.K. Koehn (Editors), Evolution of Genes and Proteins. Sinauer Associates, Sunderland, MA, pp. 165–190.
Ridgway, G.J., Sherburne, S.W. and Lewis, R.D., 1970. Polymorphism in the esterases of Atlantic herring. Trans. Am. Fish. Soc., 99: 147–151.
Ryman, N., 1983. Patterns of distribution of biochemical genetic variation in salmonids: differences between species. Aquaculture, 33: 1–21.
Shaw, C.R. and Prasad, R., 1970. Starch gel electrophoresis of enzymes – a compilation of recipes. Biochem. Genet., 4: 297–320.
Siitonen, L., 1986. Factors affecting growth in rainbow trout (*Salmo gairdneri*) stocks. Aquaculture, 57: 185–191.
Soulé, M.E., 1979. Heterozygosity and developmental stability: another look. Evolution, 33: 396–401.
Soulé, M.E., Yang, S.Y., Weiler, G.W. and Gorman, G.C., 1973. Island lizards: the genetic–phenetic variation correlation. Nature, 242: 190–192.
Sumari, O., Siitonen, L. and Linder, D., 1984. Valtakunnallinen kirjolohen rodunjalostusohjelma. (Nationwide breeding scheme for rainbow trout.) RKTL Monistettuja julkaisuja. 82 pp. (Fisheries Division of the Finnish Game and Fisheries Research Institute, Helsinki.)
Sved, J.A., Reed, T.E. and Bodmer, W.F., 1967. The number of balanced polymorphisms that can be maintained in a natural population. Genetics, 55: 469–481.
Utter, F.M., Allendorf, F.W. and Hodgins, H.O., 1973. Genetic variability and relationships in Pacific salmon and related trout based on protein variations. Syst. Zool., 22: 257–270.

Genetic Heterozygosity and Growth Rate in Louisiana Oysters (*Crassostrea virginica*)

DAVID W. FOLTZ and MARK CHATRY[1]

Department of Zoology and Physiology, Louisiana State University, Baton Rouge, LA 70803 (U.S.A.)
[1]*Louisiana Department of Wildlife and Fisheries, Lyle S. St. Amant Marine Laboratory, Post Office Box 37, Grand Isle, LA 70358 (U.S.A.)*

ABSTRACT

Foltz, D.W. and Chatry, M., 1986. Genetic heterozygosity and growth rate in Louisiana oysters (*Crassostrea virginica*). *Aquaculture*, 57: 261–269.

Heterozygosity at nine electrophoretically variable allozyme loci was studied in relation to weight and length in approximately 1600 sixteen-week-old oysters (*Crassostrea virginica*) produced from five pair crosses. There was little evidence that heterozygosity itself was important in determining growth rates within crosses, either in single-locus or in multi-locus comparisons. There was evidence that some allelic substitutions at the aminopeptidase-1 locus may have had an effect on growth. This result indicates either that some allelic substitutions within an allozyme locus affect growth and others do not, or that the effect is due to a growth-affecting gene that is linked to the allozyme locus.

INTRODUCTION

There is now abundant evidence that enzyme heterozygosity is positively correlated with fitness-related traits such as growth, viability and fertility in many different organisms [summarized by Mitton and Grant (1984) and Zouros and Foltz (1986)]. In the oyster *Crassostrea virginica* (Koehn and Shumway, 1982) and the clam *Mulinia lateralis* (Garton et al., 1984) there is a negative correlation between oxygen consumption and enzyme heterozygosity, suggesting that heterozygous individuals grow faster than more homozygous ones because heterozygotes have a higher metabolic efficiency and a lower cost of routine metabolism. These observations raise several questions. What is the biochemical basis of this apparent heterozygote superiority? Does allozyme heterozygosity directly affect growth rate or is it simply a marker for heterozygosity at linked loci affecting growth? And how (if at all) can these results be used for the genetic improvement of bivalve stocks? The last question has been discussed by Newkirk (1983) and by Gaffney and Scott (1984).

The present study addresses the possible effect on growth rate of genes linked to allozyme loci by measuring the correlation between growth and heterozygosity in five pair crosses of Louisiana oysters (*C. virginica*). If allozyme heterozygosity directly affects growth rate, then significant associations between heterozygosity and growth should appear both in samples from natural populations and in samples of progeny produced by pair crosses.

METHODS

Hatchery procedures

Broodstock oysters were collected from Barataria Bay, Louisiana in May 1984 and brought to the oyster hatchery at the Lyle S. St. Amant Marine Laboratory on Grand Terre Island. Each oyster was cleaned of fouling organisms, placed overnight in a 4-l glass jar containing 2 l of 5 μm-filtered, UV-irradiated seawater and allowed to purge itself. Spawning was induced by raising the temperature 2–4°C above ambient. Eggs from one female were collected and divided into 12 batches. All but one batch of eggs were fertilized separately by sperm from 11 different males. The last batch was left unfertilized and served as a control. Development of the unfertilized eggs would indicate either parthenogenesis or the presence of foreign sperm. Each of the 11 cultures was divided into two (A and B) or three (A, B and C) replicates 5 days after fertilization. However, crosses 1, 3, 4, 7, 10 and 11 plus one replicate of cross 5 had very low survivorship and produced few or no spat. Also, for both cross 2 and cross 8, both replicates were pooled in all analyses because small sample sizes precluded meaningful between-replicate comparisons.

The larvae were reared using the brown water culture technique of Ogle (1982). Each replicate was cultured in a 20-l polyethylene bucket containing 14 l of aerated 5 μm-filtered, UV-irradiated seawater. The water was changed every 3 days. The salinity of the water varied over time between 11 and 22‰. Each culture was provided with individually labelled clam (*Rangia*) shells as cultch material. After metamorphosis, each oyster was transferred to a common tank supplied with seawater filtered to remove foreign oyster larvae. At 16 weeks of age, the oysters were taken to Baton Rouge and processed for electrophoretic analysis.

Laboratory procedures

Each oyster (minus the left valve) was weighed on an electronic balance and its length was measured using dial calipers. In most instances, more than one oyster settled on a cultch shell. To adjust for any effect of crowding on growth rate, the number of oysters per cultch shell (which we will refer to as the "crowding index") was used as a covariate in the analysis of the size (length and weight) data. Each oyster was processed and analyzed by starch-gel electrophoresis using standard methods (e.g., Schaal and Anderson, 1974; Buroker et al., 1975). Nine allozyme loci were segregating in the five crosses: aminopeptidase-1 (Ap-1), glutamate oxaloacetate transaminase-2 (Got-2, also known as Aat-2), leucine aminopeptidase-1 (Lap-1), Lap-2, mannosephosphate isomerase (Mpi), 6-phosphogluconate dehydrogenase (6-Pgd), phosphoglucose isomerase (Pgi), phosphoglucomutase-2 (Pgm-2), and xanthine dehydrogenase (Xdh).

Within each locus, alleles were designated by numbers indicating mobility relative to the allele found to be most common in a preliminary survey of electrophoretic variation in oysters from Barataria Bay (Foltz, 1986), which was designated as "100". Negative numbers referred to cathodally migrating bands and "00" was used to designate the null (non-reactive) allele at the Lap-2 locus (Foltz, 1986).

Statistical procedure

Analyses were performed on log-transformed weight and length measurements using the Statistical Analysis System (SAS 82.4) as implemented on the System Network Computer Center's IBM 370/3081 at Louisiana State University. The partial correlation coefficient between heterozygosity and log-transformed size measurements, with the crowding index held constant, was calculated using Sokal and Rohlf's (1981, p. 656) formula. In addition to looking at the effect of multi-locus heterozygosity on growth, we were also interested in determining if there were size differences among single-locus genotypes within crosses. The number of genotypic classes occurring among the offspring of a pair cross will vary from one to four, depending on parental genotypes. The seven possible mating types are listed in Table 1, where A, B, C and D have been used to designate alleles at a hypothetical locus. The types of contrasts that are possible will vary among these mating types, and only some contrasts will be biologically meaningful. Table 1 lists the contrasts that can be interpreted in terms of additive or non-additive models of gene action. However, additive and non-additive effects of allele substitutions cannot always be separated (e.g., mating type III). Also, in some mating types (IV, VI, and VII), only additive effects can be tested. Contrasts of these sorts are easily performed using, for example, the General Linear Models (GLM) procedure in SAS.

RESULTS

In all, 1387 oysters were examined for all nine variable allozyme loci. The mean number of heterozygous loci per individual was 4.47 (range: 1–7). The corresponding values for each cross (including replicates within crosses) are reported in Table 2, which also gives the partial correlation between log-trans

TABLE 1

Single-locus mating types, progeny genotypes and contrasts among means for analysis of size data (all contrasts have 1 degree of freedom, except for the second contrast in V, which has 2)

Mating type male × female (and reciprocal)	Progeny Genotype(s) (contrast)	Possible effects leading to rejection of null hypothesis
I AA × AA	AA	None; no contrast possible
II AA × BB (BB × AA)	AB	None; no contrast possible
III AA × AB (AB × AA)	AA AB ($+1$ -1)	Additive effect of allele substitution Non-additive effect of allele substitution
IV AA × BC (BC × AA)	AB AC ($+1$ -1)	Additive effect of allele substitution
V AB × AB	AA AB BB (-1 $+2$ -1) (-1 0 $+1$)	Non-additive effect of allele substitution (a) Additive effect of allele substitution (b)
VI AB × AC (AC × AB)	AA AB AC BC ($+1$ -1 $+1$ -1) ($+1$ $+1$ -1 -1)	Additive effect of allele substitution in male gamete (a) Additive effect of allele substitution in female gamete (b)
VII AB × CD (CD × AB)	AC AD BC BD ($+1$ $+1$ -1 -1) ($+1$ -1 $+1$ -1)	Additive effect of allele substitution in male gamete (a) Additive effect of allele substitution in female gamete (b)

formed weight and heterozygosity. In two of eight instances, there was a significant positive correlation between log-transformed weight and heterozygosity. However, the results were not consistent among replicates within crosses. For cross 6, there was a significant result in replicate A, but not in B. Similarly, there was a significant positive correlation in 9B, but 9A and 9C both had (non-significant) negative correlations between log-transformed weight and heterozygosity. The weighted mean correlation between log-transformed weight and heterozygosity was 0.052. Analysis of the log-transformed lengths gave similar results but are not presented here.

Table 3 shows the outcome of single-locus tests for differences in size among

TABLE 2

Sample sizes (N), number of heterozygous loci per individual and correlation between log-transformed weight and heterozygosity (r) in five pair crosses of Louisiana oysters (Crassostrea virginica) (A, B, and C refer to replicates within crosses)

Cross	N	Number of heterozygous loci		r
		Mean	Range	
2	47	3.34	2—6	0.213
5	426	4.65	2—7	0.061
6A	162	2.95	1—4	0.203**
6B	41	2.98	1—4	0.001
8	41	5.00	3—7	0.039
9A	222	4.82	3—7	−0.008
9B	104	5.11	3—7	0.200*
9C	344	4.80	3—7	−0.052

*$P < 0.05$; **$P < 0.01$.

progeny genotypes. The only clear effect was at the Ap-1 locus, in which there was an additive effect of substituting the "127" allele for the "100" allele in the male gamete in cross 9. This result was consistent among all three replicates of that cross, with the 127/108 heterozygote having a lower mean weight than the 108/100 heterozygote. Because both progeny genotypes were heterozygous, there was no effect of heterozygosity per se. The other nominally significant results (at Got-2, Lap-2 and Mpi) involved cross 5 or 6. Because the female was heterozygous at Got-2, Lap-2, and Mpi, the results should have been consistent among crosses, because all crosses had the same female parent. The fact that inconsistent results were found at the Got-2, Lap-2 and Mpi loci suggests either that the effect was spurious or else due to an interaction between the female gamete and the male gamete. Further research is necessary to distinguish between these two alternatives.

When each cross or replicate was stratified according to time of larval settlement, there was a significant difference ($P < 0.05$) in heterozygosity between early settling larvae and late settling ones in one (6A) out of seven comparisons, with early settlers being more heterozygous than late settlers. However, a similar trend was not observed in the other replicate (6B) of this cross. A locus-by-locus test for differences in genotype frequencies between early and late settling larvae gave mostly negative results except for cross 5, in which three loci (Got-2, Pgi, and Xdh) exhibited significant differences.

DISCUSSION

The finding that in natural populations there is a positive association between level of allozyme heterozygosity and growth rate has been confirmed in several

TABLE 3

Contrasts of mean weights among progeny genotypes. Roman numerals refer to mating types as listed in Table 1 (Arabic numerals refer to the identity of alleles being contrasted, see text for further explanation; — indicates that no contrast was possible)

Cross	Locus								
	Ap-1	Got-2	Lap-1	Lap-2	Mpi	6-Pgd	Pgi	Pgm-2	Xdh
2	III 108/100	VIa 78/−100 VIb 78/−279	—	VIIa 115/108 VIIb 100/00	III 100/91	III 100/55	IV 78/58	—	—
5	—	VIa 78/−100 VIb 78/−279	—	IV 100/00*	III 100/91*	III 100/72	VIIa 112/100 VIIb 78/58	—	III 115/100
6A	—	VIa 78/−100** VIb 78/−279	—	—	Va 100/91 Vb 100/91	—	VIa 100/78 VIb 78/58	—	—
6B	—	VIa 78/−100 VIb 78/−279	—	—	Va 100/91 Vb 100/91	—	VIa 100/78 VIb 78/58	—	—
8	IV 116/100	IV 78/−279	III 100/88	—	VIa 111/100 VIb 100/91	III 100/72	VIIa 106/100 VIIb 78/58	III 100/82	—
9A	IV 127/100*	IV 78/−279	—	IV 100/00	III 100/91	III 100/72	IV 78/58	III 100/82	—
9B	IV 127/100*	IV 78/−279	—	IV 100/00	III 100/91	III 100/72	IV 78/58	III 100/82*	—
9C	IV 127/100**	IV 78/−279	—	IV 100/00	III 100/91	III 100/72	IV 78/58	III 100/82	—

*$P < 0.05$; **$P < 0.01$.

different species, although counter-examples are known [see Zouros and Foltz (1986) for a review of this literature]. In bivalves, evidence that heterozygotes have an advantage in growth rate comes from studies of samples from natural populations (Singh and Zouros, 1978; Zouros et al., 1980; Singh, 1982; Green et al., 1983; Koehn and Gaffney, 1984), from studies of ploidy manipulation (Stanley et al., 1984), and from studies of the physiological energetics of growth (Koehn and Shumway, 1982; Garton et al., 1984).

The associations between allozyme heterozygosity and growth rate observed in natural populations of bivalves are not necessarily present in hatchery populations produced from a limited number of parents. In offspring of pair crosses in *Mytilus edulis* (Beaumont et al., 1983) and *Mulinia lateralis* (Gaffney and Scott, 1984), there was no association between size and heterozygosity, although positive associations were observed in natural populations of both species. Similarly, there was little or no association between size and heterozygosity in offspring of mass spawnings in *C. virginica, Spisula solidissima, M. lateralis* (Gaffney and Scott, 1984) and *Mercenaria mercenaria* (Adamkewicz et al., 1984). The present study confirms these findings. When comparing the present study with earlier studies of heterozygosity and growth in oysters, it is important to realize that the sample sizes used here (both number of individuals and number of polymorphic loci) are as large as or larger than those employed in the earlier studies. Thus, the largely non-significant results are not due to inadequate sample sizes.

As Gaffney and Scott (1984) have emphasized, these results indicate that associations between enzyme heterozygosity and growth depend strongly on the number of parental genomes contributing to the progeny (see also Koehn and Gaffney, 1984). The failure to observe heterozygosity–growth associations in hatchery populations produced from limited numbers of parents suggests that the associations observed in natural populations are not directly due to allozyme heterozygosity itself. Although it might be argued that progeny from pair crosses have lower heterozygosity than individuals from a natural population and that this phenomenon accounts for the lack of heterozygosity–growth associations in the former, the average heterozygosity is not reduced in pair crosses or mass spawnings, although the variance in heterozygosity among loci may be increased. This result follows from the fact that the loss of heterozygosity is $1-f$, where f is the inbreeding coefficient (Crow and Kimura, 1970). If the parents have been chosen at random from the population, which is presumably the case in most hatchery matings involving wild-collected broodstock, then $f = 0$. Only if the offspring from a pair mating are themselves crossed to produce a second generation in the hatchery would there be a reduction in average heterozygosity. Even though average heterozygosity is not reduced when a limited number of parents contribute to a cross, particular alleles may easily be lost. There can be no more than four different alleles at a locus present in offspring of a pair cross, and if both parents share one or both alleles there will be even less allelic diversity among the offspring. Thus, a possible explanation for the lack of association between heterozygosity and growth in pair crosses is that some allelic substitutions at allozyme loci affect growth and others do not, and that the growth-affecting alleles were absent from the crosses which gave non-significant results. However, this model shifts attention away from heterozygosity per se and toward the effect of particular alleles. Although the results of the single-locus tests reported here (particularly for Ap-1) and some of the results of Beaumont et al. (1983), Adamkewicz et al. (1984), and Gaffney and Scott (1984) are consistent with the suggestion that only some allelic substitutions at an allozyme locus have an effect on growth rate, the possibility that the effect is due to a gene linked to the allozyme locus cannot be ruled out. Either explanation could account for the finding that correlations between allozyme heterozygosity and growth appear only sporadically in offspring of pair crosses.

ACKNOWLEDGEMENTS

We thank B. Glidewell and P. Larsen for help in the laboratory and Dr. A. Bull for assistance in the hatchery. D.W. Foltz would like to thank the director and staff of the St. Amant Marine Laboratory for their support and hospitality.

This research was supported by NSF Grant no. BSR-8407450 to D.W. Foltz. Additional support was provided by Louisiana State University and by the Louisiana Department of Wildlife and Fisheries.

REFERENCES

Adamkewicz, L., Taub, S.R. and Wall, J.R., 1984. Genetics of the clam *Mercenaria mercenaria*. II. Size and genotype. Malacologia, 25: 525–533.

Beaumont, A.R., Beveridge, C.M. and Budd, M.D., 1983. Selection and heterozygosity within single families of the mussel *Mytilus edulis* (L.). Mar. Biol. Lett., 4: 151–161.

Buroker, N.E., Hershberger, W.E. and Chew, K.K., 1975. Genetic variation in the Pacific oyster, *Crassostrea gigas*. J. Fish. Res. Board Can., 32: 2471–2477.

Crow, J.F. and Kimura, M., 1970. An Introduction to Population Genetics Theory. Harper & Row, Publishers, Inc., New York, NY, 591 pp.

Foltz, D.W., 1986. Null alleles as a possible cause of heterozygote deficiencies in the oyster *Crassostrea virginica* and other bivalves. Evolution, in press.

Gaffney, P.M. and Scott, T.M., 1984. Genetic heterozygosity and production traits in natural and hatchery populations of bivalves. Aquaculture, 42: 289–302.

Garton, D.W., Koehn, R.K. and Scott, T.M., 1984. Multiple-locus heterozygosity and the physiological energetics of growth in the coot clam, *Mulinia lateralis*, from a natural population. Genetics, 108: 445–455.

Green, R.H., Singh, S.M., Hicks, B. and McCuaig, J., 1983. An arctic intertidal population of *Macoma balthica* (Mollusca, Pelecypoda): genotypic and phenotypic components of population structure. Can. J. Fish. Aquat. Sci., 40: 1360–1371.

Koehn, R.K. and Gaffney, P.M., 1984. Genetic heterozygosity and growth rate in *Mytilus edulis*. Mar. Biol., 82: 1–7.

Koehn, R.K. and Shumway, S.E., 1982. A genetic/physiological explanation for differential growth rate among individuals of the American oyster, *Crassostrea virginica* (Gmelin). Mar. Biol. Lett., 3: 35–42.

Mitton, J.B. and Grant, M.C., 1984. Associations among protein heterozygosity, growth rate, and developmental homeostasis. Annu. Rev. Ecol. Syst., 15: 479–499.

Newkirk, G.F., 1983. Applied breeding of commercially important molluscs: a summary of discussion. Aquaculture, 33: 415–422.

Ogle, J.T., 1982. Operation of an oyster hatchery utilizing a brown water culture technique. J. Shellfish Res., 2(2): 153–156.

Schaal, B.A. and Anderson, W.W., 1974. An outline of techniques for starch gel electrophoresis of enzymes from the American oyster *Crassostrea virginica* Gmelin. Univ. Georgia Mar. Sci. Cen. Tech. Rep. Ser. No. 74-3, 17 pp., unpubl.

Singh, S.M., 1982. Enzyme heterozygosity associated with growth at different developmental stages in oysters. Can. J. Genet. Cytol., 24: 451–458.

Singh, S.M. and Zouros, E., 1978. Genetic variation associated with growth rate in the American oyster (*Crassostrea virginica*). Evolution, 32: 342–353.

Sokal, R.R. and Rohlf, F.J., 1981. Biometry, 2nd edition. Freeman, San Francisco, CA, 859 pp.

Stanley, J.G., Hidu, H. and Allen, S.K., Jr., 1984. Growth of American oysters increased by polyploidy induced by blocking meiosis I but not meiosis II. Aquaculture, 37(2): 147–155.

Zouros, E. and Foltz, D.W., 1986. The use of allelic isozyme variation for the study of heterosis. In: M.C. Rattazzi, J.G. Scandalios and G.S. Whitt (Editors), Isozymes: Current Topics in Biological and Medical Research, 13. Alan R. Liss, New York, NY.

Zouros, E., Singh, S.M. and Miles, H.E., 1980. Growth rate in oysters: an overdominant phenotype and its possible explanations. Evolution, 34: 856–867.

Triploidy Induction by Thermal Shocks in the Japanese Oyster, *Crassostrea gigas*

EDWIGE QUILLET and P.J. PANELAY

IFREMER, Centre de Brest, BP 377, 29273 Brest Cedex (France)

ABSTRACT

Quillet, E. and Panelay, P.J., 1986. Triploidy induction by thermal shocks in the Japanese oyster, *Crassostrea gigas*. *Aquaculture*, 57: 271–279.

Triploidy induction by application of thermal shocks during early development soon after fertilization was attempted in the Japanese oyster, *Crassostrea gigas*.

Experimental groups were heat-shocked at 30, 32, 35, and 38°C. Shocks of 10 min duration were applied every 5 min, from 5 min to 45 min post-fertilization. Longer shocks (20 min) were also tested at 32°C and 35°C at 5 min and 30 min post-fertilization. Karyological examination of embryos 2 h post-fertilization permitted evaluation of the treatment efficiency.

Triploid embryos were found in all treated groups, but none could be detected in control groups. Among shocks of 10 min duration, the most effective shocks were those applied 10–15 min or 35–40 min post-fertilization at 35°C and 38°C (up to 25% and 45% triploid embryos, respectively). Shocks of 20 min duration appeared more effective; up to 60% of treated embryos were found to be triploid.

Depressive effects of treatments on early survival (measured 24 h post-fertilization) could not be detected, except at higher temperatures (38°C).

INTRODUCTION

Interest in chromosomal engineering and induction of polyploidy as genetic improvement tools of marine cultured species has often been discussed (Moav, 1976; Purdom, 1983).

In fish, chemical treatments applied shortly after fertilization have already been tested: cytochalasin B in salmonids (Refstie et al., 1977; Refstie, 1981) or colchicine in goldfish (Lieder, 1964) and salmonids (Smith and Lemoine, 1979). Such treatments usually cause rather low survival rates, and poorly defined ploidy in the treated eggs (mosaic embryos are often observed).

On the other hand, physical treatments (thermal and high pressure shocks) applied shortly after fertilization, appeared to be successful. Their application usually induces very high percentages of triploids (heat shocks in salmonids: Chourrout, 1980; pressure shocks in salmonids: Benfey and Sutterlin, 1984;

Lou and Purdom, 1984) and in several cases, all triploid populations of hatched fry or juveniles were produced [by heat shocks or high pressure shocks in salmonids (Chourrout and Quillet, 1982; Chourrout, 1984) or by cold shocks in flat fish (Purdom, 1972), and in carp (Gervai et al., 1980)]. Moreover, decreased survival rates of treated groups when compared to the control are generally non-significant or low.

In molluscs, cytochalasin B has been the main agent used until now to induce polyploidy, though its negative effects on survival have often been pointed out (Stanley et al., 1981; Allen et al., 1982; Tabarini, 1984). It was therefore of interest to test the efficiency of physical treatments to induce triploidy in molluscs. In this study, induction by thermal shock was attempted in the Japanese oyster (*Crassostrea gigas*).

MATERIAL AND METHODS

Broodstock

Three- and four-year-old oysters were collected from a private oyster farm in Morbihan (France) in February 1984 and brought to our laboratory in Brest, where they were stored in 15-l tanks supplied with filtered sea water. The temperature of the water was gradually raised from the external sea water temperature of 14°C to 18°C at the rate of 0.5°C every 5 days to accelerate maturation of the oysters. Oysters were fed daily with cultured phytoplankton. A second stock, ready to spawn, was introduced in July. Experiments were performed in June and July.

Collection of gametes and insemination

For each experiment, eggs from at least two females and sperm from several males were collected by artificial pressure on the gonads, and pooled separately in filtered sea water regulated at 22–23°C. The time when sperm was added to the suspension of eggs was considered as the start of the experiment. Eggs and sperm were gently mixed together. Ten minutes later, excess sperm was rinsed from the eggs by straining on a 20-μm mesh screen. The eggs were then divided into several beakers containing filtered sea water at 22–23°C, until they were heat-shocked.

Thermal shocks

Four different temperatures were tested: 30, 32, 35, and 38°C. Most of the experiments were carried out with short (10 min) shocks but long (20 min) shocks were also applied at 32 and 35°C. Short shocks (SS) were applied every

5 min from 10 to 45 min post-fertilization. Long shocks (LS) were applied only at 5 and 30 min post-fertilization.

Larval rearing

Eggs of treated and control groups were then placed in two (three in the case of LS) replicated 15-l glass beakers. Initial concentration was adjusted to five eggs/ml, and chloramphenicol was added (8 mg/l) to the water to prevent bacterial development.

Survival rates

Counts of survivors were made in three different samples per beaker 24 h post-fertilization to estimate survival rate for each treated group, defined as the mean of survival rates observed in the replicates.

Karyological examination

Two hours after fertilization, aliquots of eggs were placed in a 0.005% colchicine solution for 1.5 h, then subjected to a hypotonic treatment (1 h in 0.2 sea water–0.8 distilled water) and fixed in three washes of ethanol/chloroform/acetic acid (3:2:1) solution. The suspension of eggs was then dropped onto a warmed (54°C) microscope slide, air-dried and stained in 4% Giemsa buffered at pH 6.8 for 10 min.

The diploid chromosome number of *Crassostrea gigas* is 20 (Ahmed and Sparks, 1967). Contrary to what is often observed in different species of bivalves, polymorphism for chromosome number was not detected by Thiriot-Quievreux (1984), but it could exist in some populations (C. Thiriot-Quievreux, personal communication, 1985). Small variation in observed chromosome number can also be due to technical reasons. Because of this, the effectiveness of treatment for inducing triploidy was estimated as follows: every metaphase spread with about 20 chromosomes was considered diploid, and those with about 30 chromosomes triploid. Counts were usually made on 25–45 embryos per treatment.

RESULTS

Ten-minute shocks (SS)

Whatever the temperature tested and the time of application, it appeared that short shocks at all tested times were effective for induction of triploidy. About 20% of the analyzed embryos were triploids (Figs. 1 and 2). All analyzed

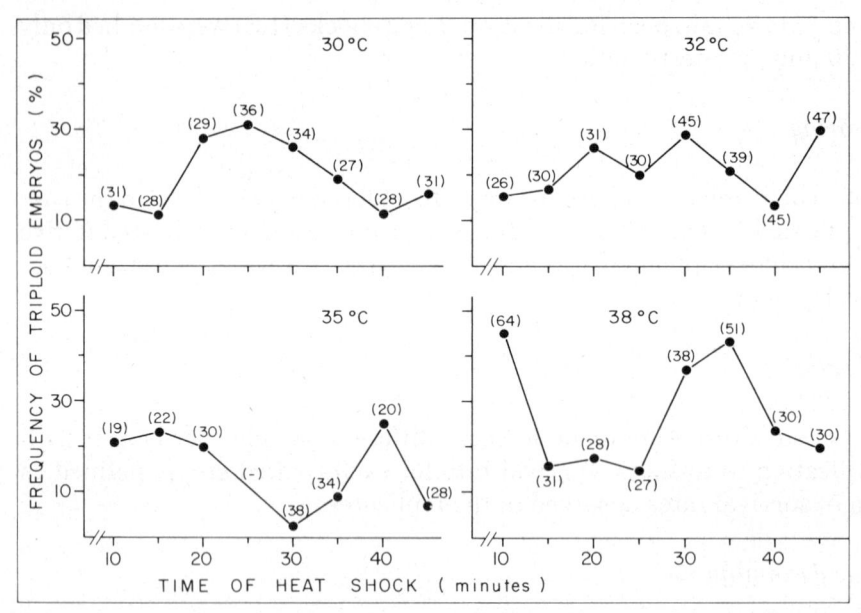

Fig. 1. Efficiency of 10-min shocks at four different temperatures in inducing triploidy. Figures in parentheses indicate numbers of embryo analyses for each shock.

control embryos (26 at 30°C, 25 at 32°C, 26 at 35°C, and 28 at 39°C) were diploids.

A more detailed analysis indicated that at 35°C, shocks applied 15 and 40 min post-fertilization were most effective, while the best results at 38°C were obtained with shocks applied 10 and 35 min post-fertilization.

Large variability in mortalities was observed between replicates of some groups, and mean survival rates of the control groups varied substantially according to the experiment (Table 1). In order to compare the effects of heat shocks on early survival, survival rates of the treated groups were expressed relative to the survival rate of the control group for each treatment (Fig. 3). It can be concluded that early survival rate is not significantly decreased by shocks up to 35°C. Higher temperatures (38°C) induce lower survival.

Twenty-minute shocks (LS)

Long shocks were only performed at temperatures of 32 and 35°C due to the observed negative effect of very high temperature on survival. When using such shocks, a higher rate of triploidy (up to 60%) was observed both at 32 and 35°C (Table 2) than that obtained with short shocks. Variation in survival rates remained high, but no significant effect of the treatment on early survival rates could be detected except for shocks applied 5 min post-fertilization at 32°C (Table 3).

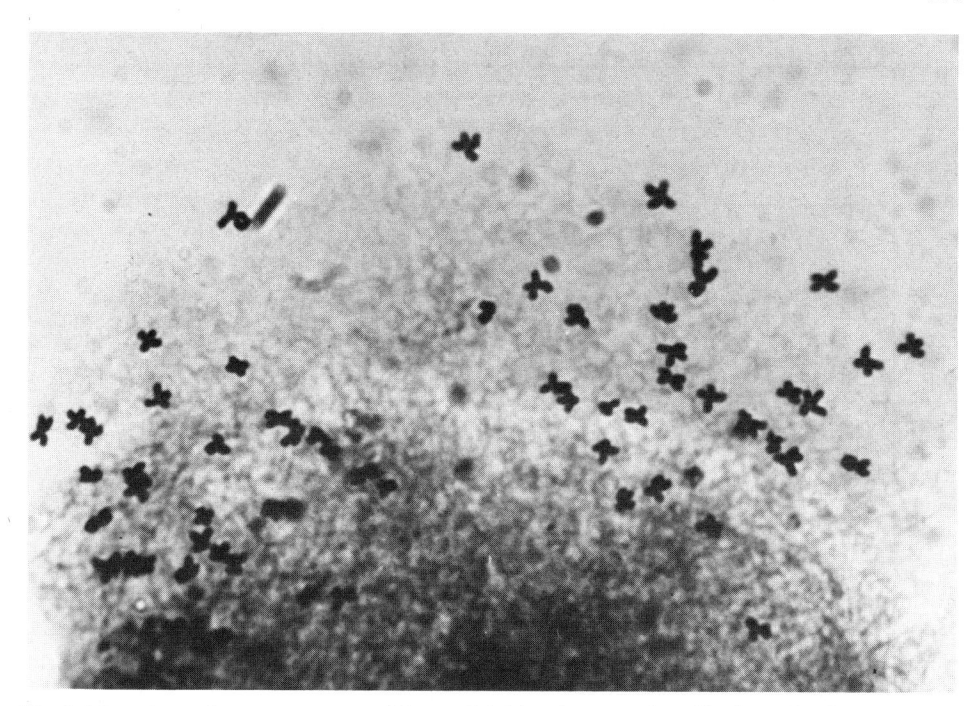

Fig. 2. Metaphase chromosome spread from triploid embryo produced by heat shock.

TABLE 1

Survival rates observed 24 h after shocks of 10- min duration as percentage of the initial concentration of eggs for each replication, R_1 and R_2

Temp. (°C)		Control	Time of initiation (min post-fertilization) of shock							
			10	15	20	25	30	35	40	45
30	R_1	35.5	44.5	49	35.5	35.5	31	40	49	42
	R_2	42	46.5	44.5	35.5	44.5	33.5	35.5	40	49
32	R_1	18	40	29	4.5	26.5	22	24.5	22	97.5
	R_2	73.5	11	40	55.5	35.5	69	6.5	—	11
35	R_1	49	72	0	53	—	59	39	7	39
	R_2	57	0	60	7	-	13	—	—	13
38	R_1	71	15.5	11	15.5	11	9	20	37.5	11
	R_2	62	0	6.5	9	22	11	11	17.5	26.5

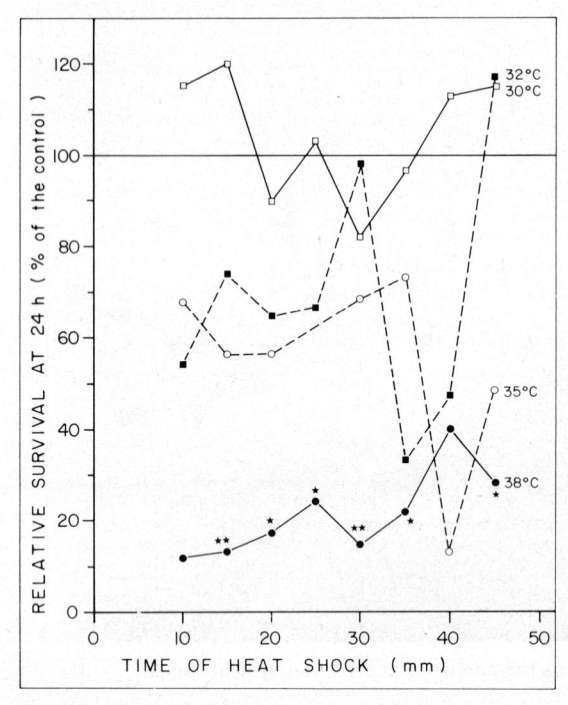

Fig. 3. Relative survival rates for the four temperature treatments (as % of the rate of the control group) at 24 h post-fertilization. Statistical analyses were performed with arc-sine transformed survival percentages. *Significantly different from the control $P < 0.05$; **significantly different from the control $P < 0.01$.

TABLE 2

Efficiency of 20-min shocks on triploidy induction rates at 24 h post-fertilization (D = number of diploid embryos observed, T = number of triploid embryos observed, N = total number of embryos observed)

Temp. (°C)		Control	5 min	30 min
32	D	25	11	17
	T	0	35	35
	N	25	46	52
35	D	27	9	22
	T	0	14	13
	N	27	23	35

TABLE 3

Effect of 20-min shocks on survival rates at 24 h post-fertilization for each replication (R_1, R_2, R_3)

Temp. (°C)	Replicate	Control	5 min	30 min
32	R1	38	9	22
	R2	27	7	22
	R3	42	7	9
35	R1	13	40	0
	R2	56	0	2
	R3	7	0	13

CONCLUSIONS AND DISCUSSION

Efficiency of the treatments

Thermal shocks are effective for induction of triploidy in molluscs, as they are in many other species. The mean yield of triploids with short shocks is moderate, and it seems that longer treatments are more effective (up to 60% triploid embryos).

Concerning the optimal time of application of shocks, no clear evidence could be drawn from our results. Nevertheless, two critical periods were identified by the effectiveness of short shocks: the period between 10 and 15 min and that between 35 and 40 min post-fertilization.

Due to the wide range of survival rates observed between replicates, which may be due to properties of the larval rearing system (use of very small beakers), an accurate estimation of the influence of high temperature on early survival rates was not possible. It can be concluded, however, that even when highly efficient in terms of triploidy induction, no major effect of heat shocks was observed on early survival, and very few abnormal larvae were observed. These results are different from those mentioned by authors using cytochalasin B, who usually report a depressive effect of treatment on survival (Stanley et al., 1981; Allen et al., 1982; Tabarini, 1984). However, our experimental design did not allow long-term rearing of larvae, so further studies will be necessary to evaluate the long-term effect of thermal shocks on survival and incidence of abnormality.

Finally, preliminary unpublished observations on the induction of triploidy by application of cold shocks indicate that these techniques could also be efficient in molluscs.

Cytological events

One of the potential benefits of triploids in aquaculture is improved growth, as observed by Stanley et al. (1984) and Tabarini (1984). The former indicate that the only triploids showing better growth than diploids are those obtained by blocking the first meiotic division, which leads to a higher level of heterozygosity than inhibition of the second meiotic division. This raises the question of the exact events of meiosis modified by the treatment in our experiments. Preliminary observations indicated that first cleavages occurred 50 min post-fertilization, and suggest that eggs had completed both meiotic divisions by that time. It is therefore possible that the two periods where short shocks were more efficient, 10–15 min and 35–40 min post-fertilization, would correspond to meiosis I and meiosis II, respectively. For this reason, 5 min and 30 min post-fertilization were the times chosen for application of long shocks. However, further observations revealed that the synchrony of development of control eggs often was very poor, even within a single batch (both single-celled and multiple-celled stages could be observed in similar proportions), so triploidy could occur by different events in different eggs of the same batch.

Sterility in triploid oysters

It does not seem that triploidy in bivalves could induce sterility similar to that observed in triploid fishes, where at least in female fishes, gonad development is totally inhibited (Lincoln, 1981a,b; Benfey and Sutterlin, 1984; Chevassus et al., 1985). Tabarini (1984) showed that there is subsequent gonad development during the maturation season in triploid bay scallop, with an increase of the gonadosomatic index, but the failure of triploid scallops to reach complete sexual maturity is accompanied by better meat quantity and quality, which is of interest for aquaculture purposes. Moreover, if triploid oysters prove to be unable to complete meiosis and to produce gametes, this advantage could be taken into consideration for reducing risks when the introduction of a new species is planned.

REFERENCES

Ahmed, M. and Sparks, A.K., 1967. A preliminary study of chromosomes of two species of oysters (*Ostrea lurida* and *Crassostrea gigas*). J. Fish. Res. Board Can., 24: 2155–2159.

Allen, S.K., Peter, S.G. and Hidu, H., 1982. Induced triploidy in the soft-shell clam. Cytogenetic and allozymic confirmation. J. Hered., 73: 421–428.

Benfey, T.J. and Sutterlin, A.M., 1984. Growth and gonadal development in triploid landlocked Atlantic salmon (*Salmo salar*). Can. J. Fish. Aquat. Sci., 41 (9): 1387–1392.

Chevassus, B., Quillet, E. and Chourrout, D., 1985. La production de truites stériles par voie génétique. Piscic. Fr., 78: 10–19.

Chourrout, D., 1980. Thermal induction of diploid, gynogenesis and triploidy in the eggs of the rainbow trout, *Salmo gairdneri* R. Reprod. Nutr. Dev., 20 (3A): 727–733.

Chourrout, D., 1984. Pressure-induced retention of second polar body and suppression of first cleavage in rainbow trout: production of all-triploids, all-tetraploids, and heterozygous and homozygous diploid gynogenetics. Aquaculture, 36: 111–126.

Chourrout, D. and Quillet, E., 1982. Induced gynogenesis in the rainbow trout: sex and survival of progenies. Production of all-triploid populations. Theor. Appl. Genet., 63: 201–205.

Gervai, J., Peter, S., Nagy, A., Horvath, L. and Csanyi, V., 1980. Induced triploidy in carp, *Cyprinus carpio* L. J. Fish Biol., 17: 667–671.

Lieder, U., 1964. Polyploidisierungsversuche bei Fischen mittels Temperaturschock und Colchizinbehandlung. Z. Fisch., 12: 247–257.

Lincoln, R.F., 1981a. Sexual maturation in triploid male plaice (*Pleuronectes platessa*) and plaice × flounder (*Platichthys flesus*) hybrids. J. Fish Biol., 19: 415–426.

Lincoln, R.F., 1981b. Sexual maturation in female triploid plaice (*Pleuronectes platessa*) and plaice × flounder (*Platichthys flesus*) hybrids. J. Fish Biol., 19: 499–507.

Lou, Y.D. and Purdom, C.E., 1984. Polyploidy induced by hydrostatic pressure in rainbow trout, *Salmo gairdneri* Richardson. J. Fish Biol., 25: 345–351.

Moav, R., 1976. Genetic improvement in aquaculture industry. In: T.V.R. Pillay and W.A. Dill (Editors), Advances in Aquaculture. Fishing New Books Ltd., Farnham, Surrey, pp. 610–622.

Purdom, C.E., 1972. Induced polyploidy in plaice (*Pleuronectes platessa*) and its hybrid with the flounder (*Platichthys flesus*). Heredity, 29: 11–24.

Purdom, C.E., 1983. Genetic engineering by the manipulation of chromosomes. Aquaculture, 33: 287–300.

Refstie, T., 1981. Tetraploid rainbow trout produced by cytochalasin B. Aquaculture, 25: 51–58.

Refstie, T., Vassvik, V. and Gjedrem, T., 1977. Induction of polyploidy in salmonids by cytochalasin B. Aquaculture, 10: 65–74.

Smith, L.T. and Lemoine, H.L., 1979. Colchicine-induced polyploidy in brook trout. Prog. Fish Cult., 41 (2): 86–88.

Stanley, J.G., Allen, S.K. and Hidu, H., 1981. Polyploidy induced in the American oyster, *Crassostrea virginica*, with cytochalasin B. Aquaculture, 23: 1–10.

Stanley, J.F., Hidu, H. and Allen, S.K., 1984. Growth of American oysters increased by polyploidy induced by blocking meiosis I but not meiosis II. Aquaculture, 37: 147–155.

Tabarini, C.L., 1984. Induced triploidy in the bay scallop, *Argopecten irradians*, and its effect on growth and gametogenesis. Aquaculture, 42: 151–160.

Thiriot-Quievreux, C., 1984. Les caryotypes de quelques Ostreidae et Mytilidae. Malacologia, 25 (2): 465–476.

Tetraploid Induction in *Oreochromis* spp.

JAMES M. MYERS

School of Fisheries, WH-10, University of Washington, Seattle, WA 98195 (U.S.A.)

ABSTRACT

Myers, J.M., 1986. Tetraploid induction in *Oreochromis* spp. *Aquaculture*, 57: 281–287.

The use of tetrapoloid broodstock in tetraploid × diploid matings has been suggested as a method of circumventing de novo triploid induction. Eggs from female tetraploids may also correct inherent developmental imbalances created by inducing polyploidy in diploid derived eggs. Tetraploid induction, using a combination of pressure and cold treatments, was attempted on two species of tilapia, *O. niloticus* and *O. mossambicus*, and their hybrid. The three genotypes were used in a full-factorial mating design under different induction regimes. Results indicated that allotetraploids were markedly more viable than autotetraploids in early development, yet both tetraploid types were subvital when compared to their diploid counterparts.

INTRODUCTION

The use of polyploid induction, especially the induction of triploidy, in aquaculture has attracted considerable attention. Triploidy provides functional sterility, because the pairing of chromosomes during meiosis I is impeded by the presence of three homologous chromosomes resulting in the uneven or aborted separation of chromosome triplets. The development of gonads in triploids is reduced or completely inhibited allowing for increased somatic growth (Thorgaard and Gall, 1979; Lincoln, 1981; Wolters et al., 1982).

Juvenile triploids exhibit inferior growth relative to diploids (Gervai et al., 1980; Utter et al., 1983). Therefore, the growth benefits of induced triploidy are not realized until after the time of maturation in diploids. The inferior growth of juvenile triploids may be partially explained by the induction method presently used in triploid production: the retention of the second polar body via physical, temperature, or chemical "shocks". The induction of polyploidy in an egg from a diploid organism may have detrimental effects on embryonic development. It has been shown that altering the nuclear–cytoplasm ratio (as occurs during polyploid induction in a diploid derived oocyte) causes the embryonic genome to activate earlier than occurs in the diploid embryo (Kafiani et al., 1969; Newport and Kirschner, 1982). Moreover, the time period from the activation of the genome to gastrulation remains unchanged such that gastrulation and early organogenesis begin with a reduced complement of progenitor cells (Chultiskava, 1970; Kobayakawa and Kubota, 1981). Therefore, the induction of triploidy results in aberrant embryonic development from which the organism is unable to compensate. The use of female tetraploids in tetraploid × diploid crosses should correct for the alteration in the nuclear–cytoplasm ratio and subsequent aberrant development created when polyploidy is induced in a diploid derived egg.

The mating of diploid and tetraploid organisms is an alternative method for inducing triploidy. Jaylet (1972) achieved 100% triploid progeny by mating tetraploid and diploid newts (*Pleurodeles waltii*). More recently, D. Chourrout (personal communication, 1984) produced triploid rainbow trout (*Salmo gairdneri*) using tetraploid males crossed with diploid females.

This study attempted to ascertain the most effective method of tetraploid induction and the effect of different parental genotypes on the induction and survival of tetraploids as a preliminary step in comparing induced triploids with those produced via tetraploid × diploid matings. For the experiment, tilapia (*Oreochromis* spp.) were selected because of their high reproductive capacity, early maturity, and commercial importance.

MATERIALS AND METHODS

Two species of tilapia, *O. mossambicus* (Peters) and *O. niloticus* (L.), and one of their hybrids, *O. mossambicus* × *O. niloticus*, were used to produce auto- (intraspecies) and allotetraploids (interspecies). Fish were raised in 220-l circular tanks with a 3 l/min flow of heated lake water (28–30°C), and fed a 40% protein diet ad libitum. Spawning and incubation procedures are outlined by Myers (1985).

Pilot experiments indicated that temperature treatments (both heat and cold) were ineffective in inducing tetraploidy. Similarly, pressure treatments alone proved incapable of inducing tetraploidy. The production of tetraploids was initially accomplished using a combination of pressure [4.8×10^4 kPa (7000 psi)] and cold treatments (7.5°C) applied for 7 min just prior to cleavage. Further experimentation indicated that an increased pressure level of 5.2×10^4 kPa (7500 psi) provided more efficient induction (Myers, 1985). Furthermore, the survival of treatment groups was greatly dependent on the slow application and removal of pressure [2.76×10^4 kPa/min (4000 psi/min)].

In addition to the optimum pressure treatment of 5.2×10^4 kPa (7500 psi) a second pressure level, 4.5×10^4 kPa (6500 psi), below the induction threshold of 4.8×10^4 kPa (7000 psi), was selected to differentiate between treatment induced mortalities and mortalities related to tetraploidy. Both treatments were used in conjunction with chilled water (7.5°C). The three genotypes were crossed in a full factorial design with two replicates at each of the treatment levels. Survival and percent induction of tetraploidy were examined at the eyed stage (36 h), and 12 h post-hatching (approximately 70 h). A minimum of 10 embryos was taken from each replicate at each sample period, and chromosome metaphase spreads for each embryo were prepared according to the procedure described by Kligerman and Bloom (1977).

Data for percent survival and induction were analyzed after being adjusted via the arcsin transformation. Analysis of variance (ANOVA) was performed using a fixed effects model.

RESULTS

Survival among the treated groups to the eyed stage was highly variable, ranging from 9.9% to 67.5%. No single genotype combination was significantly different from the rest ($P > 0.05$), nor was there any significant treatment effect ($P = 0.870$). The maternal genotype was the only factor to significantly influence survival ($P = 0.005$). A similar situation existed when survival estimates were analyzed for hatched fry. Comparison of genotype combinations within and between treatment levels using a pairwise T-test failed to reveal any significant differences. The analysis of variance (ANOVA) also indicated that neither the main effects nor the interaction effects significantly influenced survival to hatch. The overall survivals for the two pressure treatments were quite similar, 15.3 and 12.2%, for 4.5 and 5.2×10^4 kPa, respectively.

Strong influences, however, were noted when the percent tetraploidy for different treatment groups were compared. One group in the 4.5×10^4 kPa treatment level was found to contain tetraploids, and at a low incidence, 5.0%. Considerably more groups in the 5.2×10^4 kPa treatment contained tetraploids (Table 1A). The overall incidence of tetraploidy in the 5.2×10^4 kPa treatment was 5.0%, compared with 0.6% for the 4.5×10^4 kPa treatment. The highest incidence, 20.0%, was noted in the *O. mossambicus* × *O. niloticus* cross at the higher pressure. The ANOVA indicated that treatment was the only significant effect ($P = 0.014$).

Among the hatched fry the incidence of tetraploidy changed considerably in most groups. Tetraploids were found at a very low incidence in two groups at the 4.5×10^4 kPa treatment level, while eight of the nine genotype groups at the 5.2×10^4 kPa pressure treatment contained tetraploids (Table 1B). The *O. mossambicus* × *O. niloticus* crosses at 5.2×10^4 kPa continued to have the highest incidence of tetraploidy (21.6%), which was nearly five times the incidence of the next highest group. Comparisons of the change in tetraploidy from the eyed to hatch stage revealed that only the *O. mossambicus* × *O. niloticus* group did not show a decrease in tetraploid incidence (Table 2). Several significant factors did appear to influence the incidence of tetraploidy among the hatched fry (Table 3). The primary effect was that of treatment ($P = 0.009$), followed by maternal genotype ($P = 0.043$). The maternal by treatment inter-

TABLE 1

Percent tetraploidy in Oreochromid embryos at two stages of development following pressure treatment 5.2 × 10⁴ kPa (7500 psi) at 7.5°C: A, % tetraploidy and standard deviations (in parentheses) of embryos at the eyed stage; B, % tetraploidy and standard deviations (in parentheses) at hatch

Female genotype	Male genotype			
	O. niloticus	Hybrid	O. mossambicus	Mean
A				
O. niloticus	5.0 (7.1)	0 (0)	5.0 (7.1)	3.3 (5.2)
Hybrid	0 (0)	10.0 (14.1)	5.5 (7.8)	1.8 (6.0)
O. mossambicus	20.0 (14.1)	3.3 (8.1)	0 (0)	10.0 (12.0)
Mean	8.3 (11.6)		3.5 (5.4)	
B		2.8 (4.0)		
O. niloticus	0 (0)	0.5 (0.7)	4.5 (3.5)	2.4 (3.5)
Hybrid	0.5 (0.7)	0.4 (0.5)	2.7 (1.1)	1.2 (1.1)
O. mossambicus	21.6 (13.2)	1.2 (2.2)	2.9 (4.1)	8.3 (11.7)
Mean	7.4 (12.2)		3.4 (2.6)	

action was also significant ($P = 0.034$). The maternal by paternal genotype interaction and the three-way interaction between genotype and treatment showed strong but not significant influences.

The tetraploids produced were markedly subvital. Developmental abnormalities, particularly eye deformities, were highly pronounced, enabling identification of tetraploids. The yolk sac was also distinct in that it lacked the abundance of blood vessels usually found in fry. No tetraploid survived beyond

TABLE 2

Change in percent tetraploidy from the eyed stage to hatch among Oreochromid embryos treated at different pressure levels

Parental species	% Eyed stage	% Hatch	% Change[a]
4.5 × 10⁴ kPa (6500 psi) at 6.5°C			
O.n. × O.m.	10.0	0	—
5.2 × 10⁴ kPa (7500 psi) at 7.5°C			
O.n. × O.n.	10.0	0	—
O.n. × O.m.	10.0	7.1	−29.0
Hybrid × O.m.	11.1	0.7	−93.6
O.m. × O.n.	10.0	11.3	+13.0
O.m. × O.n.	30.0	31.8	+ 6.0

[a]% Change is computed as [(% hatch − % eyed)/ % eyed] × 100.

TABLE 3

Analysis of variance of tetraploid incidence among hatched Oreochromid embryos

Source of variation	df	MS	P
Main effects	5	0.006	0.015
Paternal genotype (P)	2	0.002	0.246
Maternal genotype (M)	2	0.006	0.043
Treatment (T)	1	0.008	0.009
Two-way interactions	8	0.004	0.034
P × M	4	0.004	0.052
P × T	2	0.002	0.251
M × T	2	0.006	0.039
Three-way interactions	4	0.004	0.052
P × M × T	4	0.004	0.052
Within cells	18	0.002	

7 days post-hatching. Furthermore, the development of tetraploids during the post-hatching period was slower than that seen in diploids.

DISCUSSION

The inability of temperature or pressure treatments individually to induce tetraploidy contrasts with studies by Jaylet (1972), Thorgaard et al. (1981), and Chourrout (1984). Only the combination of pressure and temperature was able to produce an observable level of induction. The application of the cold shock may have served two purposes: (1) development may have been arrested or slowed, thereby enlarging the induction window or allowing precise control of the time of pressure application; (2) cold shock may work in combination with the pressure treatment in an additive manner.

Survival differences between the two treatment levels were not significant at either the eyed or post-hatch stages. However, there was a striking difference in the level of tetraploid induction. The similarities in survival at the two treatment levels would suggest that the application of the treatment and not the induction of tetraploidy is largely responsible for mortality. No genotype combinations exhibited survival levels that significantly outperformed the rest, although there was a maternal genotype effect evident at the eyed stage. Prior experimentation revealed that there were not inherent differences in the survival of untreated crosses (Myers, 1985). However, variation within crosses due to female and incubation effects confounded the interpretation of the survival data.

The two treatments clearly defined the induction threshold for the species studied. The tetraploids observed at the 4.5×10^4 kPa treatment levels may be considered incidental occurrences, while the variability in induction at the 5.2×10^4 kPa may be due to the rapid development of tilapia eggs. The time from fertilization to first cleavage at $30\,°C$ is approximately 1 h. The induction window for tilapia occurs from 95–100% cleavage, and the asynchronous development in an egg of a few minutes may vastly alter the proportion of eggs at the appropriate cleavage stage. Despite the high variability between replicates in a cross, the *O. mossambicus* $\times$ *O. niloticus* cross consistently had a significantly higher incidence of tetraploidy at hatch, with the reciprocal cross having the second highest.

The high incidence of tetraploidy at the eyed stage in the hybrid crosses with the *O. mossambicus* maternal genotype would suggest that the *O. mossambicus* eggs are more susceptible to induction. However, the ANOVA for tetraploidy at the eyed stage did not indicate a significant maternal genotype effect. Survival differences between allo- and autotetraploids (pure and hybrid genotypes) prior to the eyed stage may have masked any *O. mossambicus* egg susceptibility to induction.

It was apparent that there were survival differences between the allo- and autotetraploids from the eyed stage to hatching. The incidence of tetraploidy was maintained only in the *O. mossambicus* $\times$ *O. niloticus* allotetraploids, while a slight decrease in percent tetraploidy was noted in the reciprocal allotetraploid. All of the other crosses either showed large declines in tetraploidy or were found to have very low incidences at hatching. The high incidence of tetraploidy in the allotetraploid groups may be due to the diverse nature of their chromosomal composition. Markert (1982) suggests that the low viability of mitotic clones may be due to the abnormally close topographical interactions of identical chromosome pairs. Topographical interactions may be necessary for the expressions of genes, and the close interaction of these chromosomes pairs may retard or inhibit normal expression. In tetraploids formed by mitotic inhibition the interaction of the two cloned pairs of chromosomes may be further complicated if the two pairs are from the same species resulting in a complex interaction between all four chromosomes. Allotetraploids may be less susceptible to these interactions because of differences between the chromosomes of the two species. Chourrout (1984), however, found that the survival of allotetraploid rainbow trout $\times$ brown trout (*Salmo trutta*) was considerably less than that of autotetraploid rainbow trout.

The inviability of both auto- and allotetraploids in this study appears to be due to developmental abnormalities. The abnormalities may be due to treatment effects rather than the chromosomal complement. Thorgaard et al. (1981) and Chourrout (1982) induced tetraploidy in rainbow trout at the time of cleavage (cytokinesis) and the embryos died prior to or soon after hatching. On the other hand, Chourrout (1984) treated rainbow trout prior to cleavage, at the time of the first metaphase (karyokinesis), and produced viable tetraploids.

Apparently significant differences exist between tetraploids induced at karyokinesis and cytokinesis. If viable tetraploid tilapia can be induced via the inhibition of karyokinesis, then producing allotetraploid tilapia may further improve the performance of these organisms.

REFERENCES

Chourrout, D., 1982. Tetraploidy induced by heat shocks in the rainbow trout. Reprod. Nutr. Dev., 22: 569–574.

Chourrout, D., 1984. Pressure-induced retention of second polar body and suppression of first cleavage in rainbow trout: production of all-triploid, all-tetraploid, and heterozygous and homozygous diploid gynogenetics. Aquaculture, 36: 111–126.

Chulitskava, E.V., 1970. Desynchronization of cell divisions in the course of egg cleavage and an attempt at experimental shift in its onset. J. Embryol. Exp. Morphol., 23: 359–374.

Gervai, J., Peter, S., Nagy, A., Horvath, L. and Csanyi, V., 1980. Induced triploidy in carp. J. Fish Biol., 17: 667–671.

Jaylet, A., 1972. Tetraploide expérimentale chez le triton (*Pleurodeles waltii*). Chromosome, 38: 173–184 (in French, trans. by C. James).

Kafiani, C.A., Timofeeva, M.J., Neyfakh, A.A., Melnikova, N.L. and Rachkus, J.A., 1969. RNA synthesis in the early development of a fish (*Misgurnus fossilis*). J. Embryol. Exp. Morphol., 21: 295–308.

Kligerman, A.D. and Bloom, S.E., 1977. Rapid chromosome preparations from solid tissues of fishes. J. Fish. Res. Board Can., 34: 266–269.

Kobayakawa, Y. and Kubota, H., 1981. Temporal pattern of cleavage and the onset of a gastrulation in amphibian embryos developed from eggs with reduced cytoplasm. J. Embryol. Exp. Morphol., 62: 83–94.

Lincoln, R.F., 1981. Sexual maturation in triploid male plaice (*Pleuronectes platessa*) and plaice × flounder (*Platichthys flesus*) hybrids. J. Fish Biol., 19: 415–426.

Markert, C., 1982. Parthenogenesis, homozygosity, and cloning in mammals. J. Hered., 73: 390–397.

Myers, J.M., 1985. An Assessment of Spawning Methodologies and the Induction of Tetraploidy in Two Oreochromid Species. M.S. thesis, Univ. of Washington, WA, 83 pp.

Newport, J.H. and Kirschner, M., 1982. A major developmental transition in early embryos. 1. Characterization and timing of cellular changes at midblastula stage. Cell, 30: 675–686.

Thorgaard, G. and Gall, G., 1979. Adult triploids in a rainbow trout family. Genetics, 93: 961–973.

Thorgaard, G.H., Jazwin, M.E. and Stier, A.R., 1981. Polyploidy induced by heat shock in rainbow trout. Trans. Am. Fish. Soc., 110: 550–565.

Utter, F., Johnson, O., Thorgaard, G. and Rabinovitch, P., 1983. Measurement and potential applications of induced triploidy in Pacific salmon. Aquaculture, 35: 125–135.

Wolters, W.R., Libey, G.S. and Chrisman, C.L., 1982. Effect of triploidy on growth and gonadal development of channel catfish. Trans. Am. Fish. Soc., 111: 102–105.

Androgenetic Rainbow Trout Produced from Inbred and Outbred Sperm Sources Show Similar Survival

PAUL D. SCHEERER, GARY H. THORGAARD, FRED W. ALLENDORF[1] and KATHY L. KNUDSEN[1]

Department of Zoology and Program in Genetics and Cell Biology, Washington State University, Pullman, WA (U.S.A.)
[1]*Department of Zoology, University of Montana, Missoula, MT (U.S.A.)*

ABSTRACT

Scheerer, P.D., Thorgaard, G.H., Allendorf, F.W. and Knudsen, K.L., 1986. Androgenetic rainbow trout produced from inbred and outbred sperm sources show similar survival. *Aquaculture*, 57: 289–298.

Androgenesis is a technique that could facilitate the rapid production of completely homozygous isogenic lines of fish. We induced diploid androgenesis in rainbow trout (*Salmo gairdneri*) by fertilizing ^{60}Co gamma-irradiated eggs with untreated sperm and then blocking the first cleavage division with hydrostatic pressure. Electrophoresis of enzymes encoded by seven loci confirmed the complete homozygosity and all-paternal inheritance of the androgenetic progeny. We used sperm from both inbred and outbred strains to determine if the survival of androgens was greater with sperm from inbred males because of a reduction in the number of deleterious recessive alleles in inbred strains. Survival of androgenetic diploids from inbred and outbred sperm sources was similar and significantly below that of outbred controls. The similar survival of androgens from inbred and outbred sperm sources may be the result of high treatment mortality associated with androgenesis.

INTRODUCTION

All-paternal inheritance (androgenesis) involves inactivation of the genetic material from the female parent and fertilization of the treated egg with normal sperm. Egg inactivation can be achieved in fish using gamma radiation (Romashow and Belyaeva, 1964; Purdom, 1969; Arai et al., 1979; Parsons and Thorgaard, 1984). Diploidization can then be restored by suppressing the first cleavage using hydrostatic pressure shock (Onozato, 1982, in Yamazaki, 1983; Parsons and Thorgaard, 1985).

Androgenesis and gynogenesis (all-maternal inheritance) potentially can facilitate the rapid production of inbred lines in fish. First-generation offspring containing two identical chromosome sets can serve as the parents for isogenic lines (Streisinger et al., 1981). Androgenesis may also prove useful as a method of gene banking by allowing strains of fish to be reconstituted from cryopreserved sperm. Sperm cryopreservation has been quite successful in fish, but cryopreservation of embryos and eggs has been unsuccessful, possibly due to their large size (Stoss, 1983).

A major obstacle to the use of androgenesis for generation of inbred lines and for gene banking is the extremely low survival expected in androgenetic offspring because of homozygosity for deleterious recessive alleles. Survivors from previous inbreeding should possess fewer harmful recessive alleles than individuals from outbred strains, and consequently might be expected to give rise to androgenetic offspring with higher survival. In this study we compared survival of androgenetic rainbow trout (*Salmo gairdneri*) generated from two inbred and two outbred sperm sources. We found no significant differences in survival among sperm sources.

MATERIALS AND METHODS

Sperm and egg sources

Rainbow trout sperm was obtained in December 1984 from two outbred and two inbred strains. The outbred strains were the Arlee (AR) strain from the Jocko River Trout Hatchery (Montana Department of Fish, Wildlife and Parks) and the Spokane (SP) Hatchery strain (Washington Department of Game). One inbred strain (HC) was from the Hot Creek Hatchery (California Department of Fish and Game) and was the result of eight generations of brother–sister mating (R. Toth, personal communication, 1984). The second inbred strain (CC) consisted of gynogenetic sex-reversed males reared at Washington State University. These partially inbred gynogenetic trout were produced in 1982 by fertilization with UV-irradiated sperm and retention of the second polar body as previously described (Thorgaard et al., 1983) from eggs from the Chambers Creek Hatchery (Washington Department of Game). The gynogenetic offspring were sex-reversed into males by adding 17-α methyltestosterone to the initial diet (Okada et al., 1979).

Eggs were obtained from eight females from the Spokane, Washington, hatchery. Gametes were transported on ice to Washington State University.

Androgenesis

Eggs from the eight females were pooled. The majority of the pooled eggs were exposed to ^{60}Co gamma radiation at the Washington State University Nuclear Radiation Center to a final dose of 3.6×10^4 R (Parsons and Thorgaard, 1984); a portion of the pooled eggs was left unirradiated. The irradiated eggs were divided into 16 treatment groups and fertilized separately with sperm from six AR males, two SP males, six HC males and two CC males. The non-irradiated eggs were also divided into 16 groups and fertilized with sperm from the same males to serve as diploid controls.

After fertilization, all egg lots were placed in an incubator at 10°C. A portion of the irradiated egg lots was removed to serve as androgenetic haploid controls; the remainder were treated with hydrostatic pressure at 345 min post-fertilization to produce androgenetic diploids. The pressure was increased steadily over a 3-min period, held at 9000 psi for 3 min (Parsons and Thorgaard, 1985), and decreased over a 2-min period.

Relative survival rates of the androgenetic treatment groups were calculated as a percentage survival of control groups (survival androgenetic group/survival control group). Survival was monitored at the eyed stage (18 days), at hatching (33 days), at the initiation of feeding (59 days), and over a period of 111 days after the initiation of feeding. At 59 days, survivors were pooled according to the sperm source and were reared in separate 5-gallon stainless steel tanks in a 10°C recirculating water system. Equal numbers of the corresponding controls were reared in identical tanks. Growth of the androgenetic trout was compared to that of the corresponding controls from day 59 through day 184.

Protein and chromosome analysis

Protein electrophoresis was used to test for levels of genetic variation in male parents and to test for homozygosity in the androgenetic offspring. Tissue extracts were prepared and horizontal starch-gel electrophoresis was accomplished as described by Allendorf et al. (1977). The male parents were found to be polymorphic at seven of 29 loci examined. The nomenclature of the polymorphic loci and allelic variants follow the system described by Allendorf et al. (1983), and are: creatine kinase, E.C.2.7.3.2., Ck1 (100,76); esterase, E.C.3.1.1.1, Est1 (100,95); isocitrate dehydrogenase, E.C.1.1.1.42, Idh2 (100,140); malate dehydrogenase, E.C.1.1.1.37, Mdh3,4 (100,83); phosphoglucomutase, E.C.2.7.5.1, Pgm2 (100,90); and superoxide dismutase, E.C.1.15.1.1, Sod1(100,152). In this paper alleles will be referred to as 1 or 2, where 1 represents the 100 allele and 2 represents the second allele. The seven HC males were screened for the 42 loci described in Ferguson et al. (1985) so that their heterozygosity could be compared with other strains of rainbow trout. Androgenetic fish that died between days 59 and 171 were stored at $-70°C$ until the samples were analyzed. Some denaturation of enzymes made it impossible to score all loci in all individuals because the fish were collected after their death that had occurred sometime within the preceeding 24 h.

Ten androgenetic diploids and five androgenetic haploids were sacrificed at day 18 (eyed stage) and chromosome peparations were made (Thorgaard et al.,

TABLE 1

Genotypes at seven protein loci of the male parents from four strains of rainbow trout.

Strain	No.	Ck1	Est1	Idh2	Mdh3,4	Pgm2	Sod1
AR	1	11	12	11	1111	11	12
	2	11	11	11	1112	11	11
	3	11	11	11	1122	11	12
	4	11	11	12	1122	11	11
	5	12	12	12	1111	11	22
	6	11	11	11	1111	11	12
CC	1	11	11	11	1111	11	12
	2	11	12	11	1111	11	11
HC	1	11	11	22	1111	22	22
	2	11	11	22	1111	22	22
	3	11	11	22	1111	22	22
	4	11	11	22	1111	22	22
	5	11	11	22	1111	22	22
	6	11	11	22	1111	22	22
	7	11	11	22	1111	22	22
SP	1	12	11	22	1122	12	11
	2	11	12	12	1112	11	12

1981). Chromosome spreads were analyzed to determine ploidy level and to check for the presence of chromosome fragments.

RESULTS

Protein analysis

The comparison of phenotypes of the male parents at seven electrophoretic loci supported the inbred origins of the CC and HC strains (Table 1). We have previously reported that gynogenetically produced rainbow trout, such as CC, have an estimated 45% reduction in overall genomic heterozygosity (Thorgaard et al., 1983).

The inbred HC fish were identically homozygous at all 42 loci examined; this is in agreement with the ancestry of eight generations of brother–sister matings. It is valid to estimate the amount of genetic variation in this strain based on only seven fish because they are full-sibs. One-half of all individuals are expected to be heterozygous at any locus segregating in this strain. Therefore, the probability of not detecting at least one heterozygote at a segregating locus is $(1/2)^7$, or 0.008. Thus, this strain has an estimated average heterozygosity

TABLE 2

Androgenetic progeny genotypes and allele frequencies in progeny and male parents at six protein loci

Locus	Strain	Progeny genotypes			Frequency of allele 1	
		11	12	22	Progeny	Parents
Ck1	AR	21	0	0	1.00	0.92
	CC	25	0	0	1.00	1.00
	HC	67	0	0	1.00	1.00
	SP	37	0	2	0.95	0.75
Est1	AR	32	0	5	0.86	0.83
	CC	21	0	1	0.95	0.75
	HC	7	0	0	1.00	1.00
	SP	6	0	2	0.75	0.75
Idh2	AR	18	0	4	0.82	0.83
Pgm2	AR	61	0	1	0.98	1.00
	CC	27	0	0	1.00	1.00
	HC	0	0	67	0.00	0.00
	SP	32	0	6	0.84	0.75
Sod1	AR	45	0	16	0.74	0.58
	CC	15	0	10	0.60	0.75
	HC	0	0	66	0.00	0.00
	SP	26	0	11	0.70	0.75

of zero. The heterozygosities of the six strains of rainbow trout examined by Ferguson et al. (1985) at these same 42 loci ranged from 0.040 to 0.080. Furthermore, Busack et al. (1979) estimated a heterozygosity of 0.087 based on 24 electophoretic protein loci for the strain from which the inbred HC was derived.

All androgenetic progeny were homozygous at the five disomic loci for which the male parents were polymorphic (Table 2). In addition, all-male inheritance was confirmed because the alleles present in the androgens reflected the genotypes of the strains from which their male parent was derived. For example, all HC progeny were homozygous for allele 2 at both Pgm2 and Sod1, just as their male parents were. The only exception is a single individual in the AR androgens that was homozygous at Pgm2 for an allele not present in the AR males. In addition, two SP androgens were homozygotes at both Pgm2 and Sod1; however, neither of the two SP male parents were segregating at both these loci (Table 1). All three of these androgens were sampled on the same collection date (4 March 1985) and we suspect that they were mislabeled. There is no other apparent explanation for these three exceptions.

TABLE 3

Androgenetic progeny genotypes and allele frequencies in progeny and male parents at Mdh3,4

Strain	Progeny phenotypes					Frequency of allele 1	
	1111	1112	1122	1222	2222	Progeny	Parents
AR	41	0	24	0	0	0.82	0.79
CC	28	0	0	0	0	1.00	1.00
HC	67	0	0	0	0	1.00	1.00
SP	23	0	15	0	2	0.76	0.63

Two loci (Mdh3,4) that share the same common allele encode the skeletal muscle form of malate dehydrogenase (MDH-B) in rainbow trout (Bailey et al., 1970). These loci are inherited disomically in females but show residual tetrasomic inheritance in males from some strains (May et al., 1982; Allendorf and Thorgaard, 1984). All androgenetic progeny had phenotypes at Mdh3,4 that were compatible with the diploidy in the androgens being the result of doubling of the sperm genome by the suppression of the first cleavage division (Table 3).

Chromosome analysis

Chromosome analyses confirmed that 12 of 12 suspected androgenetic diploid individuals were diploid ($2n = 58$ or 60) and that five of five suspected androgenetic haploids were haploid ($n = 29$ or 30). No chromosome fragments were observed among the androgenetic embryos.

Survival and growth

Survival of the androgenetic diploids from inbred and outbred sperm sources was similar and significantly below that of the diploid control lots (Table 4). Mean survival to the initiation of feeding of the control groups with SP, AR, HC and CC male parents was 0.848, 0.868, 0.837, and 0.879, respectively. No hatching was observed in any of the androgenetic haploid lots, suggesting that all individuals hatching in the experimental lots were diploid.

Survival from day 59 through 174 of the androgenetic lots produced from inbred and outbred sources was similar, and significantly below that of the control lots (Table 5). Control survival over this period in the AR, SP, HC and CC lots was 0.709, 0.737, 0.725, and 0.929, respectively.

The mean weight of the androgenetic diploids from inbred sperm sources, androgenetic diploids from outbred sperm sources, and control diploids was similar after 63 days of feeding (day 121). However, this comparison is com-

TABLE 4

Mean relative survivial through the initiation of feeding of androgenetic rainbow trout produced from inbred and outbred sperm sources (figures represent mean ± one standard deviation; ranges of relative survival are listed in parentheses)

	No. lots	No. eggs/lot[a]	Mean relative survival (%)		
			Eyed (18 days)	Hatch (33 days)	Initiation of feeding (59 days)
Inbred sources					
HC	6	888 ± 149 (675 — 1028)	40.8 ± 7.8 (33.0 — 55.3)	7.9 ± 3.3 (3.4 — 13.0)	4.8 ± 2.6 (2.1 — 8.9)
CC	2	905 ± 79 (849 — 961)	35.9 ± 6.2 (31.5 — 40.2)	7.2 ± 4.0 (4.3 — 10.0)	4.7 ± 4.1 (1.8 — 7.6)
Outbred sources					
AR	6	926 ± 111 (808 — 1115)	40.2 ± 2.9 (36.5 — 44.4)	8.8 ± 1.9 (6.3 — 11.1)	5.6 ± 2.2 (2.7 — 9.4)
SP	2	907 ± 18 (894 — 920)	38.2 — 1.8 (36.9 — 39.5)	9.5 ± 3.5 (7.0 — 11.9)	6.8 ± 2.0 (5.4 — 8.2)

[a]Sample sizes in the haploid androgenetic control groups ranged from 19–53% of the numbers in the androgenetic diploid groups. Sample sizes in the diploid control groups ranged from 20–36% of the number in the androgenetic diploid groups.

TABLE 5

Mean relative survival from the initiation of feeding (day 59) through day 170 of androgenetic rainbow trout produced from inbred and outbred sperm sources (figures represent mean ± one standard deviation; ranges of relative survival are in parentheses)

	No. lots	Mean relative survival (%)	
		Day 59–122	Day 123–170
Inbred sources			
HC	2	39.5 ± 20.2 (25.2 — 53.7)	63.3 ± 4.7 (60.6 — 66.6)
CC	1	20.0	23.1
Outbred sources			
AR	2	37.0 ± 10.7 (29.5 — 44.6)	62.8 ± 18.2 (49.9 — 75.7)
SP	1	41.7	43.8

plicated by unequal densities in the different lots because of differential mortality. At day 122 numbers and densities of fish were adjusted to be nearly equivalent in the growth lots. Comparisons at day 184 again showed similar mean weights in the androgenetic and control lots.

DISCUSSION

Our results indicate that the production of androgenetic diploid rainbow trout is feasible using sperm from a variety of sources. Survival to the initiation of feeding (relative to control) averaged 4.7–6.8%, chromosome analysis confirmed that the survivors were diploid and protein electrophoresis results were consistent with homozygosity and all-paternal inheritance. Androgenesis could be used in the rapid generation of inbred lines in fish and in the recovery of genotypes from cryopreserved sperm.

The lack of differences in survival between androgenetic diploids from inbred and outbred sperm sources that we observed is somewhat puzzling. Inbred individuals might be expected to have fewer harmful recessive alleles than outbred individuals and consequently might be expected to give rise to more viable androgenetic diploids. Chourrout (1984) has advocated that partially inbred gynogenetic diploids be used as female parents to improve survival of homozygous diploids produced by gynogenesis involving suppression of the first cleavage. Previous studies of gynogenesis and androgenesis with inbred parents have given conflicting results. Streisinger et al. (1981) found greatly improved survival among the gynogenetic progeny of homozygous diploid gynogenetic zebra fish. Gillespie and Armstrong (1981), however, found no significant improvement in survival among the androgenetic offspring of an androgenetic diploid male axolotl, suggesting that treatment effects and not the uncovering of harmful recessive alleles were responsible for most of their mortality.

Treatment effects (egg irradiation or pressure treatment) might also be responsible for the low survival in the androgenetic groups from inbred parents in our study. However, the similar survival rates of gynogenetic and androgenetic haploid rainbow trout suggest that irradiation of the egg does not have an extreme effect on early development (Parsons and Thorgaard, 1984). In this study, the survival of androgenetic haploids and pressure-treated androgenetic diploids to day 18 was similar, suggesting that the pressure treatment may not have an extremely damaging effect on early development. The egg irradiation or pressure treatments might, however, have harmful effects on later development. An alternate explanation for the uniform survival results we observed is that homozygous trout have uniformly poor viability, regardless of genotype.

To examine the contribution of genetic factors and treatment effects to the mortality in the androgenetic diploids it would be valuable to study survival of the androgenetic offspring produced from homozygous diploid androgenetic males. YY individuals appear to be viable in rainbow trout (Parsons and Thorgaard, 1985) so this may be possible if some of our homozygous androgenetic diploids survive to sexual maturity.

ACKNOWLEDGMENTS

We thank Mike Albert (Washington Department of Game), Jack Boyce (Montana Department of Fish, Wildlife and Parks), and Robert Pipkin and Dr. Graham Gall (University of California, Davis) for supplying the trout gametes. This research was supported by U.S. Depatment of Agriculture grants 82-CRSR-2-1058 (to G.H.T.) and 83-CRSR-2-2227 (to F.W.A.).

REFERENCES

Allendorf, F.W. and Thorgaard, G.H., 1984. Tetraploidy and the evolution of salmonid fishes. In: B.J. Turner (Editor), Evolutionary Genetics of Fishes. Plenum, New York, NY, pp. 1-53.

Allendorf, F.W., Mitchell, N., Ryman, N. and Stahl, G., 1977. Isozyme loci in brown trout (*Salmo trutta*): detection and interpretation from population data. Hereditas, 86: 179-190.

Allendorf, F.W., Knudsen, K.L. and Leary, R.F., 1983. The adaptive significance of differences in the tissue-specific expression of a phosphoglucomutase gene in rainbow trout. Proc. Natl. Acad. Sci. U.S.A., 80: 1397-1400.

Arai, K., Onozato, H. and Yamazaki, F., 1979. Artificial androgenesis induced with gamma irradiation in masu salmon, *Oncorhynchus masou*. Bull. Fac. Fish., Hokkaido Univ., 30: 181-186.

Bailey, G.S., Wilson, A.C., Halver, J.E. and Johnson, C.L., 1970. Multiple forms of supernatant malate dehydrogenase in salmonid fishes. J. Biol. Chem., 245: 5927-5940.

Busack, C.A., Halliburton, R. and Gall, G.A.E., 1979. Electrophoretic variation and differentiation in four strains of domesticated rainbow trout. Can. J. Genet. Cytol., 21: 81-94.

Chourrout, D., 1984. Pressure-induced retention of second polar body and suppression of first cleavage in rainbow trout: production of all-triploids, all-tetraploids and heterozygous and homozygous gynogenetics. Aquaculture, 36: 111-126.

Ferguson, M.M., Danzmann, R.G. and Allendorf, F.W., 1985. Developmental divergence among hatchery strains of rainbow trout (*Salmo gairdneri*). I. Pure strains. Can. J. Genet. Cytol., 27: 289-297.

Gillespie, L.L. and Armstrong, J.B., 1981. Suppression of the first cleavage in the Mexican axolotl (*Ambysotma mexicanum*) by heat shock or hydrostatic pressure. J. Exp. Zool., 218: 441-445.

May, B., Wright, J.E. and Johnson, K.R., 1982. Joint segregation of biochemical loci in Salmonidae. III. Linkage associations in Salmonidae including data from rainbow trout (*Salmo gairdneri*). Biochem. Genet., 20: 29-39.

Okada, H., Matumoto, H. and Yamazaki, F., 1979. Functional masculinization of genetic females in rainbow trout. Bull. Jpn. Soc. Sci. Fish., 45: 413-419.

Onozato, H., 1982. Artificial induction of androgenetic diploid embryos in salmonids. Abstr. Annu. Meet. Jpn. Soc. Sci. Fish, p. 142 (in Japanese).

Parsons, J.E. and Thorgaard, G.H., 1984. Induced androgenesis in rainbow trout. J. Exp. Zool., 231: 407-412.

Parsons, J.E. and Thorgaard, G.H., 1985. Production of androgenetic diploid rainbow trout. J. Hered., 76: 177-181.

Purdom, C.E., 1969. Radiation-induced gynogenesis and androgenesis in fish. Heredity, 24: 431–444.

Romashov, D.D. and Belyaeva, V.N., 1964. Cytology of radiation gynogenesis and androgenesis in loach (*Misgurnus fossilis* L.). Dokl. Akad. Nauk S.S.S.R., 157: 964–967.

Stoss, J., 1983. Fish gamete preservation and spermatozoan physiology. In: W.S. Hoar, D.J. Randall and E.M. Donaldson (Editors), Fish Physiology, Vol. 9B. Academic Press, New York, NY, pp. 305–350.

Streisinger, G., Walker, C., Dower, N., Knauber, D. and Singer, F., 1981. Production of clones of homozygous diploid zebra fish (*Brachydanio rerio*). Nature, 291: 293–296.

Thorgaard, G.H., Jazwin, M.E. and Stier, A.R., 1981. Polyploidy induced by heat shock in rainbow trout. Trans. Am. Fish. Soc., 110: 546–550.

Thorgaard, G.H., Allendorf, F.W. and Knudsen, K.L., 1983. Gene-centromere mapping in rainbow trout: high interference over long map distances. Genetics, 103: 771–783.

Yamazaki, F., 1983. Sex control and manipulation in fish. Aquaculture, 33: 329–354.

Commercial Methods for the Control of Sexual Maturation in Rainbow Trout (*Salmo gairdneri* R.)

VICTOR J. BYE and RICHARD F. LINCOLN

Ministry of Agriculture, Fisheries and Food, Directorate of Fisheries Research, Fisheries Laboratory, Lowestoft, Suffolk NR33 0HT (Great Britain)

ABSTRACT

Bye, V.J. and Lincoln, R.F., 1986. Commercial methods for the control of sexual maturation in rainbow trout (*Salmo gairdneri* R.). *Aquaculture*, 57: 299–309.

Over half of the production of United Kingdom trout farms is accounted for by female-only stocks, and the use of sterile trout, particularly for sport fisheries and saltwater culture, is increasing. The practical sex control techniques by which these are produced were developed at this laboratory. They are suitable for use on technically unsophisticated farms, require no specialized equipment and are aesthetically acceptable to the consumer. The production of entirely female stocks requires phenotypically male broodstock of female genotype to fertilize eggs from untreated females. The sex-inversed males are produced by feeding fry for 700°C days with food containing 3 mg kg^{-1} 17α methyltestosterone. The majority of the males lack sperm ducts, requiring the semen to be removed surgically, but reduced duration androgenization increases the proportion of males with intact ducts. No fish destined for human consumption are treated with hormones. Sterile trout are produced by the induction of triploidy in female eggs. Eggs are heatshocked at 28°C for 10 min, commencing 40 min after fertilization. Commercially acceptable levels of survival and sterility are achieved. This paper presents details of the techniques developed and considers the problems of applying genetic methods to commercial aquaculture.

INTRODUCTION

For the majority of fish which are cultured commercially for food, the control of sexual development is economically desirable, to produce either monosex or sterile stocks. The technique which is currently most commonly employed is the control of gender by the administration of synthetic sex steroids to sexually undifferentiated fish. The production of sterile fish is less common but can be achieved either by induced triploidy or by the administration of high doses of sex steroids. The extensive literature relating to control of fish sexual development has been reviewed both for the hormonal (Hunter and Donaldson, 1983) and chromosome engineering methods (Purdom, 1983; Thorgaard, 1983). Research will undoubtedly continue to refine these techniques and extend them to a wider range of species but there must be a parallel development to translate them from the research laboratory into commercial practice. The use of sex control is widespread in the United Kingdom trout farming industry. This is a result of the policy of the Fisheries Directorate of the Ministry of Agriculture, Fisheries and Food which actively involves commercial fish farms at all stages in its programme of research and development. In this paper we describe the methods developed and their transfer to commercial farms.

The fish farming industry in the United Kingdom produces mainly salmonids with an estimated output for 1984 of 12 000 t of rainbow trout, 250 t of brown trout and 4000 t of Atlantic salmon. The latter are grown mainly in sea cages off the coast of Scotland. The majority of rainbow trout is produced in fresh water for marketing at 250–300 g but over the past 5 years the production of larger fish of 500 g–3 kg has increased. Recently a few seawater farms for rainbow trout have been established and we expect this sector of the industry to expand because of the intensifying competition for good quality fresh water. All the brown trout and at least 2000 t of rainbow trout are grown for the sporting market to stock rivers, lakes and reservoirs for anglers. The minimum size for this market is 500 g and there is an unsatisfied demand for fish heavier than 2 kg. Angling is an economically important part of fish farming because of the high first-sale price of the fish.

In 1974, when we first became involved in Research and Development in support of fish farming, it was apparent that sexual development in rainbow trout was an economic burden for the industry. A substantial proportion of males become sexually mature before they reach market size and, apart from their poor growth and food conversion efficiency, they are virtually unsaleable because of their unpleasant appearance and flesh quality. Maturity in females was initially less of a problem because it never occurred in fish less than 2 years old. But with the increasing demand for large fish both for food and sport and the expansion of saltwater cultivation, where maturation is often lethal, there is now a need to control sexual development in females.

METHODS AND RESULTS

Sex ratio control

We commenced our research by identifying the options for sex control and deciding that the production of entirely female stocks by sex steroid treatment and of sterile fish by either high dose steroid or induced polyploidy were the most promising lines to follow. Initial experiments established the parameters for steroid-induced sex inversion: these included hormone species, method of administration, concentration, time and duration of treatment. Masculinization was accomplished relatively easily by the administration of 17α methyltestosterone at 3 mg kg^{-1} in the diet of first feeding fry for 1000°C days. Feeding was two to four times an hour for 12 h each day. Feminization was more difficult and, although it was established that the dietary administration of 17β estradiol at 30 mg kg^{-1} for 30 days from first feeding was the most effective treatment, the degree of feminization was low and unpredictable. We surmised that effective feminization required the maintenance of an adequate concentration of estradiol in the blood of the fry throughout the period of gonadal differentiation and that this could only be achieved by uninterrupted feeding for at least 18 h each day. This was obviously impractical on commercial farms particularly when combined with the higher mortality of estrogenized fish compared with controls or those receiving androgen. In addition, we sought to avoid hormone treatment of fish ultimately destined for human consumption. Although individual hormone doses were very low (less than 5 μg/fish) and all residues would be eliminated before the fish were eaten (Johnstone et al., 1978; Fagerlund and Dye, 1979) we wished to preclude consumer resistance and avoid the necessity for farms to use large amounts of hormone in order to treat all their fish.

At this time, it had been demonstrated that gynogenetic trout were all female, indicating a homogametic female (XX) and heterogametic male (XY) sex determining system. This provided an opportunity for making entirely female stocks which were not hormone treated. By masculinizing female fry, we could create functional homogametic males whose semen, when used to fertilize the ova from untreated females, would produce entirely female offspring since the spermatozoa contained no Y chromosomes. A potential difficulty was our inability to separate the XX males from the natural XY males at maturity since karyotyping to identify genetic gender was problematic (Thorgaard, 1977). It was therefore necessary to conduct progeny tests. At maturity, the androgenized stock, although almost entirely male with well developed secondary sexual characteristics (dark colour, kype, etc.) was comprised of approximately half which could be stripped of milt in the normal way and half which were unstrippable. We found that although the latter group had well developed testes the afferent ducts were absent or occluded (Fig. 1). Progeny test results revealed that the offspring of the ducted males included equal numbers of males and females but those from the ductless group were entirely female. This facilitated the identification of the homogametic males at the initial masculinization, and in subsequent generations by ensuring that the only fry which were androgenized were from female-only progenies; we could produce male broodstocks which were all homogametic, the Y chromosome having been eliminated from the stock.

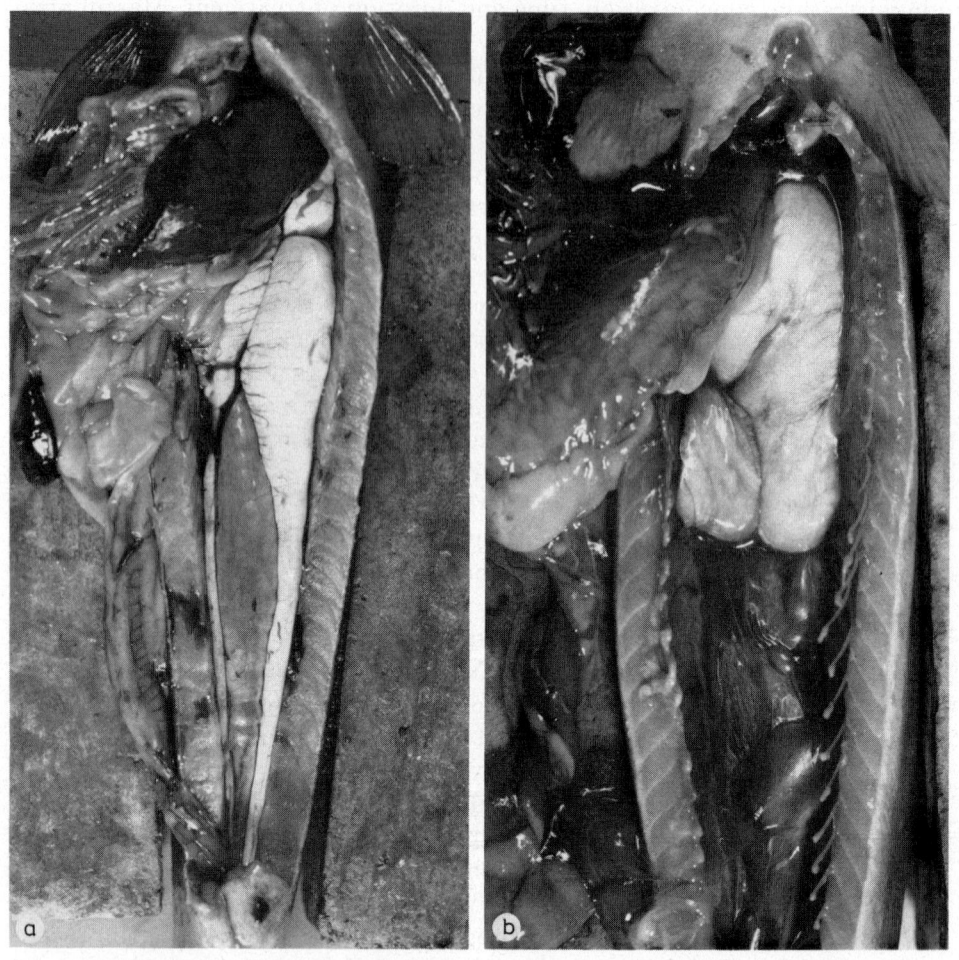

Fig. 1. Surgically exposed testes of mature male rainbow trout demonstrating development of sperm ducts. (a) Normal heterogametic male with semen-filled ducts connecting testes to vent. (b) Steroid-inversed homogametic male showing well developed testes without connecting ducts to vent.

Industrial trials

Commercial exploitation started in 1979 on broodstock farms in the south of England and was in two stages. In the first masculinization trial, the food was hormone treated by farm staff under our supervision and the fry hand-fed throughout the treatment period. The fish were tested the following spring and found to be all males. When they matured at 2 years of age, we supervised the surgical removal of the ductless testes, the extraction of the semen and its use for the fertilization of eggs, modifying our methods as necessary to suit conditions and practices on the farm. Concurrently with the first masculinizations, we supplied interested commercial farms with surplus homogametic males from our laboratory experiments. When they matured the farmers used them, with our assistance, to produce female-only ova, some of which they masculinized to produce homogametic male broodstock for future use by themselves and other farmers.

The first commercially supplied, all-female trout ova were advertised in the trade press in the summer of 1980 and became available in January 1981. Interest spread rapidly throughout the industry, particularly when the customers of the egg and fry farms began demanding female-only material. There were relatively few broodstock farms in the United Kingdom and so, despite the increased interest, we were able to provide assistance to our existing collaborators and to those new farmers wishing to join the trial. Over the first 3 years of the commercial launch of the 'Lowestoft technique', we attempted to make two visits each year to the collaborating farms. One visit was made during the spawning season, to help with aspects of the semen collection and fertilization procedures and the masculinization of the fry. A second visit was made 4–6 months later to monitor sex ratios in 2–5 g fingerlings for both masculinization and female-only production. The farmers were shown how to dissect the immature gonads from the fingerlings and examine them, as a fresh squash preparation, under the microscope in order to determine gender. At this stage female fish are readily identified by the presence of oocytes or the characteristic lamellar structure of the ovary while the testes are small, string-like and packed with undifferentiated spermatogonia (Fig. 2). In addition to the practical demonstrations, the farmers were provided with explanatory leaflets and detailed instruction sheets for each stage of the process. We also published articles in trade magazines (Bye and Lincoln, 1979, 1981) and lectured at conferences (Bye, 1982; Bye and Purdom, 1982) and training courses.

In most countries the use of steroids is rigorously controlled and in the United Kingdom this is covered by the Medicines Acts (Great Britain, Parliament, 1968, 1971). To comply with this legislation, we ensured that the masculinization was supervised by a veterinarian and we provided methyltestosterone without charge.

Problems in industrial trials

Few problems were encountered by farmers in preparing and using the androgenized diet for masculinizing their trout. Originally we supplied vials holding 3 mg of 17α methyltestosterone which was first dissolved in a few drops of benzyl alcohol and then dispersed in ethyl alcohol before addition to 1 kg of diet. However, because of Customs and Excise restrictions, farmers found ethyl alcohol expensive and difficult to obtain so an alternative carrier, iso-propyl alcohol (propan-2-ol), was identified. This was not toxic, was easily

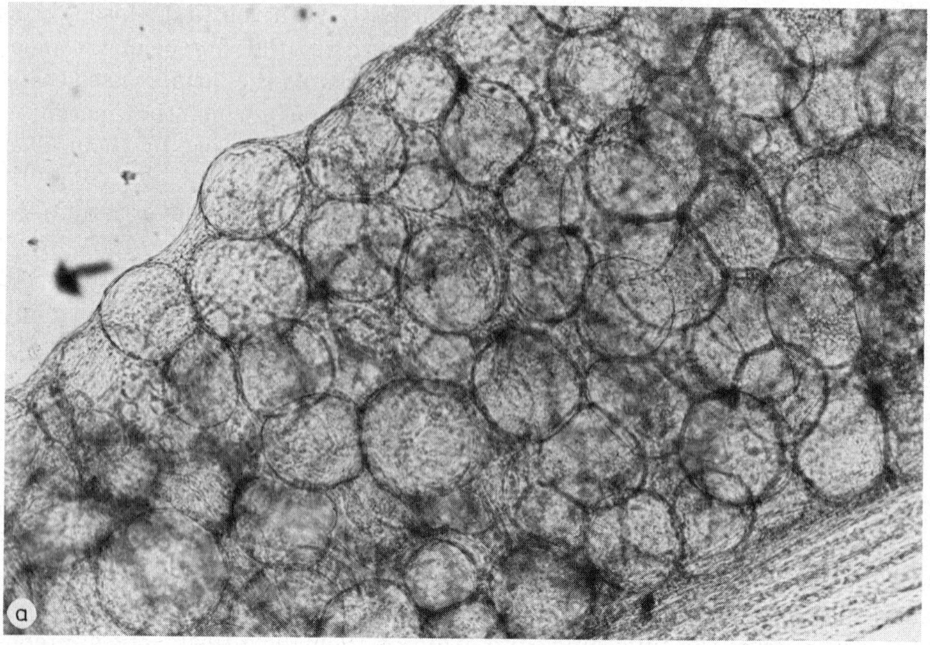
a

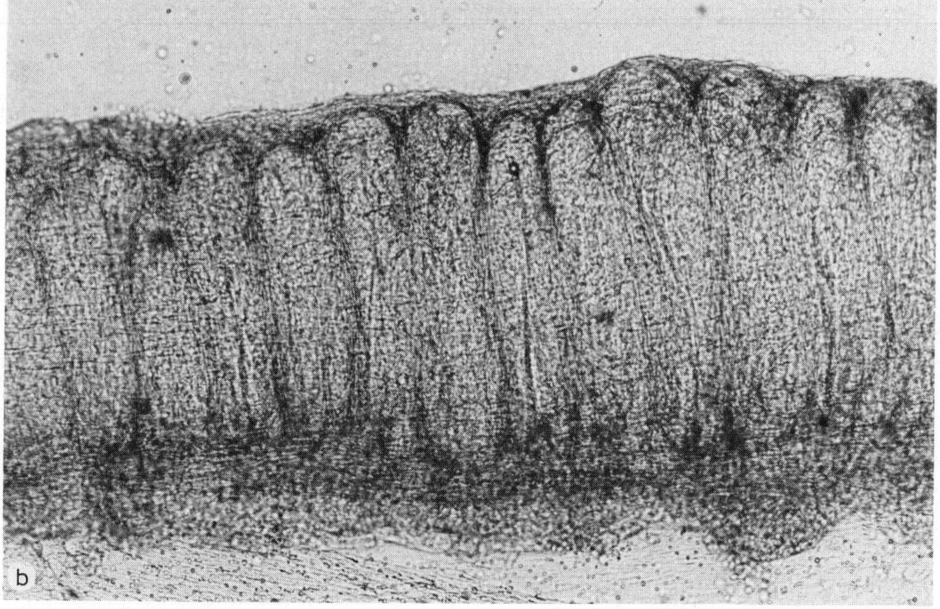
b

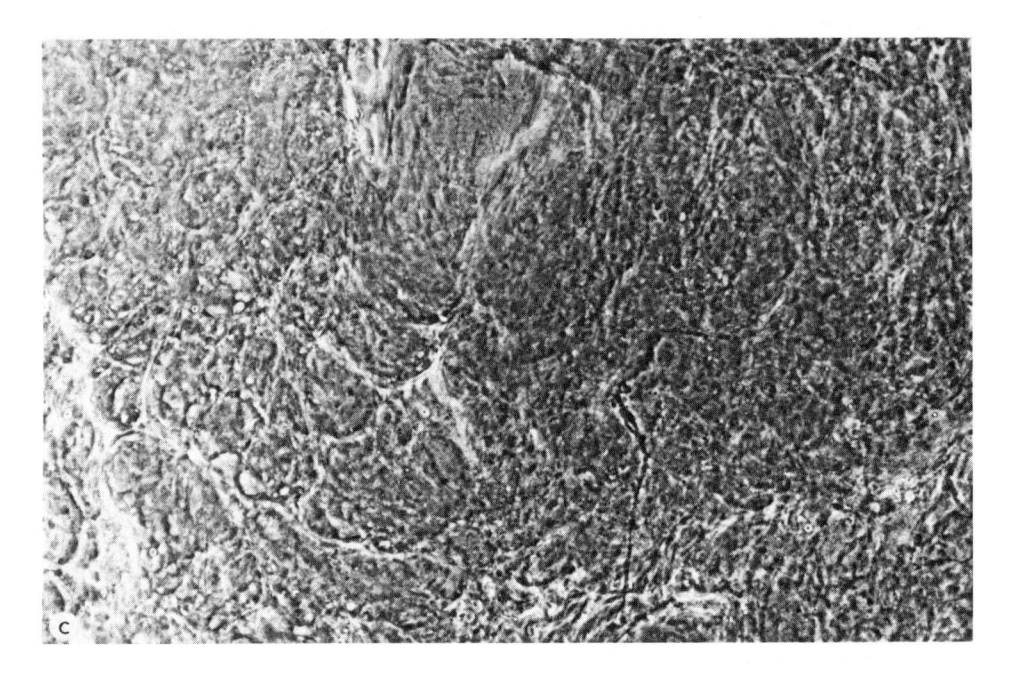

Fig. 2. Micrographs of wet-mount gonad squashes from 5-month-old rainbow trout. (a) Ovary from diploid showing oocytes. (b) Ovary from triploid showing typical lamellar structure and absence of oocytes. Ovaries from very immature diploids have the same appearance. (c) Testis from diploid. Triploid testes have the same appearance.

available from pharmacies and was exempt from Excise duty. Because methyltestosterone dissolved easily in iso-propyl alcohol, the initial benzyl alcohol stage could be eliminated. In order to ensure an even distribution throughout the diet, 3 mg of hormone was dissolved in 200 ml of alcohol and thoroughly mixed into 1 kg of fry food. This was spread in a thin layer and allowed to dry overnight in a warm room. As the alcohol evaporated the hormone remained dissolved in the lipid component of the diet. When the food was completely dry it was sealed into polythene bags and stored in a freezer or refrigerator. Sufficient food for 2 days feeding was removed at a time.

The absence of sperm ducts in the sex-inversed males created difficulties for hatchery staff at spawning time. Although this feature permitted the separation of homogametic and normal males when fry of both sexes had been masculinized, it prevented the farmer assessing whether individual males were ripe without first killing them. This was very wasteful, particularly during the early part of the season, since even in fish with well-developed secondary sexual characteristics the testes could be firm and contain little liquid semen. Attempts to use these testes resulted in poor rates of fertilization. Although farmers were shown how to check for activation of spermatozoa with a microscope, most found this difficult and preferred to trust to luck. Even with ripe ductless testes, the thick concentrated semen was difficult to extract and use so we formulated a diluent (S.M. Baynes, personal communication, 1980) which could extract the spermatozoa from the minced testes without activating them. This enabled us to make an artificial semen of similar consistency and spermatozoan concentration to that from normally strippable males which was usually used within a few hours but could be stored under oxygen at 4°C for 2–4 days.

Although the salt mixture for the diluent was supplied to farmers it was never extensively used and farmers quickly became adept at using the concentrated semen, many reporting that fertilization rates were higher than when the normally stripped semen was used. Nevertheless, the difficulty of identifying ripe testes without killing the fish remains a problem, although an observation made in 1982 has provided a potential solution. The original 1000°C days duration of masculinization delayed or prevented sexual development in a significant proportion of fish and so was shortened to 700°C days to reduce the number of steriles. At maturation it was apparent both in the laboratory and on commercial farms that although only female fry had been masculinized some of the mature sex-inversed fish were strippable. When tested, these ducted males achieved excellent fertilization rates and produced only female progeny. Current trials are delimiting the duration of treatment necessary to achieve a high degree of masculinization combined with the presence of ducts. Durations as short as 450°C days induce complete masculinization in fish assessed at 1 year, and treatments down to 150°C days are under investigation.

The Lowestoft female-only procedure has been readily accepted by the United Kingdom trout farming industry, stimulated to a considerable extent by the marked preference for female fish in the sporting sector. A full survey of the industry is planned but we estimate that at least half of the total United Kingdom trout production is achieve by the female-only method. As farmers continue to become aware of advantages of the absence of males, even on table farms, we expect production to become almost exclusively female.

Sterility

The potential benefits of sterility on the growth of farmed fish has long been realized, particularly for salmonids. A variety of methods for inducing sterility have been investigated in recent years (Hunter and Donaldson, 1983) but none has proved commercially practical or has been acceptable to the consumer. The only method that is suitable for commercial use and unlikely to arouse consumer resistance involves the production of triploid fish. We have been studying polyploidy in marine flatfish (Purdom, 1972; Lincoln, 1981a, b) but reliable techniques for inducing triploidy in salmonids are recent (Chourrout, 1980; Lincoln and Scott, 1983; Lou and Purdom, 1984). However, observations on spontaneous rainbow trout triploids found in hatchery stocks (Thorgaard and

Gall, 1979; Lincoln and Bye, 1980) indicated that although both sexes were genetically sterile only female triploids showed the complete suppression of gonadal development essential to avoid the deleterious characteristics associated with maturation. It was clear that the combination of our female-only method with induced triploidy provided an opportunity for producing entirely sterile stocks in an aesthetically acceptable way. The first batch of all-female triploid trout was produced in 1980 and subsequently confirmed as sterile (Lincoln and Scott, 1983, 1984). Although the parameters of the induction process required further investigation, the interest from the industry persuaded us to commence commercial trials concurrently with those on the female-only method.

Industrial trials

During the 1981/1982 hatchery season, small batches of female ova were heat-shocked at 28°C for 10 min commencing 40 min after fertilization. In these early trials, although triploidy was usually in excess of 90%, survival was low, ranging from 20 to 40% at swim up. Subsequent improvements in the method, in particular the heat-shocking of large volumes of eggs and the control of temperature, increased survival without significantly reducing the degree of triploidy. Currently, several farms are heat-shocking around 500×10^3 ova per year and one claims in excess of 4×10^6 for the 1984/1985 season. Our recent experiments indicate that the conditions we used originally for the induction of triploidy do not require great change, although there is still a need to scale-up the process to meet the requirements of industry. In order to encourage industrial acceptance of triploidy, the benefits of sterility have been described in popular articles (Lincoln and Bye, 1980, 1984a), at conferences and on farm visits.

The advantages of sterility have only been assessed on a small scale. Small batches of marked triploids have been released into lakes and reservoirs for sporting purposes and the response of anglers has been very favourable. The few farms which are growing triploids to produce large table-fish have found their growth rate, conversion efficiency and flesh quality entirely acceptable. Controlled, blind taste trials have indicated a significant preference for sterile triploid trout when tested against diploids of similar size. Under laboratory conditions trials in sea water have demonstrated that sterile triploids exhibit significantly better survival and growth over the spawning season when compared to mature diploids (Lincoln and Bye, 1984b). Commercial trials in sea cages commenced in 1984 in South Wales.

DISCUSSION

Although improvements are still possible, the methods we have developed make the production of entirely female and sterile trout stocks a practical proposition for the fish farming industry in the United Kingdom. Our future role will be to refine the methods, encourage a wider exploitation and deal with individual difficulties. A few farmers consider that for the production of 300 g trout, entirely male populations would be an advantage because of superior early growth rate. Although there is no convincing evidence for this supposition, we are investigating the possibility of producing all-male stocks by androgenesis or by the production of heterogametic (XY) females by estrogenization followed by the use of YY males.

The transfer of new technology to industry is as important and almost as difficult as the original research on which it is based. In our experience, a few aspects are of particular significance. Obviously the development must meet a definable need of economic consequence and the potential financial benefit must be stressed in publicity aimed at clearly identified customers. Trials at commercial sites make it possible to tailor new techniques to suit industrial conditions and identify problems not apparent in the laboratory. The possible competitive advantage enjoyed by farms cooperating in these trials will usually encourage others to follow suit until the majority of the industry is involved. This of course assumes that the new technology fulfils its promise. During the early stages, it is essential to provide adequate follow-up with advice and site visits to deal with the difficulties which inevitably will arise. New methods require considerable adjustments in attitude and practice for farmers who become discouraged relatively easily without constant support. It is important to ensure that novel techniques become firmly established in at least part of the industry within 3 years of the initial publicity, otherwise the impetus will be lost and the techniques gradually abandoned.

The decision of whether or not to patent methods depends primarily on their novelty but, at least for state-financed organizations, the effect this will have on their adoption by the industry is also important. We considered that patenting our methods and introducing charges for their use would inhibit adoption and ultimately be detrimental to the developing industry which we were trying to support. The fate of the auto-immune castration method patented by Aberdeen University in consort with the United Kingdom National Research Development Corporation and apparently untested since (Anon., 1980) justifies our decision.

REFERENCES

Anon., 1980. Sexless salmon technique in danger. Fish Farmer, 3 (5): 15.
Bye, V.J., 1982. Methods for the production of female and sterile trout. In: M.J. Bulleid (Editor), Commercial Trout Farming. Janssen Services, Chislehurst, Great Britain, pp. 75–85.
Bye, V.J. and Lincoln, R.F., 1979. Male trout, an expensive luxury. Fish Farmer, 2 (3): 19–20.
Bye, V.J. and Lincoln, R.F., 1981. Get rid of the males and let the females prosper. Fish Farmer, 4 (8): 22–24.

Bye, V.J. and Purdom, C.E., 1982. Selecting the fish of the future. Fish Farmer, 5 (5): 6-8.

Chourrout, D., 1980. Thermal induction of diploid gynogenesis and triploidy in the eggs of rainbow trout (*Salmo gairdneri* Richardson). Reprod. Nutr. Dev., 20: 727-733.

Fagerlund, U.H.M. and Dye, H.M., 1979. Depletion of radioactivity from yearling coho salmon (*Oncorhynchus kisutch*) after extended ingestion of anabolically effective doses of 17α methyltestosterone-1,2-3H. Aquaculture, 18: 303-315.

Hunter, G.A. and Donaldson, E.M., 1983. Hormonal sex control and its application to fish culture. In: W.S. Hoar, D.J. Randall and E.M. Donaldson (Editors), Fish Physiology, IXB. Academic Press, New York, NY, London, pp. 223-303.

Johnstone, R., Simpson, T.H. and Youngson, A.F., 1978. Sex reversal in salmonid culture. Aquaculture, 13: 115-134.

Lincoln, R.F., 1981a. Sexual maturation in triploid male plaice (*Pleuronectes platessa*) and plaice × flounder (*Platichthys flesus*) hybrids. J. Fish Biol., 19: 415-426.

Lincoln, R.F., 1981b. Sexual maturation in female triploid plaice (*Pleuronectes platessa*) and plaice × flounder (*Platichthys flesus*) hybrids. J. Fish Biol., 19: 499-507.

Lincoln, R.F. and Bye, V.J., 1980. Triploid rainbows: nature's mistakes may be the fish of the future. Fish Farmer, 3 (4): 46.

Lincoln, R.F. and Bye, V.J., 1984a. Triploid rainbows show commercial potential. Fish Farmer, 7 (5): 30-32.

Lincoln, R.F. and Bye, V.J., 1984b. The growth and survival of triploid rainbow trout (*Salmo gairdneri*) in sea water over the spawning season. International Council for the Exploration of the Sea, Council Meeting 1984/F:6, 6 pp. mimeo.

Lincoln, R.F. and Scott, A.P., 1983. Production of all-female triploid rainbow trout. Aquaculture, 30: 375-380.

Lincoln, R.F. and Scott, A.P., 1984. Sexual maturation in triploid rainbow trout, *Salmo gairdneri* Richardson. J. Fish Biol., 25: 385-392.

Lou, Y.D. and Purdom, C.E., 1984. Polyploidy induced by hydrostatic pressure in rainbow trout, *Salmo gairdneri* Richardson, J. Fish Biol., 25: 345-351.

Medicines Act 1968. Medicines (Applications for Product Licences and Clinical Trial and Animal Test Certificates) Regulations 1971. Great Britain, Parliament, HMSO, London.

Purdom, C.E., 1972. Induced polyploidy in plaice (*Pleuronectes platessa*) and its hybrid with the flounder (*Platichthys flesus*). Heredity, 29: 11-24.

Purdom, C.E., 1983. Genetic engineering by the manipulation of chromosomes. Aquaculture, 33: 287-300.

Thorgaard, G.H., 1977. Heteromorphic sex chromosomes in male rainbow trout. Science, N.Y., 196: 900-902.

Thorgaard, G.H., 1983. Chromosome set manipulation and sex control in fish. In: W.S. Hoar, D.J. Randall and E.M. Donaldson (Editors), Fish Physiology, IXB, Academic Press, New York, NY, London, pp. 405-434.

Thorgaard, G.H. and Gall, G.A.E., 1979. Adult triploids in a rainbow trout family. Genetics, 93: 961-973.

Broodstock Development for Monosex Production of Grass Carp

WILLIAM L. SHELTON

Zoology Department, University of Oklahoma, Norman, OK 73019 (U.S.A.)

ABSTRACT

Shelton, W.L., 1986. Broodstock development for monosex production of grass carp. *Aquaculture*, 57: 311–319.

Monosex female grass carp production has been accomplished with a breeding program that involves functionally sex-inverted males as broodstock. Between 1977 and 1982, 102 phenotypic males were produced by treating genetic females with methyltestosterone; 16 have sired progeny. Only females have been found among 1339 offspring sexed from 3 year classes.

INTRODUCTION

Broodstock management is a cornerstone of fish culture since the general condition of the stock at spawning and their genetic composition are primary quality determinants of the progeny. The development of specifically selected traits further elevates the importance of broodstock management. The present discussion concerns the development of broodstock grass carp (*Ctenopharyngodon idella*) that produce monosex progeny. This protocol is primarily applicable to other exotic fishes rather than as a developmental solution to a previous introduction. Grass carp importation should serve as an example of the problems than can result when introductions are permitted without adequate means of control. Protocol development for aquatic introductions is being considered on a global basis (Welcomme, 1986) as well as for the United States (Kohler and Stanley, 1984) and was the theme of a recent symposium (Kohler, 1986). The grass carp stocking controversy has stimulated developments that should be useable for future exotic fish introductions where evaluations can proceed without an irreversible commitment to potential naturalization (Shelton, 1986); the protocol discussed in this paper is one option among currently available ones.

MATERIALS AND METHODS

Induced spawning

Fishes that do not spawn under hatchery conditions must be hormonally induced to ovulate; their eggs need artificial incubation and their young must have appropriate care. Grass carp were induced to spawn using the techniques described by Bailey and Boyd (1971). Stripped eggs were inseminated, then incubated in flow-through jars (8 l) at about 30 000 eggs per jar. One-day-old fry were stocked in prepared nursing ponds according to techniques described by Rothbard (1982). Growth was managed through variable stocking densities (Shelton et al., 1981; Rosenblatt, 1982).

Gynogenesis and normal fertilization

Gynogenesis is essentially "self-fertilization" of eggs by retention of the second polar body, without genetic contribution of the male genome (Purdom, 1983). Retention of the second polar body is necessary to restore diploidy in eggs stimulated to develop by radiation-treated sperm of common carp (*Cyprinus carpio*). Induced gynogenesis for grass carp was reported by Stanley (1976a) and was the basis for demonstrating female homogamety (Stanley, 1976b). The yield of diploids can be enhanced by an appropriately timed thermal shock as described by Chourrout (1982). During the initial period of study, some eggs ($\sim 50\ 000$) from each ovulated female were fertilized with grass carp milt; the remainder were induced to develop through insemination with irradiated common carp milt. Some trials with various thermal shocks were used in attempts to increase the yield of diploids; however, our primary goal was to produce a pool of all-female fish for androgen treatment, not to optimize gynogenesis.

Sex inversion

Steroid-induced sex inversion directs the development of the functional phenotypic sex of fishes but does not alter their genotypes; an effective steroid hormone must be administered throughout the period of gonadal differentiation (GD) (Yamamoto, 1969). The most practicable means of administering an efficacious level of hormone to fish is through addition to the diet. However, oral treatment for grass carp has not been effective (Stanley and Thomas, 1978; Jensen, 1979). Consequently, an alternative mode of administration was developed in the present series of studies (Jensen et al., 1983; Boney et al., 1984). The androgenic hormone methyltestosterone (MT) was hand-packed (5 or 12 mg) into a segment of silastic (dimethylpolysiloxane) tubing, the ends were sealed and the carrier was implanted in the body cavity through a small abdominal incision. A liquid phase carrier was also tested for use with small fish. Treatment was initiated primarily using age and size as indicators of the period of GD (Jensen and Shelton, 1983). Following implantation, fish were stocked at variable rates in ponds to control the rate of growth and consequently the relative levels of circulating androgen. Diffusion of the MT from the implant is a function of implant dimensions and various environmental conditions, especially temperature (Boney, 1982). Effectiveness of treatment in relation to size at implantation and stocking rate was evaluated through histological examination of the gonads; however, some fish from selected treatments were not killed and these were used in breeding studies.

Progeny evaluation

Functionally sex-inverted individuals were used in progeny tests by inseminating eggs ovulated from non-steroid treated females. Some trials were conducted through artificial fertilization while in other trials, spawnings was in circular tanks as described by Rottmann and Shireman (1979). Progeny were nursed in ponds and their sex was determined by gonadal examination at one or more years of age at lengths greater than 200 mm (TL).

RESULTS

This breeding program involves a number of phases and critical procedures within each. Basically, the program is initiated with gynogenetically derived females that are then androgen-treated to produce functional phenotypic males (Fig. 1A); these sex-inverted males (genetic females) are bred with normal females to produce all-female progeny (Fig. 1B). Some of these progeny are androgen-treated to produce replacement male broodstock. Major events are discussed below, somewhat in the chronology of development, although considerable overlap and re-examination of various procedures occurred.

Gonadal development

Successful sex inversion through exogenous hormone treatment requires exposure of individuals during the period of GD. The most practicable treatment period for grass carp was determined through histological examination of the developmental chronology of male and female gonads. The GD in grass carp occurs in two distinct stages, anatomical and cytological differentiation. Anatomical differentiation is initiated between 50 and 60 mm (TL) and permits identification of the two gonad types. Cytological differentiation, during which primary germ cells develop, begins at about 120 mm and is completed between 180 and 200 mm. An appropriate level of androgen must be administered throughout this second phase of GD to induce functional sex inversion.

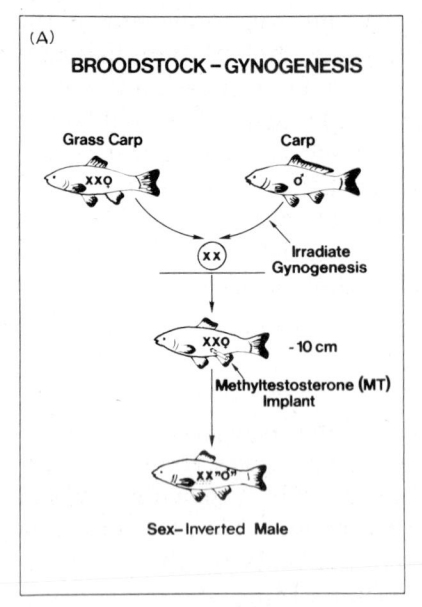

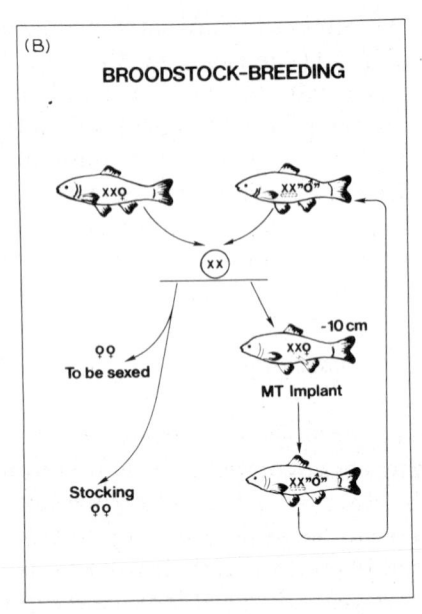

Fig. 1. A, Broodstock production by sex inversion of gynogenetic grass carp. B, Progeny testing of sex-inverted male grass carp and second-generation broodstock production.

Androgen-carrier development

Delivery of androgen to effect sex inversion is most efficiently accomplished by adding an androgen to the diet, as the desired level of input is easily maintained throughout the period of GD by periodic adjustments in feeding level appropriate to growth. Since sex inversion of grass carp by adding androgenic hormone to the diet had been unsuccessful, the concept of a slow-release androgen implant was explored. A prototype was developed that could be accommodated in the body cavity of gonadally undifferentiated grass carp. This implant, containing 12 mg of MT, was found to be suitable (Jensen, 1979). A smaller implant (5 mg) and a liquid phase carrier were tested so as to permit insertion into slightly smaller fish. The liquid carrier released MT too rapidly and was unsuitable; however, both implants were effective (Boney et al., 1984). The in vivo release rate from the 5-mg implant was studied under controlled temperatures (Boney, 1982). Fig. 2, derived from that study, characterizes the changing release pattern in relation to temperature. Following insertion, the rate of MT diffusion was high but diminished to a more stabile level within 30–40 days. The variable release rate complicates controlled delivery of physiologially effective levels during GD. Therefore, the rate of growth in relation to the release pattern from the implant is a critical consideration.

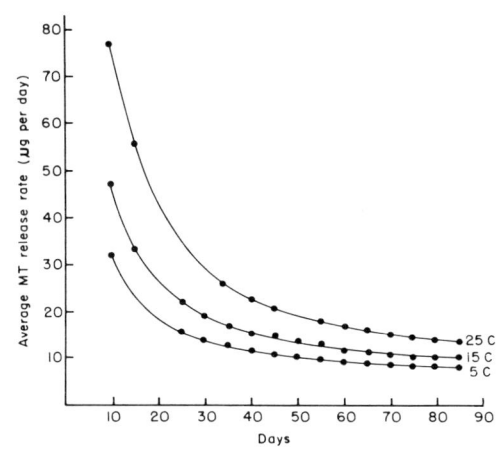

Fig. 2. Change in release rate of methyltestosterone from 5-mg implants in grass carp (Boney, 1982).

Growth management

Growth management actually begins prior to androgen treatment, since fish of an appropriate size are needed by mid-growing season. During the initial post-hatching period, fry are nursed at high stocking densities (1–2 million/ha), then during a secondary nursing phase, stocking rate can be reduced (30 000–50 000/ha) so as to obtain fish of appropriate size for androgen treatment (Shelton et al., 1982). The total nursing period is beteween 60 and 90 days.

In the preliminary study trials grass carp were stocked at low rates (< 100/ha) after androgen implantation; thus, growth was maximized during the initial high androgen-release period. However, growth management was subsequently considered as a more efficient means of controlling the relative androgen exposure. Based on an initial characterization of density-related growth for grass carp by Shelton et al. (1981) and these preliminary data, the scope of stocking was narrowed to a range appropriate for androgen treatment (Rosenblatt, 1982; Fig. 3). Stocking rate during androgen treatment is the primary means of controlling exposure through the period of GD within the first growing season.

Androgen treatment

With the foregoing considerations, the most appropriate treatment regime can be characterized. A complete discussion of various treatments is included in Jensen et al. (1983) and Boney et al. (1984), but most of these data are based on histological evaluations; few fish were retained alive. Table 1 summarizes the most successful treatments in these studies and includes data on fish

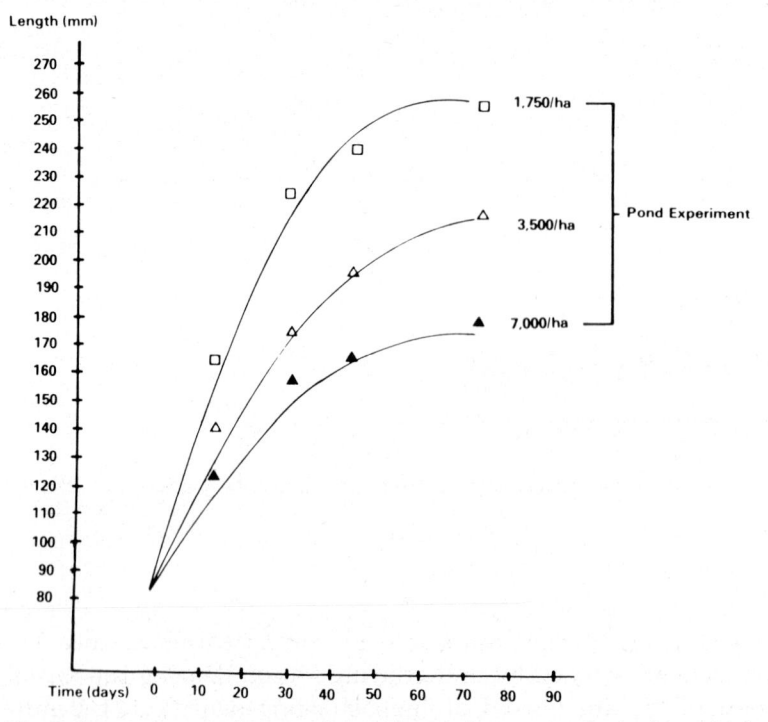

Fig. 3. Composite growth curves for pond-reared grass carp beginning at 52 days of age (modified from Rosenblatt, 1982).

TABLE 1

Androgen treatment of genetic female grass carp, 1978-1982

Female source		Treatment (no.)				
Year	Mode[a]	Length (mm)	Age (days)	Implant[b] (no.)	Survival[c] (No.)	Sex-inverted[d] (No.)
1977	G	120–135	55	51	39	6/8
1978	G	100–165	75	11	7	2/3
1979	G	110–130	55	22	17	10/15
1980	B	65–75	55	81	61	12/61[e]
1981	B	85–110	50	60	29	27/29
1982	B	95–105	75	107	86	45/54

[a]Gynogenetic (G); breeding (B).
[b]Twelve-mg implant used in 1977–1978; 5-mg implant used in 1979–1982.
[c]Survived treatment period.
[d]Number sex-inverted of those evaluated at maturity.
[e]Includes about one-half that were treated with the liquid carrier.

TABLE 2

Progeny tests of grass carp broodstock, 1980-1982

Broodstock		Progeny sexed (no.)	
Test	No.	Female	Male
Monosex breeding			
Females	9		
Sex-inverted males	18[a]	1,339[b]	0
Normal fertilization			
Females	16[c]		
Normal males	35	421	403
Gynogenesis			
Females	16[c]	158[d]	0

[a]Sixteen of 102 sex-inverted males with progeny.
[b]Includes progeny of eight sex-inverted males tank spawned.
[c]Same females used in gynogenesis and normal fertilization.
[d]Stanley and Thomas (1978) sexed 350 gynogenetic grass carp; all females.

retained as potential broodstock. Fish specifically designated for broodstock development were initially obtained through gynogenesis. In the course of our program, approximately 1200 diploid gynogenetic offspring were produced. Survival was sufficient to permit the development of initial generations of sex-inverted broodstock. Survival during androgen treatment was usually reasonable ($>50\%$) and functional sex inversion was greater than 60%. Six treatment years have been evaluated and 102 sex-inverted broodstock males have been produced. Treatment appears to have been most effective when initiated at a fish size of 85–110 mm (TL) and an age of 50–75 days; stocking rate during androgen treatment should be between 2000 and 7000/ha. This regime considers a spawning time of mid-May, allows a nursing period of 2–3 months to bring fish to the suggested treatment size and a treatment period of about 60 days so as to attain a size of 180–200 mm by the end of the growing season.

Breeding

Progeny testing of sex-inverted males is the ultimate feasibility test for the overall management program. Progeny were obtained from sex-inverted males during spawning in 3 years (1980–1982). Not all of the 102 sex-inverted males have been progeny tested, but 16 produced offspring that have been evaluated (Table 2). Milt from these males inseminated eggs from nine females producing over 500 000 swim-up fry, of which 92 000 survived to fingerling size (>10 cm). Progeny from each year class were grown (>20 cm) and sexed by gonadal examination. A total of 1339 progeny have been sexed (640 from 1980; 526 from 1981; 173 from 1982), all were females.

Normal sex ratio data were obtained from samples of progeny of 16 females and 35 males; the sex ratio for 824 grass carp was 1:1 (Table 2). To provide more data on the genetic mechanism of sex determination for grass carp, 158 gynogenetic offspring were sexed; all were females. In addition, all 350 gynogenetic grass carp sexed by Stanley and Thomas (1978) were females.

The trauma associated with stripping sex-inverted males during induced spawning should be an important consideration in the practical application of this type of breeding program. Tank spawning would reduce the stress on broodstock if sex-inverted males would court females and fertilize the eggs. In 1982, eight sex-inverted males tank spawned with three females siring 26 000 viable offspring.

CONCLUSIONS

Reproductive limitation should be a component of the evaluation of an exotic fish; thus, thorough and safe testing can be conducted within the country so as to permit a rational decision as to the limits of its utilization (Shelton, in press). Monosex production, as described for the grass carp is one alternative that may be considered in relation to other developments such as triploid production; either will offer security from unwanted naturalization during preliminary evaluation.

ACKNOWLEDGMENTS

These studies were funded by a series of grants from the Office of Water Research and Technology, U.S. Department of the Interior through the Water Resources Research Institute of Auburn University, Alabama. The dedication of three graduate students, S. Boney, G. Jensen and E. Rosenblatt is sincerely appreciated.

REFERENCES

Bailey, W.M. and Boyd, R.L., 1971. A preliminary report on spawning and rearing grass carp (*Ctenopharyngodon idella*) in Arkansas. Proc. Southeast. Assoc. Game Fish Comm., 24: 560–569.
Boney, S.E., 1982. Monosexing Grass Carp by Sex Reversal and Breeding. Doctoral dissertation, Auburn University, AL, 77 pp.
Boney, S.E., Shelton, W.L., Yang, S.L. and Wilken, L.O., 1984. Sex reversal and breeding of grass carp. Trans. Am. Fish. Soc., 113: 348–353.
Chourrout, D., 1982. Gynogenesis caused by ultraviolet irradiation of salmonid sperm. J. Exp. Zool., 223: 175–181.
Jensen, G.L., 1979. Administration of Methyltestosterone to Direct Gonadal Sex Differentiation of Female Grass Carp. Doctoral dissertation, Auburn University, AL, 202 pp.

Jensen, G.L. and Shelton, W.L., 1983. Gonadal differentiation in relation to sex control of grass carp, *Ctenopharyngodon idella* (Pisces: Cyprinidae). Copeia, 1983: 749–755.

Jensen, G.L., Shelton, W.L., Yang, S.L. and Wilken, L.O., 1983. Sex reversal of gynogenetic grass carp by implantation of methyltestosterone. Trans. Am. Fish. Soc., 112: 79–85.

Kohler, C.C., 1986. Strategies for reducing risks from the introduction of aquatic organisms. Fisheries, 11 (2): 2–3.

Kohler, C.C. and Stanley, J.G., 1984. A suggested protocol for evaluating proposed exotic fish introductions in the United States. In: W.R. Courtenay and J.R. Stauffner (Editors), Distribution, Biology and Management of Exotic Fishes. The John Hopkins University Press, Baltimore, pp. 387–406.

Purdom, C.E., 1983. Genetic engineering by manipulation of chromosomes. In: N.P. Wilkins and E.M. Gosling (Editors), Genetics in Aquaculture. Aquaculture, 33: 287–300.

Rosenblatt, E.M., 1982. Effect of Stocking Density on Growth and Sex Reversal of Methyltestosterone-treated Female Grass Carp. M.S. thesis, Auburn University, AL, 44 pp.

Rothbard, S., 1982. Induced reproduction in cultivated cyprinids – the common carp and the group of Chinese carps. II. The rearing of larvae and the primary nursing of fry. Bamidgeh, 34: 20–32.

Rottmann, R.W. and Shireman, J.V., 1979. Tank spawning of grass carp. Aquaculture, 17: 257–267.

Shelton, W.L., 1986. An aquaculture perspective on strategies for reducing risks from introduced aquatic organisms. Fisheries, 11 (2): 16–19.

Shelton, W.L. (in press). Reproductive control of exotic fishes – a primary requisite for utilization in management. In: R.H. Stroud (Editor), Proc. Symp. Fish Culture in Fisheries Management, Lake Ozark, Missouri, 31 March–3 April 1985. Am. Fish. Soc. (Publ.), pp. 427–434.

Shelton, W.L., Smitherman, R.O. and Jensen, G.L., 1981. Density related growth of grass carp, *Ctenopharyngodon idella* (Val.), in managed small impoundments in Alabama. J. Fish Biol., 18: 45–51.

Shelton, W.L., Boney, S.E. and Rosenblatt, E.M., 1982. Monosexing grass carp by sex reversal and breeding. In: T.O. Robson (Compiler), Second International Symposium on Herbivorous Fish, Proceedings. European Weed Research Society, Wageningen, The Netherlands, pp. 184–194.

Stanley, J.G., 1976a. Production of hybrid, androgenetic, and gynogenetic grass carp and carp. Trans. Am. Fish. Soc., 105: 10–16.

Stanley, J.G., 1976b. Female homogamety in grass carp (*Ctenopharyngodon idella*) determined by gynogenesis. J. Fish. Res. Board Can., 33: 1372–1374.

Stanley, J.G. and Thomas, A.E., 1978. Absence of sex reversal in unisex grass carp fed methyltestosterone. In: R.O. Smitherman, W.L. Shelton and J.H. Grover (Editors), Culture of Exotic Fishes Symposium, Proceedings. Fish Culture Section, Am. Fish. Soc., Auburn University, AL, pp. 194–199.

Welcomme, R.L., 1986. International measures for the control of introductions of aquatic organisms. Fisheries, 11 (2): 4–9.

Yamamoto, T., 1969. Sex differentiation. In: W.S. Hoar and D.J. Randall (Editors), Fish Physiology, Vol. 3. Academic Press, New York, NY, pp. 117–175.

Color, Growth and Maturation in Ploidy-Manipulated Fancy Carp

N. TANIGUCHI, A. KIJIMA, T. TAMURA, K. TAKEGAMI and I. YAMASAKI

Department of Cultural Fisheries, Faculty of Agriculture, Kochi University, Nankoku, Kochi 783 (Japan)

ABSTRACT

Taniguchi, N., Kijima, A., Tamura, T., Takegami, K. and Yamasaki, I., 1986. Color, growth and maturation in ploidy-manipulated fancy carp. *Aquaculture*, 57: 321–328.

Ploidy manipulation was used to examine the effects of triploidizatin and gynogenetic diploidization in fancy carp, a race of *Cyprinus carpio*. Ultraviolet dosages of at least 9000 erg/mm^2 were required to inactivate sperm genetically. An interval of 10–12 min before the start of 60-min cold treatments was required to retain the second polar body in fertilized eggs. At optimum conditions, survival rates of gynogenetic diploids were 80% to the eyed embryo stage and 25% to hatching; corresponding values in triploids were 73% and 32%, respectively. The highest success rates in inducing gynogenetic diploids and triploids were 100% and 68.9%, respectively. Triploids with the Japanese strain as the female and European strain as the male parent grew slower than control diploids when they were reared in the same pond for 20 months. However, triploids using the Japanese strain as both parents grew larger in 7 months than control diploids. Triploids at 20 months of age had less developed gonads and more fat around the digestive tract than diploid fish. Male-specific color spots were smaller in triploids than in control diploids. The typical red and white fancy carp colors were found in the gynogenetic diploids in almost the same frequency as in the control diploids. Recombination between gene and centromere was observed at the *Gpi-2* locus in gynogenetic diploids.

INTRODUCTION

In anticipation of the immediate applicability of chromosome technology in the field of fish farming, ploidy manipulation has been tried in various fish species (Purdom, 1983). The carp, being very important in freshwater fish farming around the world, has been the target for ploidy manipulation for many years. The conditions necessary for induction of triploidy and gynogenetic diploidy in carp have already been investigated (Makino and Ojima, 1943; Cherfas, 1975; Stanley, 1976; Nagy et al., 1978; Ojima and Makino, 1978; Gervai et al., 1980; Ueno, 1984). However, few data on the effects of diploidy manipulation on various characters of carp are available.

Fancy carp are very popular as pets in Japan because of their beautiful combinations of red, white, black, gold and platinum body colors. Fancy carp were developed from common carp (*Cyprinus carpio*) in the northeast part of Japan, Niigata Prefecture, about 100 years ago. Several color varieties now exist, such as red and white ("Kouhaku" in Japanese), tricolor ("Taisho sanke"), showa-tricolor ("Showa sanke"), gold ("Oogon"), and others. These color varieties are not fixed genetically, and the genetic models for red, white, black and brilliant coloration are not yet resolved. These colors, however, are useful genetic markers for ploidy manipulation studies. In this study, ploidy manipulation was performed to examine the optimal conditions to induce high percentages of triploids and gynogenetic diploids with high survival rates, and to determine the effects of the induced conditions on growth, maturation, and coloration.

MATERIALS AND METHODS

Examination of optimal condition to induce gynogenetic diploid and triploid

Female breeders used were "Kouhaku" (red and white), while male breeders were "Karasu" (black) or "Taisho sanke" (red, white and black) which have melanophore marker for genetic activity of sperm. Eggs and sperm were collected by stripping. Fertilization was performed by mixing eggs with semen diluted by physiological saline solution (7.98 g NaCl, 0.02 g $NaHCO_3$ in 1 l of distilled water).

To inactivate sperm genetically, sperm was diluted 100 times using physiological saline solution and exposed to 75 erg mm^{-2} s^{-1} ultraviolet ray on the surface of the diluted sperm solution in a 1-mm layer, following the method of Onozato and Yamaha (1983). UV doses examined ranged from zero to 13 500 erg/mm^2 (0–180 s), and the effect was evaluated based on the appearance of the haploid syndrome in the embryos and the absence of melanophores, a marker for the genetic material of male breeders. The development rate was recorded at 12 h post-fertilization. The survival rate was observed in eyed embryo stage (4 days post-fertilization).

The best time for initiation of cold treatment (0–0.5°C for 60 min) was determined by laying fertilized eggs on a nylon fiber mesh placed in 20°C water and allowing an elapsed time of 0–15 min before the start of the cold shock. Eggs collected from breeders were divided into two lots, one for fertilization by UV-treated sperm to induce gynogenetic diploidy, and the other for fertilization by normal sperm to induce triploidy.

Evaluation of the effects of ploidy manipulation

Female parents used were "Kouhaku" in both 1983 and 1984 and the male parents were "Matsuba-oogon" (brilliant with black spots) in 1983 and "Taisho sanke" in 1984. The success rate for induction of triploidy was determined by analysis of blood cell size (Ueno, 1984) and the rate for gynogenetic diploidy by checking the appearance of the color marker (brilliant) of male breeders in 3-month-old juveniles.

Growth rates of triploid and diploid fish were compared by mixing the two groups 2 months after hatching and rearing in the same pond for 16 months in 1983 and 7 months in 1984. Fish were first fed with Moina and then with small pellets. Fork lengths and body weight were measured at 9 and 14 months of age in 1983 and 4 and 7 months of age in 1984. Gonads of 20-month-old fish were observed and weighed to determine degree of sexual development.

Coloration (red, white, black and brilliant on the body) of each individual was recorded. Brilliant means brightness on the body surface of golden or silvery colored fish. Area of red and black spots on the surface of the body was determined by measuring free-hand drawings of the spots with an image analyzer (Shimazu Co Ltd.). Starch-gel electrophoretic methods used for detection of biochemical markers were the same as those of Taniguchi and Okada (1980).

RESULTS AND DISCUSSION

Determination of conditions for inducing triploidy and gynogenetic diploidy

Fig. 1 shows the survival rates of embryos and the frequency of haploid embryos or melanophore at 1 day before hatching, after the eggs were fertilized by sperm irradiated under varying UV dosages. The V-shaped pattern of the observed survival rate is well known as the Hertwig effect. The haploid syndrome characterized by abnormal retina and short tail (Onozato and Yamaha, 1983) was considered diagnostic for haploid embryos. The frequency of haploid embryos increased with higher doses of UV, from 0% in the control to 100% at 6420 erg/mm^2. Based on the occurrence of haploid embryos and survival rate of embryos, a dosage of 9630 erg/mm^2 (90 s) was considered necessary to inactivate sperm genetically.

Fig. 2. shows the survival rate of embryos at 1 day before hatching, when eggs, inseminated by irradiated sperm or normal sperm, were cold-shocked (0–0.5°C) for 60 min at selected times (0–15 min) post-fertilization. Survival of embryos was relatively high in groups shocked at 2.5, 10 and 12.5 min post-fertilization and very low in all other groups. At optimal conditions for induction, the survival rates were 80% to the eyed embryo stage and 25% to the newly hatched larval stage, for gynogenetic diploids, and 73% and 32%, respectively, for triploids. The incidence of haploids was very high in the control (unshocked) eggs fertilized with irradiated sperm, but drastically decreased in shocked groups. Melanophores did not appear in the embryos except those shocked at 0 min and the intact control groups. These fluctuating patterns of

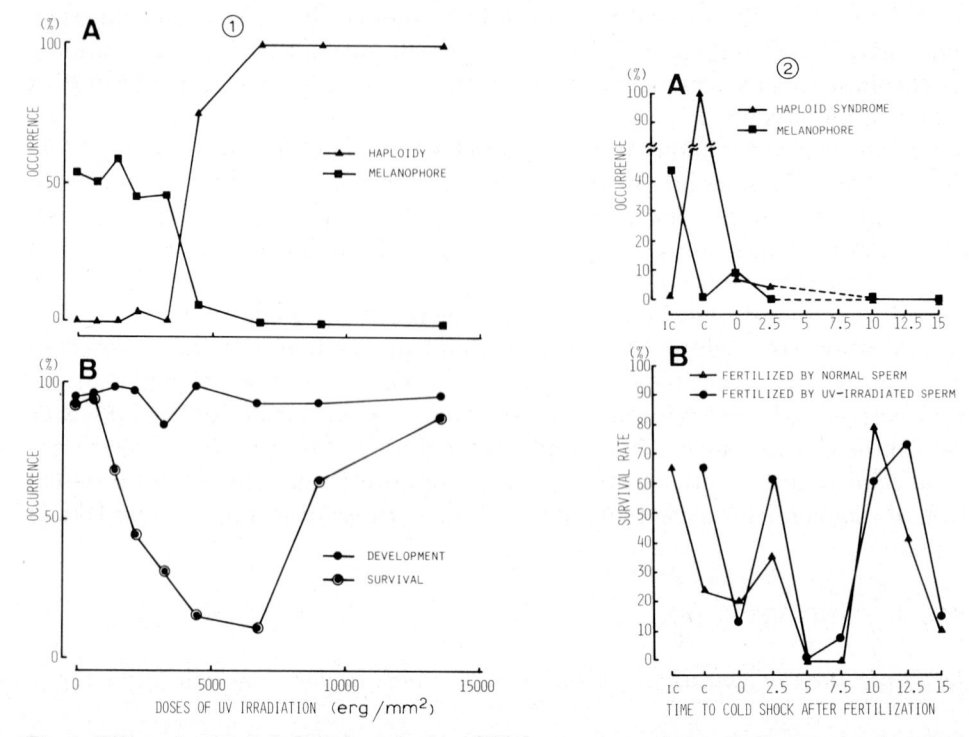

Fig. 1. Effect of sperm irradiation at various UV dosages on embryos. A, frequency of haploid syndrome and melanophore marker. B, Development rate (12 h after fertilization) and survival rate (eyed embryo stage).

Fig. 2. Effect of various time delays post-fertilization before cold shock. A, Incidence of haploidy and melanophore markers in embryos fertilized by UV-irradiated sperm. B, Survival rate of embryos fertilized by normal sperm and by UV-irradiated sperm. IC means intact control for embryos fertilized by UV-irradiated sperm.

survival rates and haploid incidence suggest that second polar body expulsion in the fertilized eggs is arrested by cold treatment at 2.5, 10 and 12.5 min post-fertilization.

The time to start cold shock after fertilization and the duration of cold shock seemed to be the most important factors in retaining the second polar body. Our results were very similar to those of Nagy et al., (1978).

Estimation of induction efficiency

The erythrocytes of triploids were clearly larger than those of diploids. Brilliant, a color marker for sperm, appeared in all control offspring, but not in the gynogenetic diploids. Based on erythrocyte size and absence of the color marker, the success rates for inducing triploids and gynogenetic diploids were 83.3%

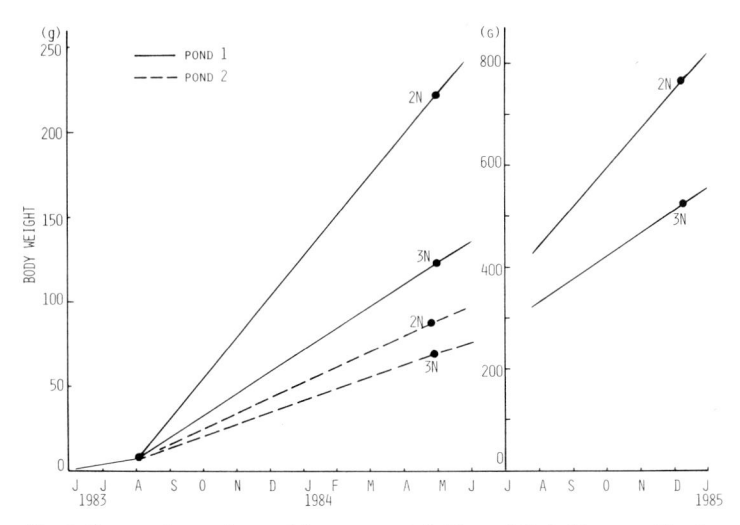

Fig. 3. Comparison of growth between triploids and diploids reared in the same pond for 16 months in the 1983 experiment. The fish of ponds 1 and 2 were placed in a bigger pond in the latter part of the experiment. The 3N and 2N were identified by the size of red blood cells collected by tuberculin syringe without killing the fish.

and 48.6%, respectively, in 1983, and 68.6% and 100% in 1984. Chromosome counts of fish presumed to be triploids based on erythrocyte size were about $2N = 150$, although the precise number could not be determined. Counts in fish presumed to be gynogenetic diploids were about $2N = 100$.

Growth of triploids

The average fork length of diploids was significantly larger than that of triploids in both rearing ponds at 9 and 16 months of age in the 1983 experiment (Fig. 3). In the 1984 experiment, however, triploids were larger than control diploids at 7 months of age (Fig. 4). The difference in the results of 1983 and 1984 may be caused by genetic differences in the parents. In 1983, native "Kouhaku" females and European "Matsuba oogon" males were used and the European fish is known to have a faster growth rate than the native one. Two native strains, "Kouhaku" and "Taisho-sanke", were used as parents in 1984. The effect of growth of European fancy carp may be superior to the effects of triploidy.

Purdom (1983) stated that triploids may grow faster than diploids by avoiding the process of gametogenesis. Gervai et al. (1980) showed that triploid carp do not grow faster than diploids during the juvenile stage. The results of the present study in 1984 showed that the triploids may be superior to the diploids in growth even during the juvenile stage.

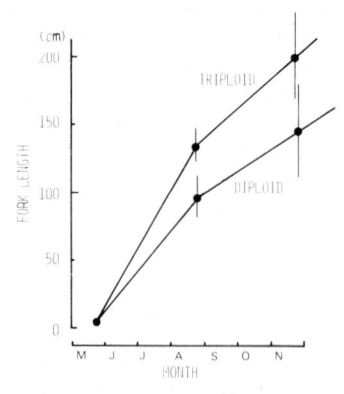

Fig. 4. Comparison of growth between triploids and diploids reared in the same pond for 7 months in the 1984 experiment.

Gonadal development

Although the spawning season was approaching, the triploids had less developed gonads and much more fat around the digestive tract than the diploids. Table 1 shows average gonad weight and gonadosomatic indices for triploids and diploids at 20 months of age. Both values for triploids were distinctly smaller than those for diploids.

Body color

Table 2 shows the incidence of body color in parents and 6-month-old offspring. Triploids had a lower incidence of red spots and a higher incidence of white area than diploids. Black spots and brilliant, markers of genetic material of the male parent, did not appear in gynogenetic diploids, which demonstrates evidence of 100% success in inducing gynogenesis.

There was no difference between gynogenetic and control diploids in the frequency distribution of red color area (Table 3). This suggests that the genes for red (R) and white (W) have no allelic relationship.

TABLE 1

Maturation of 20-month-old triploid of fancy carp in the 1983 experiment

Ploidy	Sex	No. of fish	Fork length (cm)	Body wt. (g)	Gonad wt. (g)	Gonadosomatic index
3N	—	3	33.0±1.1	711.0±29.4	1.29± 0.80	0.18±0.11
2N	Male	3	30.6±1.3	628.6±77.8	20.87±20.42	3.11±2.72
2N	Female	3	34.6±0.8	825.5±66.1	10.76± 6.80	1.35±0.89

TABLE 2

Development of body color in triploid and gynogenetic diploid progeny at 6 months after hatching; numbers represent percentage of fish with respective color

Parents			Progeny				
Sex	Name	Color	Ploidy	Black	Red	White	Brilliant
Female	Kohaku	White, red	2N	58.8	57.6	86.7	0
Male	Sanke	White, red, black	3N	60.0	20.0	100	0
Female	Kohaku	White, red	2N	55.8	88.3	42.9	100
Male	Oogon	White, red, black, brilliant	G-2N[a]	0	54.5	100	0

[a]Gynogenetic diploids.

The area of black spots was lower in triploids compared with the diploids (Table 3). This phenomenon may be due to the male parent's allele (B) being suppressed as a result of the additive effect of the female parent's allele (b) which negatively affects black color development.

Genotype frequency of biochemical markers for gynogenetic diploid

Table 4 shows the genotypes of glucose phosphate isomerase (Gpi-2) and sarcoplasmic protein (Sp-2) for the parents and gynogenetic diploid progenies. The A allele of Sp-2, a male genetic marker, did not appear in gynogenetic diploids. The BB genotype of Gpi-2, which was not found in diploids, was frequently observed in gynogenetic diploids as a result of retention of the second polar body in the fertilized eggs.

The AB genotype of Gpi-2 was also observed in gynogenetic diploids, which may be a result of recombination between gene and centromere. Although the rate of recombination between locus and centromere has already been shown for a few loci of carp (Nagy et al., 1978, 1979), more recombination data are

TABLE 3

Area of body surface covered by red and black color spots for gynogenetic diploids and triploids with diploid using female "Kohaku" (white and red colored) and male "Taisho-sanke" (white, red and black colored)

Color	Ploidy	Area of body surface with color spots (%)										
		0	0.1–10	10.1–20	20.1–30	30.1–40	40.1–50	50.1–60	60.1–70	70.1–80	80.1–90	90.1–100
Red	G-2N	45.5	27.3	0.0	4.5	9.1	0.0	0.0	4.5	0.0	4.5	4.5
	2N[a]	45.7	14.3	11.4	2.9	0.0	2.9	2.9	2.9	8.6	5.7	2.9
Black	3N	40.0	60.0	0	0	0	0	0	0	0	0	0
	2N	41.2	9.4	7.0	15.3	12.9	8.2	4.7	1.2	0	0	0

[a]Offspring with black spots in the control diploid were eliminated.

TABLE 4

Observed frequency (%) of genotypes of biochemical markers in gynogenetic diploids and control diploid, and the genotypes expected for their parents (number of fish examined in parentheses)

Loci	Parents		Progeny			
	Female	Male	Ploidy	AA	AB	BB
Gpi-2	AB	AA	2N (20)	55	45	0
			G-2N (10)	60	10	30
Sp-2	BB	AB	2N (20)	0	60	40
			G-2N (10)	0	0	100

needed to ascertain the efficiency of gynogenetic diploidization making inbred lines.

REFERENCES

Cherfas, N.B., 1975. Investigations on radiation-induced gynogenesis in carp. I. Experiments on the mass production of diploid gynogenetic offspring. Genetika, 11: 78–86.

Gervai, J., Peter, S., Nagy, A., Horvath, L. and Csanyi, V., 1980. Induced triploidy in carp, *Cyprinus carpio* L. J. Fish Biol., 17: 667–671.

Makino, S. and Ojima, Y., 1943. Formation of the diploid egg nucleus due to suppression of the second maturation division, induced by refrigeration of fertilized eggs of the carp, *Cyprinus carpio*. Cytologia, 13: 55–60.

Nagy, A., Rajki, K., Horvath, L. and Csanyi, V., 1978. Investigation on carp, *Cyprinus carpio* L., gynogenesis. J. Fish Biol., 13: 215–224.

Nagy, A., Rakji, K., Bakos, J. and Csanyi, V., 1979. Genetic analysis in carp (*Cyprinus carpio*). Heredity, 43: 35–40.

Ojima, Y. and Makino, S., 1978. Triploidy induced by cold shock in fertilized eggs of the carp. Proc. Jpn. Acad., 54: 359–362.

Onozato, H. and Yamaha, E., 1983. Induction of gynogenesis with ultraviolet ray in four species of salmoniformes. Bull. Jpn. Soc. Sci. Fish., 49: 693–699.

Purdom, C.E., 1983. Genetic engineering by the manipulation of chromosomes. Aquaculture, 33: 287–300.

Stanley, J.G., 1976. Production of hybrid, androgenetic, and gynogenetic grass carp and carp. Trans. Am. Fish. Soc., 1: 10–16.

Taniguchi, N. and Okada, Y., 1980. Genetic study on the biochemical polymorphism in red sea bream. Bull. Jpn. Soc. Sci. Fish., 46: 437–443.

Ueno, K., 1984. Induction of triploid carp and their haematological characteristics. Jpn. J. Genet., 59: 585–591.

Comparative Growth and Development of Diploid and Triploid Coho Salmon, *Oncorhynchus kisutch*

ORLAY W. JOHNSON[1,2], WALTON W. DICKHOFF[1,2] and FRED M. UTTER[1]

[1]*Northwest and Alaska Fisheries Center, National Marine Fisheries Service, National Oceanic and Atmospheric Administration, 2725 Montlake Boulevard East, Seattle, WA 98112 (U.S.A.)*
[2]*School of Fisheries, University of Washington, Seattle, WA 98195 (U.S.A.)*

ABSTRACT

Johnson, O.W., Dickhoff, W.W. and Utter, F.M., 1986. Comparative growth and development of diploid and triploid coho salmon, *Oncorhynchus kisutch*. *Aquaculture*, 57: 329–336.

This paper compares growth and gonadal development of triploid and diploid coho salmon in three treatment groups. The comparisons were made in fresh and seawater from the time of smoltification at 18 months to the onset of sexual maturity at 30 months. No differences were detected in the ability of triploids to smoltify and to successfully adapt to seawater under normal conditions. Likewise, no differences were observed in growth parameters (length, body weight or condition factor).

Gonadal development in both sexes was severely retarded in all triploid groups at 30 months of age. The average gonadosomatic index (GSI) of triploid males was 35.7% of diploids, but overall male gonadal development was not far enough advanced to determine if spermiation was altered in the triploids. Triploid females showed an almost complete blockage of gonadal development and oocyte maturation. The average GSI of triploid females was 11.8% that of diploid females. Vitellogenin was undetectable in triploid females, but was present in diploid females; a correlated reduction in the hepato-somatic index was observed in triploid relative to diploid females. However, we did not detect any significant differences between levels of estradiol in the plasma of triploids and diploids at this time.

INTRODUCTION

Triploidy can be readily induced in many teleost groups resulting in blocked or retarded gonadal development and sterility (Thorgaard, 1983). This capability has considerable value in fish culture where blockage of gonadal development diverts energies towards greater somatic growth and precludes undesirable reproduction.

The potential benefits of induced triploidy to a fish cultural program may, however, be offset by unacceptable side-effects such as increased levels of mortality (Utter et al., 1983), modified migratory behavior (Liley, 1969; Youngson and Simpson, 1984), and infertile matings caused when sterile individuals exhibit reproductive behavior. Performance between triploids and diploids appears to vary among closely related species (B. Chevassus, personal communication, 1985; E. Quillet, personal communication, 1985; T.J. Benfey, personal communication, 1985, unpublished abstracts, 2nd Symp. Genetics in Aquaculture). Thus it is necessary to accumulate adequate comparative data before induction of triploidy is routinely implemented in a particular species.

The coho salmon is a major fish cultural species throughout North America. Comparative performance of diploid and triploid coho salmon in fresh water through the age of 18 months (Utter et al., 1983) indicated reduced overall growth of triploids, and preferential survival of diploids from time of hatching in one of two sibling lots.

The purpose of this paper is to extend these initial comparative observations through the onset of sexual maturation of diploid individuals. We present comparative observations from fresh- and saltwater reared groups on growth, survival, gonadal development and blood chemistry. These data are compared to those of similar studies in other salmonid species (Benfey and Sutterlin, 1984; Lincoln and Scott, 1984; Solar et al., 1984), and their relevance to the use of induced triploidy as a tool in salmonid culture is discussed.

MATERIALS AND METHODS

Triploid induction and rearing

In November 1982, gametes of University of Washington coho salmon were treated by elevated temperature as described in Utter et al. (1983) to produce triploid treatment lots. Fry were reared at NMFS Northwest and Alaska Fisheries Center, Seattle, WA in fiberglass hatchery troughs. At 5 months fingerling fish were transferred to pools (3 m diameter, 20 cm depth) with a density of 5000 fish per pool. Two experimental ploidy groups were created at this time. Group I (50% triploid) consisted of equal numbers of fish from a 100% triploid and a 100% diploid control group. Group II consisted of all heat-treated fish with a triploid-to-diploid ratio of 40:60. The triploid fish in Group I were identified by clipping the adipose fins. The two groups were kept in separate pools.

At smoltification 1000 fish each from Groups I and II were transferred to two saltwater netpens (3 m × 5 m square and 10 m deep) at the NMFS research station on Puget Sound near Manchester, WA. Prior to saltwater entry, the fish were acclimated over a 5-day period to increasing concentrations of saltwater. Five hundred fish from Group I were not transferred to saltwater but kept in one of the freshwater pools (designated Group III).

Ploidy determination

Ploidy was monitored monthly by fluorescence of red blood cell nuclei using an ICP flow cytometer as described in Utter et al. (1983).

Saltwater challenge

At the time of saltwater entry approximately 20 fish from each treatment group were subjected to a saltwater challenge for 24 and 48 h as described in Clarke and Blackburn (1977).

Growth measurements

Growth as measured by weight and fork length was recorded monthly for 100 fish in Group I and once every 3 months for 100 fish in Groups II and III. The 3-month interval was selected to prevent shock-related growth depression. Also at these times 20 fish per group were killed and the following growth parameters measured:

gonado-somatic index $(GSI) = $ (gonad weight/fish weight) $\times 100$;
hepato-somatic index $(HSI) = $ (liver weight/fish weight) $\times 100$;
condition factor $(K) = $ (gutted fish weight/length^3) $\times 100$.

Statistically significant differences in these parametric measurements between and among treatment groups were determined using one-way analysis of variance.

Hormone determination

Blood samples were collected in the 30th month (May 1985) in heparinized tubes and plasma was separated from the red blood cells via centrifugation. Samples were immediately assayed for 17-β estradiol by direct radioimmunoassay technique (Korenman et al., 1974; Sower and Schreck, 1982) and for the presence of vitellogenin by immunodiffusion technique (Utter and Ridgway, 1967).

Gonadal morphology and histology

Each fish used for growth studies was also monitored for sex determination and gonadal development. The gonad was observed, gonadal development estimated, and when possible, the sex was determined. The gonad was then immediately excised, weighed and fixed in Bouins solution. After 48–72 h pieces of gonad were dehydrated, embedded in paraffin and sectioned (to a thickness of 8 μm) following standard procedures (Humason, 1972). The sections were stained with Gill's hemotoxylin and eosin, and then observed histologically to re-determine sex and the stages of gonadal development.

RESULTS

Growth

Length and weight comparison among diploid or triploid groups sampled during a particular month revealed no significant differences among groups.

These data were therefore pooled for comparisons between diploid and triploid groups. No significant differences were detected in fork length, body weight, gut weight or condition factor (K) in any of these monthly comparisons. Accumulation of fat bodies was observed, however, around the intestine of all triploid females by the 20th month.

Saltwater entry and survival

No statistical differences were detected between either individual or pooled diploid and triploid mean plasma sodium levels in Groups I, II, and III following saltwater challenges. The mean for diploid fish was 174.70 (± 1.7 SD) and for triploid fish, 174.13 (± 2.3 SD). Seawater challenges resulted in no mortalities in any group.

Based on these high survivals and similar sodium concentrations in challenged fish, transfers to saltwater of the majority of fish in Groups I and II were made. No mortality occurred within 48 h of full saltwater entry in either group. Survival within these groups and in Group III, the freshwater controls, during the subsequent year was good ($>95\%$). However, ploidy measurements of 50 cumulative mortalities in Group I indicated that 36 were triploids; this is significantly higher than expected since only 50% of the fish in Group I were triploid.

Gonadal development

Gross morphology of triploid and diploid testes at 30 months revealed a reduction in size of triploid testes. The GSI for triploid males (0.05 ± 0.01 SD, $N=26$) was significantly smaller ($P<0.05$) than the GSI for diploid males (0.14 ± 0.10 SD, $N=31$). However, the triploid and diploid testes were indistinguishable histologically. Spermatids were not observed in either group.

Ovarian development revealed more pronounced differences between triploids and diploids. The diploid ovaries appeared filled with hundreds of oocytes, many of which were seen to be advanced as far as vitellogenesis when viewed under the microscope (Baker, 1982). Gross observation of 16 triploid ovaries revealed six which resembled immature testes and appeared to have no oocyte development. Microscopic observation revealed these ovaries contained a few immature oocytes surrounded by small oogonial-like cells and extensive connective tissue. Examination of the 10 remaining triploid ovaries revealed similar histological structures, but each of these ovaries also had five to 12 vitellogenic oocytes per ovary (compared to hundreds in the diploid ovaries). The mean GSI was significantly higher ($P<0.05$) for diploid females (0.68 ± 0.04 SD, $N=25$) compared to triploid females (0.08 ± 0.01 SD, $N=16$).

In triploids the reduced development of gonads made it difficult to determine sex on the basis of external gonadal morphology. Reliable determination of sex of triploid fish with undifferentiated gonads required histological examination. Without such examinations, we misclassified many females. Sex ratios were 50:50 when determined by histological examination.

Secondary indications of sexual maturation

The presence of vitellogenin and levels of 17-β estradiol were monitored as secondary indications of sexual development. Estradiol levels in female triploid ($N=3$) and diploid ($N=6$) fish assayed at 30 months of age were near basal levels (around 200 pg/ml) and revealed no significant differences between ploidy groups (Table 1). The presence of vitellogenin, however, was detected in all of the diploid ($N=5$) and in none of the triploid ($N=5$) females assayed at 30 months.

A second indication of vitellogenin production is the enlargement of the liver that is a normal observed result of estrogen stimulation. At 30 months there was a significant difference ($P<0.05$) between the size of female diploid and triploid livers as indicated by the HSI. The mean HSI in three triploid females was 0.88 (± 0.16 SD) and in six diploid females (1.77 ± 0.04 SD) (Table 1). There were no significant differences between males at any age or between females prior to 30 months.

DISCUSSION

Growth and condition factor

The equivalent growth of diploids and triploids contrasts with greater diploid growth in coho salmon reported earlier (Utter et al., 1983). The major difference between these two studies is that the fish in the present investiga-

TABLE 1

Estradiol levels, hepato-somatic index and gonadosomatic index in triploid and diploid coho salmon at 30 months of age (May 1985)

Groups	17-β Estradiol (pg/ml $\pm$ SE) (N)	Hepato-somatic index $\pm$ SE (N)	Gonado-somatic index $\pm$ SE (N)
Males			
Triploid	101.1$\pm$32.2 (14)	1.12$\pm$0.08 (26)	0.05$\pm$0.01 (26)*
Diploid	64.3$\pm$17.0 (4)	0.95$\pm$0.16 (31)	0.14$\pm$0.10 (31)*
Females			
Triploid	247.4$\pm$169.1 (3)	0.88$\pm$0.16 (16)*	0.08$\pm$0.01 (16)*
Diploid	235.6$\pm$113.3 (6)	1.77$\pm$0.04 (25)*	0.68$\pm$0.06 (25)*

*$P<0.05$.

tion were reared under the optimal conditions of reduced densities and rigorous feeding schedules. The underlying bases of these different genotype–environment interactions remain largely unknown. However, commercial applications of induced triploidy will undoubtedly benefit from similar reductions in stress.

The absence of significant differences between length, body weight, gut weight, and condition factor in triploid and diploid coho salmon differs from reports comparing diploid and triploid Atlantic salmon and rainbow trout. In Atlantic salmon at a similar stage of maturity, triploids were consistently longer than diploids and thus had a lower condition factor (Benfey and Sutterlin, 1984). In mature rainbow trout (Lincoln and Scott, 1984) no differences between fork length and dressed body weight were observed but condition factor and gut weight were both significantly higher in the triploids. Our fish have not yet reached sexual maturity, but the accumulation of fat bodies we observed may indicate that gut weight will increase as sexual maturity approaches.

Smoltification

A previous observation of the ability of triploid coho salmon to adapt to seawater initiated a more careful examination. In August of 1982 a mixed group of 20-month-old diploid (51%) and triploid (49%) fish (1981 brood year) were transferred from fresh water directly to seawater. A heavy mortality occurred within 24 h of the transfer. Ploidy measurements were obtained from each of the 102 dead fish, and all but two were triploid. This one-sided mortality suggests a serious problem with osmoregulation in seawater associated with the triploid condition.

Differential mortality of triploids upon saltwater entry was not found during the present series of experiments. This suggests that environmental factors may play a significant role in seawater acclimation of triploids (Mahnken, 1973; Clarke and Shelbourn, 1977). In the saltwater transfer where there was excessive mortality among triploid coho salmon, seawater entry was in August, 2–3 months after the usual time of natural seaward migration or hatchery release. In the present experiment seawater entry was in early June, coincident with the usual time of seaward migration. The somewhat higher proportion of triploids among the mortalities observed in this study occurred throughout the year in both seawater and fresh water and was therefore not associated with the transition to seawater. It is clear that triploid coho salmon can adapt and grow in seawater. However, it is also apparent that such fish, under certain conditions, are more susceptible to the stress of saltwater challenge. These conditions need to be more precisely identified through further investigations.

Gonadal development

Gonadal development of triploids up to this stage of maturity is similar to that reported for most other fish species (reviewed in Thorgaard, 1983) including Atlantic salmon (Benfey and Sutterlin, 1984). However, these findings differ from similar studies of rainbow trout where no oocyte development has been reported in triploid females (Lincoln and Scott, 1984; Solar et al., 1984). We detected several vitellogenic oocytes (Baker, 1982) in 10 of 16 triploid females sampled and pre-vitellogenic oocytes (Baker, 1982) in all the triploid females.

Hormone production

Reduced oocyte development in triploids might be expected to result in significantly reduced estrogen levels, as follicular cell layers surrounding the oocyte are the main source of estrogen production (Nagahama, 1983) and the diploid ovary has a much greater mass of this steroidogenic tissue. Reductions in the level of estrogen in triploids would also be predicted from observations in triploid females of reduced liver weights and the lack of vitellogenin in the blood plasma. At this time liver growth is stimulated by increasing estrogen levels and there is increased production of vitellogenin, the lipophosphoprotein that acts as a substrate for yolk deposition in teleosts (Ng and Idler, 1983).

The similar levels of $17\text{-}\beta$ estradiol between triploids and diploids at this stage may be attributed to several factors. The sampling was performed early in sexual maturation before the great increase in steroidogenic tissue associated with development of oocytes and increases in plasma estradiol. There may be a lag in estrogen production until the rates of synthesis of androstenedione and/or testosterone are built up in the ovarian thecal layer to provide precursors for conversion to $17\text{-}\beta$ estradiol in the granulosa layer (Nagahama, 1983). Estrogen produced at this early stage may be immediately metabolized by the ovary or the liver, and thus be undetectable in the peripheral blood plasma. If this is so, we might expect a surge of $17\text{-}\beta$ estradiol production to be detected within the subsequent months.

ACKNOWLEDGMENTS

This work was supported by Washington Sea Grant (Projects R/A 44 and 42). The authors would like to thank Marcia Kamin and Ron Salor for their excellence in fish culture, Mike Shen and Ann Adams for technical expertise on the flow cytometer, and Penny Swanson and Susan Drohan for assistance with the hormone assays.

REFERENCES

Baker, T.G., 1982. Oogenesis and ovulation. In: C.R. Austin and R.V. Short (Editors), Reproduction in Mammals, 2nd edition. Book 1: Germ Cells and Fertilization. Press Syndicate of the University of Cambridge, Cambridge, pp. 17-45.

Benfey, T.J. and Sutterlin, A.M., 1984. Growth and gonadal development in triploid landlocked Atlantic salmon (Salmo salar). Can. J. Fish. Aquat. Sci., 41: 1387-1392.

Clarke, W.C. and Blackburn, J., 1977. A seawater challenge test to measure smolting of juvenile salmon. Fish. Mar. Serv. Tech. Rep., 705: 11 pp.

Clarke, W.C. and Shelbourn, J.E., 1977. Effect of temperature, photoperiod and salinity on growth and smolting of underyearling coho salmon. Am. Zool., 17(4): 957-964.

Humason, G.L., 1972. Animal Tissue Techniques, 3rd edition. W.H. Freeman and Co., San Francisco, CA, 641 pp.

Korenman, S.G., Stevens, R.H., Carpenter, L.A., Robb, M., Niswender, G.D. and Sherman, B.M., 1974. Estradiol radioimmunoassay without chromatography: procedure, validation and normal values. J. Clin. Endocrinol. Metab., 38: 718-720.

Liley, N.R., 1969. Hormones and reproductive behavior in fishes. In: W.S. Hoar and D.J. Randall (Editors), Fish Physiology, Vol. III. Academic Press, New York, NY, pp. 73-160.

Lincoln, R.F. and Scott, A.P., 1984. Sexual maturation in triploid rainbow trout, Salmo gairdneri Richardson. J. Fish Biol., 25: 385-392.

Mahnken, C.V.W., 1973. The size of coho salmon and time of entry into sea water. Part I. Effects of growth and condition index. Proc. Annu. Northwest Fish Cult. Conf., 24: 30.

Nagahama, Y., 1983. The functional morphology of teleost gonads. In: W.S. Hoar, D.J. Randall and E.M. Donaldson (Editors), Fish Physiology, Vol. IXA. Academic Press, New York, NY, pp. 223-275.

Ng, T.B. and Idler, D.R., 1983. Yolk formation and differentiation in teleost fishes. In: W.S. Hoar, D.J. Randall and E.M. Donaldson (Editors), Fish Physiology, Vol. IXA. Academic Press, New York, NY, pp. 373-397.

Solar, I.I., Donaldson, E.M. and Hunter, G.A., 1984. Induction of triploidy in rainbow trout (Salmo gairdneri Richardson) by heat shock and investigation of early growth. Aquaculture, 42: 57-67.

Sower, S.A. and Schreck, C.B., 1982. Steroid and thyroid hormones during sexual maturation of coho salmon (Oncorhynchus kisutch) in seawater or fresh water. Gen. Comp. Endocrinol., 47: 42-53.

Thorgaard, G.H., 1983. Chromosome set manipulation and sex control in fish. In: W.S. Hoar, D.J. Randall and E.M. Donaldson (Editors), Fish Physiology, Vol. IXB. Academic Press, New York, NY, pp. 405-434.

Utter, F.M. and Ridgway, G.J., 1967. A serologically detected serum factor associated with maturity in the English sole, Parophrys vetuius and the Pacific halibut, Hippoglossus stenolepis. Fish. Bull., 6: 47-48.

Utter, F.M., Johnson, O.W., Thorgaard, G.H. and Rabinovitch, F.S., 1983. Measurement and potential applications of induced triploidy in Pacific salmon. Aquaculture, 35: 125-135.

Youngson, A.F. and Simpson, T.H., 1984. The downstream changes in serum thyroxine levels during smolting in captive and wild Atlantic salmon, Salmo salar L. J. Fish Biol., 24: 29-39.

Increased Resistance of Triploid Rainbow Trout × Coho Salmon Hybrids to Infectious Hematopoietic Necrosis Virus

JAMES E. PARSONS[1], ROBERT A. BUSCH[1], GARY H. THORGAARD[2] and PAUL D. SCHEERER[2]

[1]*Clear Springs Trout Co., Buhl, ID 83316 (U.S.A.)*
[2]*Department of Zoology and Program in Genetics and Cell Biology, Washington State University, Pullman, WA (U.S.A.)*

ABSTRACT

Parsons, J.E., Busch, R.A., Thorgaard, G.H. and Scheerer, P.D., 1986. Increased resistance of triploid rainbow trout × coho salmon hybrids to infectious hematopoietic necrosis virus. *Aquaculture*, 57: 337–343.

Diploid and heat-induced triploid treatment groups of rainbow trout (*Salmo gairdneri*), coho salmon (*Oncorhynchus kisutch*), and reciprocal hybrid crosses were produced and monitored for survival and growth through early life stages. Survival to the fry stage showed a significant increase in the triploid rainbow trout female × coho salmon male hybrids relative to the diploid hybrid group. At a mean weight of 0.65 g all groups were administered a standardized static bath challenge with Leong Type 2 virulent IHN virus. The triploid rainbow trout female × coho salmon male hybrids showed a significant increase in IHN resistance when compared to pure-species rainbow trout groups. Early growth results suggest a poorer growth rate for the triploid hybrid.

Triploid hybridization improved viability of the hybrid cross, increased relative resistance to IHN virus challenge, and may provide a useful method for the interspecific transfer of desired characters into commercial fish stocks.

INTRODUCTION

Interspecific hybridization has often been proposed as a method for combining desired characters in fish (Chevassus, 1979, 1983; Scheerer and Thorgaard, 1983). Such hybrids frequently demonstrate reduced survival (Chevassus, 1979). However, Chevassus et al. (1983) and Scheerer and Thorgaard (1983) demonstrated that induced triploid interspecific hybrids often survive significantly better than their diploid counterparts. This increased survival has made it feasible to test interspecific hybrid performance in aquaculture situations.

Infectious hematopoietic necrosis (IHN) disease is a serious viral disease of salmonid fishes, often causing severe mortalities in hatchery-raised trout and salmon in the United States, Canada, and Japan (Pilcher and Fryer, 1980; Groberg and Fryer, 1983; Leong, 1984). Recent IHN virus epizootics have resulted in severe economic losses for the Idaho commercial foodfish rainbow trout (*Salmo gairdneri*) industry (Busch, 1983). The virus has been reported to cause mortalities in most salmonid species. However, no natural mortalities from IHN virus have been reported in coho salmon (*Oncorhynchus kisutch*). Laboratory challenge with IHN virus by injection (Parisot and Pelnar, 1962), or waterborne exposure (Wingfield et al., 1969; Wingfield and Chan, 1970), failed to induce mortality in coho salmon, suggesting a natural resistance. A cell line developed from coho salmon embryos (CSE-119) also showed a lack of sensitivity to IHN virus infection (Lannan et al., 1984), indicating that this resistance may occur at a cellular level. Chen (1984) suggested that interspecific hybrids between coho salmon and rainbow trout may exhibit a decreased susceptibility to IHN virus infection. However, attempts at producing such a cross have met with little or no success (Chevassus, 1979), until Chevassus et al. (1983) showed that triploid hybrids between rainbow trout females and coho salmon males showed a significant increase in early survival over the diploid hybrid.

In the present study we have compared early growth and survival of reciprocal diploid and triploid hybrids between rainbow trout and coho salmon with their pure-species counterparts. Surviving hybrid and pure species lots were experimentally infected with a waterborne challenge of IHN virus and monitored for mortality, recovery, and post-infection growth. Our results show the rainbow trout female×coho salmon male triploid hybrid to be a successful cross that demonstrates a significant increase in relative resistance to IHN virus infection when compared to the rainbow trout pure-species controls.

MATERIALS AND METHODS

Coho salmon eggs and sperm were obtained from Oregon Aqua Foods (Springfield, OR) in December 1984. Rainbow trout eggs and sperm were obtained from the Clear Springs Trout Company broodstock operations (Buhl, ID) at the same time. All gametes were transported on ice to the Clear Springs Trout Company Research Laboratory (Buhl, ID). Gametes from six female and 10 male coho salmon were pooled, as were gametes from four female and five male rainbow trout. All possible pure-line and reciprocal matings were attempted. A portion of each egg lot was heat-shocked 10 min post-fertilization by immersion in a 27°C water bath for 20 min to induce triploidy (modification of Chourrout, 1980). All groups were placed in individual upwelling incubator systems and held at 14.5°C through all remaining procedures.

Chromosome preparations were made from a sample of 10- to 12-day-old embryos from each treatment group (Thorgaard et al., 1981). Counts of chromosomes from five metaphase spreads per individual were examined to deter-

TABLE 1

Percent survival and proportion of triploidy in experimental lots

Cross[a] ($♀ \times ♂$)	No. of eggs	Survival (%)			Proportion of $3N$ karyotypes
		To eye	To hatch	To swim-up	
R×R					
Heat shock	3338	35.8	34.2	32.4	6/6
Control	2174	38.8	33.2	32.0	0/5
R×C					
Heat shock	4531	35.4	22.4	5.9	8/8
Control	1787	0.6	0.0	0.0	—[b]
C×C					
Heat shock	1824	26.6	0.2	0.1	3/5
Control	1893	63.7	10.8	9.2	0/3
C×R					
Heat shock	1010	0.3	0.0	0.0	0/3
Control	2285	0.4	0.1	0.1	—[b]

[a]R = *Salmo gairdneri*; C = *Oncorhynchus kisutch*.
[b]No viable embryos obtained for chromosome preparations.

mine ploidy. Survival to eyed stage, hatching, and swim-up were recorded for each group.

At a mean weight of 0.65 g, replicate samples (50–200 fish per replicate) of each treatment group were administered a standardized 60-min static bath challenge with 2.23×10^4 TCID$_{50}$ units/ml of virulent Leong Type 2 IHN virus. All mortalities were removed and counted for 28 days. Periodic post-mortality examinations were conducted to confirm the cause of mortality.

Survivors from each replicate were pooled 28 days after challenge, placed into separate circular tanks, and fed for 60 days. Growth was monitored during this period.

RESULTS

Early survival

The triploid rainbow trout female × coho salmon male hybrid group (RC-3) showed a highly significant ($P<0.01$) increase in survival to the eyed stage, hatching, and swim-up over the diploid (RC-2) hybrid group (Table 1). The high losses in the RC-3 group between hatch and swim-up were primarily attributed to crippled fry. There was no significant difference in the survival of the diploid and triploid pure-species rainbow trout groups (RR-2, RR-3) (Table 1). The relatively low overall survival in each of these groups suggests

TABLE 2

Comparison of susceptability of experimental lots to IHN virus

Group[a]	No. challenged	Number of replicates	Challenge[b] survival (%)	Significance
RR-3	1000	7	16.2	n.s.
RR-2	807	5	19.1	**
RC-3	354	4	79.4	n.s.
CC-2	160	2	89.4	

[a]RR-2 = rainbow × rainbow $2n$; RR-3 = rainbow × rainbow $3n$; RC-3 = rainbow × coho $3n$; CC-2 = coho × coho $2n$.
[b]Combined average of replicate survivals.
n.s. $P > 0.05$; **$P < 0.01$.

poor initial gamete quality which may have detrimentally affected the hybrids as well. Chromosome preparations from the RR-3 and RC-3 groups showed 100% induction of triploidy.

High mortality occurred in both the diploid and triploid pure-species coho salmon groups (CC-2, CC-3) at the time of hatching (Table 1). It is possible that the incubation temperature of 14.5°C is too high for successful hatching of these groups, as it seemed that the developed sac fry were unable to emerge from the egg shell. The coho salmon female × rainbow trout male hybrid and coho pure-species heat-shocked groups showed varying proportions of triploid progeny (Table 1).

IHN Challenge

Total mortality and mortality patterns varied considerably among the groups challenged with IHN virus (Table 2, Fig. 1). No significant differences ($P > 0.05$) in IHN challenge survival were observed between the RR-2 and RR-3 groups (16.2% and 19.1%, respectively) or between the RC-3 and CC-2 groups (79.4% and 89.4%, respectively). The RC-3 and CC-2 groups did, however, show a highly significant ($P < 0.01$) increase in survival following IHN infection when compared to both pure-species rainbow trout groups.

Post-challenge growth

Post-challenge growth of the hybrid group was slower than that of the three pure-species groups (Table 3). Some of this difference can be attributed to the extreme size variation observed in the hybrid lot as expressed by the high coefficient of variation for 60-day weight. The RR-3 group experienced continued mortality during the post-challenge period. Distinct differences in adaptation

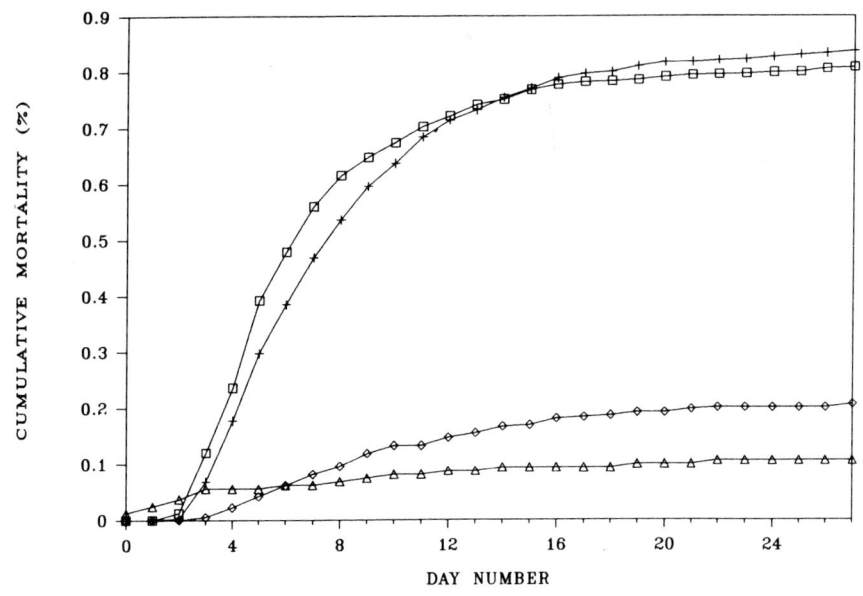

Fig. 1. Cumulative daily frequencies of mortalities in experimental groups during 28-day IHN virus challenge. $+$ = RR-3; $\square$ = RR-2; $\diamond$ = RC-3; $\triangle$ = CC-2.

to culture conditions were noted between the hybrid lot and the more aggressive pure-species lots during the grow-out period.

DISCUSSION

The results clearly demonstrate the increased resistance of the rainbow trout × coho salmon triploid hybrids to IHN virus. This study suggests a genetic basis for resistance to IHN virus in coho salmon that can be transferred to an interspecific hybrid involving rainbow trout. However, no conclusions as to

TABLE 3

Post-challenge growth and survival of experimental lots

Group	Post-challenge survival (%)	Weight[a] (g)			CV day 60 (%)[b]	Growth rate (% body wt./day)
		Day 1	Day 30	Day 60		
RR-2	79.6	2.2	8.5	20.7	41.3	3.81
RR-3	60.0	1.7	6.6	14.4	37.1	3.63
CC-2	81.3	2.3	8.3	19.8	40.9	3.65
RC-3	87.6	1.9	5.9	12.2	66.6	3.15

[a]Day number indicates days post-challenge (i.e., day 1 begins 29 days after challenge initiation).
[b]CV = Coefficient of variation of weight at day 60 [CV = (SD/$\bar{x}$) × 100].

the number or nature of the resistance genes can be drawn from the experimental design.

The higher proportion of triploid embryos observed in crosses using rainbow trout eggs when compared to those using eggs of coho salmon may possibly be explained by the smaller size of the rainbow trout egg, as the higher surface area-to-volume ratio may have permitted more efficient heating and cooling (Scheerer and Thorgaard, 1983).

Although the high initial egg and fry mortalities and the poor post-challenge growth of the triploid hybrid may preclude its commercial application, the potential for interspecific triploid hybrids can clearly be seen. Induction of triploidy in hybrids may provide a tool for transferring desired characters (i.e., disease resistance, saltwater tolerance, hardiness, rapid growth, etc.) from one species to another in crosses that have historically produced few or no offspring.

ACKNOWLEDGEMENTS

The authors wish to acknowledge Robert Anderson and Oregon Aqua Foods for providing coho salmon eggs and sperm for this study. We especially thank Andy Morton and Max Woolley for their invaluable technical assistance on this project. Work at Washington State University was supported by NSF grant PCM8108787 and USDA grant CRSR82-1058.

REFERENCES

Busch, R.A., 1983. Viral disease considerations in the commercial trout industry in Idaho. In: J.C. Leong and T.Y. Barila (Editors), Workshop on Viral Diseases of Salmonid Fishes in the Columbia River Basin, 7–8 October 1982, Portland, OR. Spec. Publ., Bonneville Power Admin., pp. 84–100.
Chen, M.S., 1984. Comparative Susceptibility of Salmonid Fish and Salmonid Fish Cell Lines to Five Isolates of Infectious Hematopoietic Necrosis Virus, and Biological Properties of Two Plaque Variants. Ph.D. dissertation, Oregon State University, 124 pp.
Chevassus, B., 1979. Hybridization in salmonids: results and perspectives. Aquaculture, 17: 113–128.
Chevassus, B., 1983. Hybridization in fish. Aquaculture, 33: 245–262.
Chevassus, B., Guyomard, R., Chourrout, D. and Quillet, E., 1983. Production of viable hybrids in salmonids by triploidization. Genet. Sel. Evol., 15: 519–532.
Chourrout, D., 1980. Thermal induction of diploid gynogenesis and triploidy in the eggs of rainbow trout (*Salmo gairdneri* Richardson). Reprod. Nutr. Dev., 20: 727–733.
Groberg, W.J. and Fryer, J.L., 1983. Increased Occurrences of Infectious Hematopoietic Necrosis Virus at Columbia River Basin Hatcheries; 1980–1982. Oregon State University Sea Grant College Program. Corvallis, OR, pp. 1–16.
Lannan, C.N., Winton, J.R. and Fryer, J.L., 1984. Fish cell lines: establishment and characterization of nine cell lines from salmonids. In Vitro, 20: 671–676.
Leong, J.C., 1984. Rapid diagnosis of IHN infection in salmon and steelhead trout. Bonneville Power Admin., Portland, OR, Proj. No. 82-4, 145 pp.

Parisot, T.J. and Pelnar, T., 1962. An interim report on Sacramento River chinook disease: a virus-like disease of chinook salmon. Prog. Fish Cult., 24: 51–55.

Pilcher, K.S. and Fryer, J.L., 1980. The viral diseases of fish: a review through 1978. Part 1: diseases of proven viral etiology. Crit. Rev. Microbiol., 7: 287–364.

Scheerer, P.D. and Thorgaard, G.H., 1983. Increased survival in salmonid hybrids by induced triploidy. Can. J. Fish. Aquat. Sci., 40: 2040–2044.

Thorgaard, G.H., Jazwin, M.E. and Stier, A.R., 1981. Polyploidy induced by heat shock in rainbow trout. Trans. Am. Fish. Soc., 110: 546–550.

Wingfield, W.H. and Chan, L.D., 1970. Studies on the Sacramento River chinook disease and its causative agent. In: S.F. Snieszko (Editor), A Symposium of Diseases of Fishes and Shellfishes. Am. Fish. Soc. Spec. Publ., 5: 307–318.

Wingfield, W.H., Fryer, J.L. and Pilcher, K.S., 1969. Properties of the sockeye salmon virus (Oregon strain). Proc. Soc. Exp. Biol. Med., 130: 1055–1059.

Female Brown Trout × Atlantic Salmon Hybrids Produce Gynogens and Triploids when Backcrossed to Male Atlantic Salmon

KENNETH R. JOHNSON[1] and JAMES E. WRIGHT

Department of Biology, The Pennsylvania State University, University Park, PA 16802 (U.S.A.)

[1]Present address: Department of Zoology, Washington State University, Pullman, WA 99164-4220, U.S.A.

ABSTRACT

Johnson, K.R. and Wright, J.E., 1986. Female brown trout × Atlantic salmon hybrids produce gynogens and triploids when backcrossed to male Atlantic salmon. *Aquaculture*, 57: 345–358.

We examined protein electromorphs and karyotypes of progeny produced from interspecific mass spawnings between female brown trout (*Salmo trutta*) and male Atlantic salmon (*S. salar*). Species-specific allelic differences for the loci Gpi-1, Gpi-3, Ldh-5, Mpi, and Pgm-2 enabled assessment of species contributions in the hybrids. All of the F_1 hybrids examined showed equal expression of both parental alleles at all loci, as would be expected for codominant alleles. However, unexpected results were obtained for backcross progeny. Seventy of 177 backcross individuals had electrophoretic phenotypes identical to F_1 phenotypes; 107 appeared to be triploid based on isozyme dosage relationships. All triploids expressed one copy of brown trout alleles and two copies of Atlantic salmon alleles at all loci examined. Karyotypes of four F_1 and nine backcross hybrids were consistent with the ploidy levels and species contributions deduced from the electrophoretic phenotypes. In addition, a novel hybrid generated by crossing a female F_1 hybrid to a male brook trout (*Salvelinus fontinalis*) appeared to contain and express the haploid genomes of all three species. The disparate parental chromosome complements of the F_1 hybrids (brown trout $2n = 80$, Atlantic salmon $2n = 54$–60) most likely led to a disruption of meiosis. Meiotic mechanisms that could account for the unusual backcross progeny are discussed. The type of inheritance exhibited by these F_1 hybrids has not been reported before in salmonid fish but appears similar to that reported in female hybrids of other vertebrate species. These results show the importance of using genetic markers to determine parental contributions and gene segregation in species hybrids.

INTRODUCTON

Interspecific hybridization experiments in fish often have been more concerned with the evaluation of survival and performance than with the identification of the genomic composition of the resultant progeny. Chevassus (1983) reviewed methods that have been used to identify hybrid genomes and some of the possible types of progeny that could result from heterospecific insemination. In this paper, we present the results of our electrophoretic and chromosomal analyses of hybrids between brown trout (*Salmo trutta*) and Atlantic salmon (*Salmo salar*).

Naturally occurring F_1 hybrids between Atlantic salmon and brown trout have been identified by protein electrophoresis in Europe (Payne et al., 1972; Solomon and Child, 1978) and in North America (Beland et al., 1981); these F_1 hybrids were found at low frequencies and there was no evidence of introgression. Artificial crosses between Atlantic salmon and brown trout have produced viable F_1 hybrids (Alm, 1955; Piggins, 1971; Refstie and Gjedrem, 1975; Blanc and Chevassus, 1979). Protein electrophoresis (Nyman, 1970; Nygren et al., 1975; Cross and O'Rourke, 1978; Vuorinen and Piironen, 1984) and cytological analysis (Nygren et al., 1972, 1975) have been used to verify the F_1 status of progeny from artificial crosses.

The fertility of F_1 hybrids and the viability of F_2 and backcross hybrids are still objects of controversy (Chevassus, 1979). Piggins (1971 cited in Cross and O'Rourke, 1978) reported viable F_2, F_3, and backcross hybrids. An electrophoretic investigation of hybrids obtained from Piggins (Nyman, 1970) confirmed the hybrid nature of five F_1 fish and reported that all Atlantic salmon genes were suppressed in five F_2 fish examined. However, Nygren et al. (1972) reported that the F_1 fish they obtained from Piggins had karyotypes that were identical to brown trout. A group of F_2 hybrids from the same stock showed only brown trout hemoglobins and appeared to contribute only trout genes to F_3 and backcross progeny (Cross and O'Rourke, 1978). These results suggest that at least some of the purported F_1 hybrids produced by Piggins were actually brown trout.

Atlantic salmon have a diploid chromosome number of 54–58 and a chromosome arm number of 72–76 (Rees, 1967; Roberts, 1970; Nygren et al., 1972). Brown trout have a diploid chromosome number of 80 and a chromosome arm number of 100 (Nygren et al., 1971). The disparate karyotypes of these two species is thought to cause meiotic pairing difficulties in the F_1 hybrid, resulting in low viability of gametes. A high number of univalents were observed in metaphase spermatocytes of an F_1 hybrid (Nygren et al., 1972), apparently supporting this hypothesis. A backcross of a cytologically confirmed F_1 hybrid male to a female Atlantic salmon produced only four progeny that survived to 11 months (Nygren et al., 1975).

In 1980, it came to our attention that personnel of the Quinebaug Valley Hatchery, Central Village, Connecticut, had produced hundreds of F_1 and backcross hybrids by mass spawnings between brown trout and Atlantic salmon. We were somewhat surprised by the reported success of the backcrosses because of the aforementioned meiotic pairing disruptions and consequent poor gametic viability expected of F_1 hybrids. We examined 132 F_1 and 177 backcross hybrids for electrophoretic variation and examined the karyotypes of four F_1 and nine backcross hybrids to corroborate our genetic interpretations based on the electrophoretic phenotypes.

Hybrids between closely related salmonid species whose karyotypes differ by, at most, one or two centric fusions have shown normal Mendelian segregation as determined from isozyme analysis of F_2 and backcross progeny [*Salvelinus fontinalis* × *S. namaycush* (May et al., 1980), *S. fontinalis* × *S. alpinus* and *Salmo gairdneri* × *Salmo clarki utah* (Johnson, 1984; Johnson and Wright, 1984)]. However, the female brown trout × Atlantic salmon hybrids that we examined in this study exhibited an aberrant type of inheritance that has not been reported before in salmonids, but appears similar to that reported to occur in female hybrids of other fish species (Poeciliidae — Schultz, 1969, 1980; Cimino, 1972; Cyprinidae — Ojima et al., 1975; Cherfas and Ilyasova, 1980; Cherfas et al., 1981; Oriziatidae — Iwamatsu et al., 1984).

MATERIALS AND METHODS

Spawnings

Interspecific mass spawnings to produce F_1 hybrids were made between female brown trout and male Atlantic salmon by personnel of the Quinebaug Valley Hatchery, Central Village, Connecticut, in the fall of 1979 and again in the fall of 1981. Concurrently, female F_1 hybrids produced from mass spawnings in 1976 (female brown trout × male Atlantic salmon) were backcrossed to male Atlantic salmon. The same female F_1 hybrids were used to make the backcrosses in both 1979 and 1981. Several females and several males were used in each of the crosses, but exact numbers were not recorded. In addition, one female F_1 hybrid was crossed to a male brook trout (*Salvelinus fontinalis*) in 1981. Fertility rates of parents and survival and performance of progeny were not experimentally assessed; however, thousands of viable progeny were produced from the mass spawnings.

Sample collection

Progeny samples from the 1979 spawnings were collected in November 1980. Eighty F_1 and 87 backcross hybrids were frozen and transferred for electrophoretic analysis. Five Atlantic salmon, five F_1 hybrids, and 10 backcross hybrids were transported live for cytological analysis. All laboratory analyses were carried out at The Pennsylvania State University.

Progeny samples from the 1981 spawnings were collected in July 1982. Fifty-two F_1 hybrids and 90 backcross hybrids were frozen and transferred for electrophoretic analysis. Two hybrids (the only survivors) from the cross of an F_1 hybrid with a brook trout were transported live. One of these hybrids died in

August 1981, and was immediately analyzed for electrophoretic variation. The other, a female, was sacrificed for cytological analysis in the fall of 1984.

Electrophoretic analysis

Horizontal starch-gel electrophoresis and genic nomenclature followed that detailed in May et al. (1979). Enzymes were visualized by the staining recipes listed in Allendorf et al. (1977). Phenotypes of loci coding for muscle forms of glucosephosphate isomerase (Gpi-1,2,3) and lactate dehydrogenase (Ldh-1), and liver forms of lactate dehydrogenase (Ldh-4) and superoxide dismutase (Sod) were resolved using the buffer system described by Ridgway et al. (1970). Phenotypes of the locus coding for mannosephosphate isomerase (Mpi) and an additional lactate dehydrogenase locus (Ldh-5) were resolved from eye tissue using the buffer system described by Markert and Faulhaber (1965). Phenotypes of a phosphoglucomutase locus (Pgm-2) were resolved from muscle tiussue using the buffer system described by Clayton and Tretiak (1972).

Alleles at each locus were identified by the numbers 1, 2, 3, or 4 so that genotype designations would not be cumbersome. Alleles whose products had the same electrophoretic mobility were assumed identical and given the same numeric designation. The electrophoretic mobilities of allelic homomers relative to those of rainbow trout (*Salmo gairdneri*) are given in Table 1.

Chromosome analysis

Mitotic metaphase preparations were obtained from anterior kidney cells by the technique described by Gold (1974). Mitotic spreads were photographed, enlarged, and karyotyped. Chromosomes were counted and classified as metacentric or acrocentric.

RESULTS

The interspecific crosses between brown trout and Atlantic salmon were made by mass spawnings; consequently, numbers of fish contributing to the crosses and their individual genotypes were not known. However, species-specific allelic differences between brown trout and Atlantic salmon for the loci Gpi-1, Gpi-3, Ldh-5, and Pgm-2 (Table 1) enabled assessment of species contributions in the interspecific crosses.

All of the F_1 hybrids collected in 1980 and in 1982 showed equal expression of both parental alleles at all loci examined (Table 1, Fig. 1), as would be expected for codominant alleles. However, unexpected results were obtained for backcross hybrids.

Most (70) of the 87 backcross hybrids collected in 1980 had diploid phenotypes that were identical to F_1 phenotypes. Seventeen of the 87 backcross

TABLE 1

Genotypic frequencies of interspecific hybrids examined in this study based on electrophoretic phenotypes
BT = Brown trout, AS = Atlantic salmon, ST = brook trout, F_1 = BT × AS, BC = (BT × AS) × AS, and BAS = (BT × AS) × ST

Locus	Allele	Mobility[a]	Species specificity[b]	Genotype	Genotypic frequencies				
					1980 Collection		1982 Collection		
					F_1	BC	F_1	BC	BAS
Gpi-1	1	40	BT	12	80	70	52	0	0
	2	100	AS, ST	122	0	17	0	90	1
Gpi-2	1	62	BT, AS	11 or 111	80	87	52	90	0
	2	100	ST	112	0	0	0	0	1
Gpi-3	1	94	BT	13	80	70	52	1*	0
	3	110	AS, ST	133	0	17	0	89	1
Ldh-1	1	100	BT, AS	11 or 111	80	87	52	90	0
	2	−100	ST	112	0	0	0	0	1
Ldh-4	1	100	BT, AS	11 or 111	80	87	52	90	0
	2	32	ST	112	0	0	0	0	1
Ldh-5	1	97	BT	12	80	70	52	0	0
	2	100	AS, ST	122	0	17	0	90	1
Mpi	1	95	BT	12	4	15	42	1*	0
	2	100	BT, AS, ST	122	0	1	0	87	1
				22 or 222	1	1	10	2	0
Pgm-2	1	−150	BT	12	1	8	48	0	0
	2	−300	AS	13	4	5	4	0	0
	3	−340	AS	122	0	1	0	39	0
	4	−100	ST	123	0	1	0	21	0
				124	0	0	0	0	1
Sod	1	190	BT, AS	11 or 111	80	87	52	90	0
	2	240	ST	112	0	0	0	0	1

[a]Electrophoretic mobility of allelic homomers relative to the mobility of the most common allelic homomer of rainbow trout (arbitrarily designated 100); a negative sign indicates migration towards the cathode.
[b]Species-specific alleles as determined from analysis of 550 brown trout, 260 Atlantic salmon, and 280 brook trout from North America (Johnson, 1984; Stoneking et al., 1981).
*This individual appeared diploid for the linked loci Gpi-3 and Mpi, but triploid for all other loci.

hybrids were triploid based on isozyme dosage relationships (Table 1, Fig. 1). All of the triploids expressed one copy of brown trout alleles and two copies of Atlantic salmon alleles for all loci examined.

Ploidy levels and species contributions were corroborated by karyotypes made on four F_1 and nine backcross hybrids collected in 1980. Brown trout have a diploid chromosome number ($2n$) of 80 and an arm number (NF) of 100 (Nygren et al., 1971). The Atlantic salmon we examined had $2n = 54–56$ and NF = 72–74.

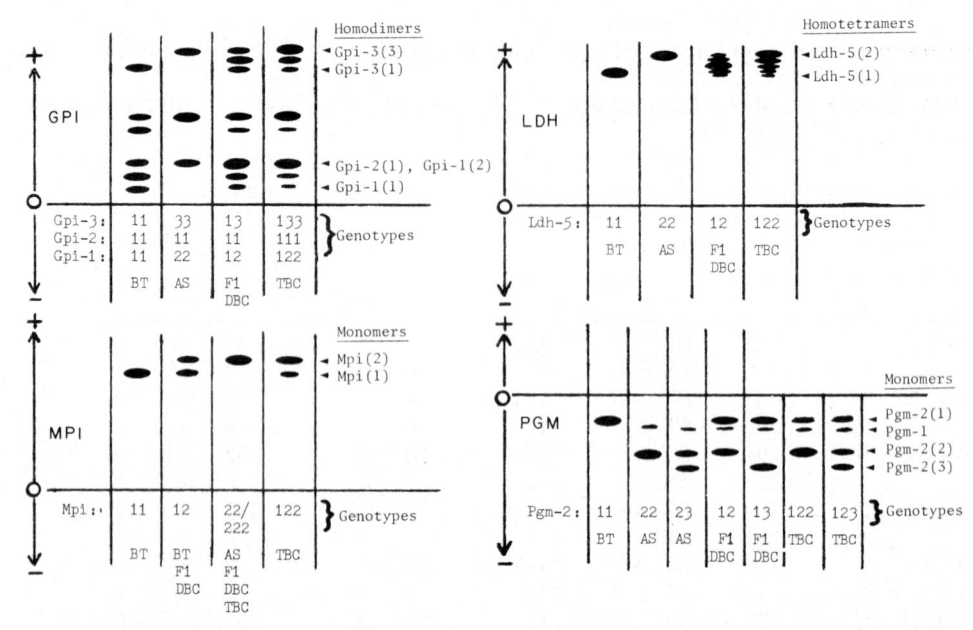

Fig. 1. Electrophoretic banding patterns and proposed genotypes of brown trout (BT); Atlantic salmon (AS); their F_1 hybrid (F_1); and backcross of the F_1 hybrid to Atlantic salmon, both diploid (DBC) and triploid (TBC), for the loci Gpi-1, Gpi-3, Ldh-5, Mpi, and Pgm-2. The proposed genotype for each locus is given below its electrophoretic phenotype. Homomers are designated by the loci and alleles (in parentheses) that code for them.

Therefore, F_1 hybrids would be expected to have $2n=67$ or 68 and $NF=86$ or 87, values very close to those we observed in the four F_1 hybrids and seven of the nine backcross hybrids that we examined ($2n=67-70$ and $NF=85-88$). Two backcross hybrids had karyotypes ($2n=95$ or 96 and $NF=121$ or 123) identical to that expected from a triploid hybrid containing two haploid sets of Atlantic salmon chromosomes and one haploid set of brown trout chromosomes. In all cases, genomic compositions determined from karyotypes agreed with those determined from electrophoretic phenotypes.

All of the 90 backcross individuals collected in 1982 were triploids based on their electrophoretic phenotypes and they also expressed one copy of brown trout alleles and two copies of Atlantic salmon alleles at all loci examined (Table 1, Figs. 1–3). However, one exceptional individual appeared to be diploid for the loci Gpi-3 and Mpi, but triploid for all other loci examined. Gpi-3 and Mpi are linked in brown trout (Johnson and Wright, 1984) which indicates that at least one Atlantic salmon chromosome was lost in this otherwise triploid individual.

One of the two individuals collected in 1982 that resulted from the cross between a female brown trout × Atlantic salmon F_1 and a male brook trout was

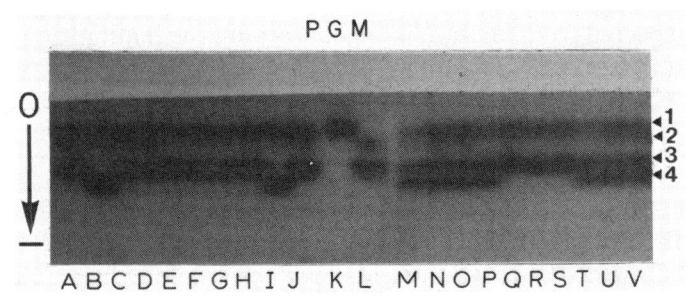

Fig. 2. Electrophoretic phenotypes of PGM from muscle tissue of brown trout (lane K), Atlantic salmon (lane L), F₁ hybrids (lanes A-J), and backcross hybrids (lanes M-V). Positions of monomers are indicated by the numbers at the right side of the figure. Allele 1 (from brown trout) of Pgm-2 codes for the monomer at position 1. Alleles 2 and 3 (from Atlantic salmon) of Pgm-2 code for the monomers at positions 3 and 4, respectively. The monomer at position 2 is the product of another locus, Pgm-1. Proposed Pgm-2 genotypes are 11 (lane K), 22 (lane L), 12 (lanes A, C-H, J), 13 (lanes B, I), 122 (lanes Q-S), and 123 (lanes M-P, T-V).

triploid based on isozyme analysis. Alleles from all three parental species were codominantly expressed in the hybrid (Table 1, Fig. 3); thus, this individual expressed the haploid genomes of all three species. Cytological examination of the second such individual determined that it also was triploid. While high chromosome numbers and poor spreads made definitive analysis impossible, counts of 104–107 chromosomes were found and metacentrics and submeta-

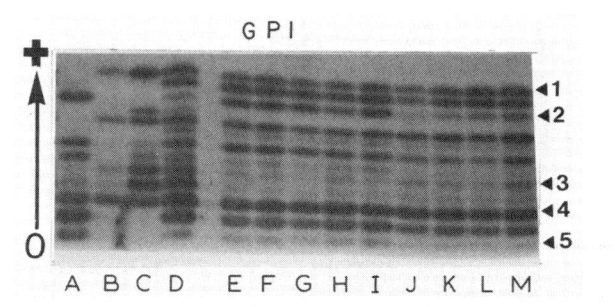

Fig. 3. Electrophoretic phenotypes of GPI from muscle tissue of brown trout (lane A), Atlantic salmon (lane B), brook trout (lane C), (brown trout × Atlantic salmon) × brook trout hybrid (lane D), brown trout × Atlantic salmon F₁ hybrids (lanes E-I), and backcross of F₁ hybrids to Atlantic salmon (lanes J-M). Positions of homodimers are indicated by the numbers at the right side of the figure. Allele 3 (from Atlantic salmon or brook trout) of Gpi-3 codes for the homodimers at position 1. Allele 1 (from brown trout) of Gpi-3 codes for the homodimers at position 2. Allele 2 (from brook trout) of Gpi-2 codes for the homodimers at position 3. Allele 1 (from brown trout or Atlantic salmon) of Gpi-2 and allele 2 (from Atlantic salmon or brook trout) of Gpi-1 code for the homodimers at position 4. Allele 1 (from brown trout) of Gpi-1 codes for the homodimers at position 5. Proposed Gpi-1 genotypes are 11 (lane A), 22 (lanes B, C), 12 (lanes E-I), and 122 (lanes D, J-M). Proposed Gpi-2 genotypes are 11 (lanes A, B, E-F), 22 (lane C), and 112 (lane D). Proposed Gpi-3 genotypes are 11 (lane A), 22 (lanes B, C), 12 (lanes E-I), and 122 (lanes D, J-M).

centrics equaled that expected from an individual containing the haploid genomes of all three parental species.

DISCUSSION

The disparate chromosome complements of the F_1 hybrids most likely caused some type of meiotic disruption. The karyotypes and electrophoretic phenotypes of their progeny suggest that the F_1 hybrids produced diploid eggs that either developed gynogenetically or, in most cases, formed triploid zygotes with paternal sperm sources. The genotypes of these diploid eggs appeared to be identical to the genotypes of the F_1 hybrids. We considered and discuss below five possible explanations for these aberrant results — premeiotic endomitosis, ameiotic egg maturation, meiotic misdivisions, hybridogenesis, and misclassification.

Premeiotic endomitosis

We believe the most likely explanation for our results is that a premeiotic endomitosis occurred in some of the oogonial cells of the F_1 hybrids. The doubling of the chromosome number coupled with meiotic pairing of the duplicated chromosomes (sister homologues) would produce diploid ova, identical in genotype to the F_1 hybrid, regardless of any crossovers. (Recombination between identical chromosomes would produce no new combination.) This type of oogenesis has been reported to occur in three gynogenetic triploid *Poeciliopsis* interspecific hybrids and one gynogenetic diploid and two gynogentic triploid *Poecilia* interspecific hybrids (Schultz 1980). This mechanisms has also been reported in gynogentic triploid salamanders and parthenogenetic triploid reptiles (Bogart, 1980).

An investigation of diplotene chromosomes from *Xenopus* hybrid oocytes provided direct, cytological evidence for this type of oogenesis. Muller (1977) found mixed populations of diploid and tetraploid oocytes in diploid *Xenopus* interspecific hybrids. He observed few bivalents in diploid oocytes, but complete bivalent pairing in tetraploid oocytes and concluded that pairing in the tetraploid oocytes must be restricted to identical sister homologues.

Tetraploid spermatocytes are common in Atlantic salmon (Nygren et al., 1968), brown trout (Nygren et al., 1971), and F_1 hybrids between these two species (Nygren et al., 1972). In males, these tetraploid cells showed the same types of multivalent pairing configurations as diploid cells (Nygren et al., 1968, 1971, 1972). Therefore, in male F_1 hybrids, meiotic pairing disruptions would probably occur in tetraploid as well as diploid cells. Indeed, Nygren et al. (1972) observed many multivalents and univalents in tetraploid as well as diploid spermatocytes. These meiotic disruptions would result in poor gametic viability as was reported in a male F_1 hybrid (Nygren et al., 1975). However, a different chromosomal pairing mechanism may occur in female F_1 hybrids.

Meiosis in female brown trout × Atlantic salmon hybrids has not been examined. Meiotic studies of other salmonid species and hybrids have shown that multivalent pairing occurs frequently in males but rarely, if at all, in females (Wright et al., 1983a). If tetraploid populations of oocytes occur in brown trout × Atlantic salmon F_1 hybrids and if pairing in these cells is restricted to bivalents between sister chromosomes, then balanced, diploid gametes would be produced that would contain the chromosome sets of both species.

Ameiotic egg maturation

The progeny phenotypes that we observed could also have been produced if egg maturation had proceeded without meiosis. Ameiotic egg maturation has been proposed to occur in female F_1 hybrids between *Carassius auratus* and *Cyprinus carpio* (Cherfas and Ilyasova, 1980; Cherfas et al., 1981) and in female F_1 hybrids between *Oryzias latipes* and *O. celebensis* (Iwamatsu et al., 1984). However, the only cytological evidence for this mechanism comes from a study of the gynogentic triploid carp *Carassius auratus gibelio*. Cherfas (1966 cited in Kirpichnikov, 1981) reported that, in females of these triploid carp, chromosome pairing, crossing-over, and the reductional division of meiosis were all aborted. The equational division of meiosis (now equivalent to a mitotic division) proceeded and, after extrusion of a single polar body, an egg would be produced whose genotype was identical to the genotype of the triploid female.

Premeiotic endomitosis and ameiotic egg maturation would both result in identical gametes and progeny genotypes. Cytological examinations of F_1 hybrid oocytes are needed to distinguish between these two mechanisms.

Meiotic misdivisions

Diploid eggs could also be produced if normal chromosome pairing and crossing-over were followed by a repression of either the first or the second meiotic division. However, based on gene-centromere recombination frequency estimates obtained by our laboratory, we were able to eliminate the possibility that meiotic misdivisions occurred in the brown trout × Atlantic salmon hybrids. If the first meiotic division was repressed, then no crossovers could have occurred between any of the marked loci and their centromeres since all gametes appeared to be heterozygous at these loci. Evidence from female brown trout and *Salvelinus* hybrids (Table 2) that crossovers occur frequently between the centromere and the loci Mpi and Gpi-3 implies that repression of the first meiotic division did not occur in the brown trout × Atlantic salmon F_1 hybrids. Repression of the first meiotic division has been reported in the frog *Rana sphenocephala* (Wright et al., 1976).

TABLE 2

Gene-centromere crossover frequencies for the loci Mpi, Gpi-3, Ldh-5, and Pgm-2 estimated from heterozygote frequencies of diploid gynogens (development activated by UV-irradiated sperm, second polar body retained by heat shock according to Chourrout, 1980)

Heterozygous locus	Species or hybrid	No. of females[a]	No. of gynogens	No. of heterozygous gynogens	Gene-centromere crossover frequency (2SE)[b]
Mpi	Brown trout	2	151	145	0.960 (0.032)
Gpi-3	F_1 Splake hybrid[c]	1	29	28	0.966 (0.067)
Ldh-5	Rainbow trout	3	288	2	0.007 (0.010)
Pgm-2	Rainbow trout	2	237	22	0.093 (0.038)
Pgm-2	F_1 Cutbow hybrid[d]	6	111	14	0.126 (0.063)

[a]Females were combined to obtain better estimates since genotypic frequencies of their gynogens were homogeneous.
[b]Assuming complete interference (see Thorgaard et al., 1983).
[c]Female *Salvelinus fontinalis* × male *Salvelinus namaycush*.
[d]Female *Salmo gairdneri* × male *Salmo clarki utah*.

If the second meiotic division was repressed, then a single crossover must have occurred between every marked locus and its centromere since all gametes appeared to be heterozygous at these loci. However, gene-centromere crossovers rarely occur for Ldh-5 and Pgm-2 in females of other salmonid species or hybrids (Table 2) and few crossovers probably occur between these two loci and their centromeres in brown trout or Atlantic salmon since most locus arrangements appear to be conserved among salmonid species (Johnson and Wright, 1984). Repression of the second meiotic division has been shown to be the mechanism responsible for the occurrence of occasional triploid progeny in *Salmo gairdneri* (Wright et al., 1983b). In trout, retention of the second polar body can occur spontaneously in over-ripe eggs (our unpublished data), or it can be induced artificially by heat or pressure treatments (Chourrout, 1980; Thorgaard et al., 1981, 1983).

Hybridogenesis

A mechanism termed "hybridogenesis" (Schultz, 1969) has been reported to occur in females of three diploid *Poeciliopsis* interspecific hybrids (Schultz, 1980) and in the hybrid frog *Rana esculenta* (Bogart, 1980). In this mechanism, there is a preferential loss of the entire paternal genome following a premeiotic endomitosis in hybrid oocytes (Cimino, 1972). Hybridogenesis could have produced the diploid, but not the triploid, backcross progeny phenotypes that we observed in this study. We were not able to distinguish paternal from grandpaternal alleles (both were Atlantic salmon) and thus could not determine if the diploid backcross progeny were gynogens or hybridogens.

Misclassification

Another possible explanation for the aberrant backcross segregation we observed is that the supposed F_1 hybrids used in the backcrosses to Atlantic salmon actually were brown trout. This type of misclassification has confounded the results of previous studies (Nyman, 1970; Nygren et al., 1972; Cross and O'Rourke, 1978). In this study, we were not able to examine the female F_1 hybrids used to make the backcrosses. However, they must have been true F_1 hybrids because they produced hundreds of triploid backcross progeny that all expressed two copies of Atlantic salmon alleles and only one copy of brown trout alleles for each locus examined. Likewise, each of the two triploid backcross karyotypes examined appeared to have two sets of Atlantic salmon chromosomes and only one set of brown trout chromosomes. The hybrid resulting from the cross made between an F_1 hybrid and a brook trout expressed brown trout, Atlantic salmon, and brook trout alleles; clearly, its female parent must have been a brown trout × Atlantic salmon hybrid.

CONCLUSION

Brown trout and Atlantic salmon are closely related species based on morphology (Scott and Crossman, 1973) and protein similarities (Johnson, 1984), but they have very different karyotypes (Nygren et al., 1971, 1972). The viability of F_1 hybrids and their codominant expression of alleles suggest that there are no major regulatory incompatibilities between these two species. However, even though some F_1 hybrids are fertile, meiotic pairing disruptions appear to prevent introgression. Male F_1 hybrids appear to have very poor gametic viability (Nygren et al., 1975). The female F_1 hybrids we examined produced many viable gametes, but these appeared to contain the haploid genomes of both parental species. Thus, efforts to transfer desirable traits of one species to the other by successive backcrossing will most likely fail. Likewise, there appears to be little need for concern that widespread introductions of brown trout will lead to genic introgression with indigenous Atlantic salmon in North America. If female brown trout × Atlantic salmon F_1 hybrids consistently produce many unreduced eggs, then activation of these diploid eggs by irradiated sperm could be used to generate F_1 hybrid clones that would be useful for a number of genetic studies.

The possibilities of androgenesis, gynogenesis, hybridogenesis, polyploidy, skewed sex ratios, and aberrant gene segregation should always be considered in the analysis of interspecific crosses because of potential chromosome pairing disruptions and incompatibilities of regulatory mechanisms (Chevassus, 1983). Our results show the importance of using karyotypes and well-defined genetic markers such as isozymes to determine parental contributions and gene segregation in species hybrids. This study also illustrates how information from gene mapping can be used to analyze inheritance results.

The entire Salmonidae family is thought to be of tetraploid origin (Ohno et al., 1969; Wright et al., 1983a). Hybridization and gynogenesis are thought to be important steps in the evolution of polyploid species in other vertebrate groups (Bogart, 1980; Schultz, 1980). The results of this study show that at least one salmonid interspecific hybrid produces unreduced eggs and suggests that hybridization may have played a role in the evolution of salmonid tetraploidy.

ACKNOWLEDGEMENTS

This work was supported by National Science Foundation Grant DEB-8120149. We would like to thank Michael Vernesoni and David Sumner of the Fish and Wildlife Division of the Connecticut Department of Environmental Protection for supplying the fish samples. We thank Dr. Vincent Ringrose of Hartford, Connecticut, for rearing the novel $F_1 \times$ brook trout hybrids to fingerling size. We also thank Mary Delany for the chromosome cytology results shown, and Bernie May for aid in electrophoretic analyses and interpretations.

REFERENCES

Allendorf, F.W., Mitchell, N., Ryman, N. and Stahl, G., 1977. Isozyme loci in brown trout (*Salmo trutta* L.): detection and interpretation from population data. Hereditas, 86: 179–190.

Alm, G., 1955. Artificial hybridization between different species of the salmon family. Rep. Int. Freshwater Res., Drottingholm, 36: 13–59.

Beland, K.F., Roberts, F.L. and Saunders, R.L., 1981. Evidence of *Salmo salar* × *Salmo trutta* hybridization in a North American river. Can. J. Fish. Aquat. Sci., 38: 552–554.

Blanc, J.M. and Chevassus, B., 1979. Interspecific hybridization of salmonid fish. I. Hatching and survival up to the 15th day after hatching in F_1 generation hybrids. Aquaculture, 18: 21–34.

Bogart, J.P., 1980. Evolutionary implications of polyploidy in amphibians and reptiles. In: W.H. Lewis (Editor), Polyploidy: Biological Relevance. Plenum Press, New York, NY, pp. 341–378.

Cherfas, N.B., 1966. Natural triploidy in the females of the unisexual variety of the silver crucian carp (*C. auratus gibelio* Bloch). Genetika (Moscow), 2(5): 16–24.

Cherfas, N.B. and Ilyasova, V.A., 1980. Induced gynogenesis in silver crucian carp and carp hybrids. Genetika (Moscow), 16(7): 1260–1269.

Cherfas, N.B., Gomelsky, B.I., Emeljanova, O.V. and Rekoubratsky, A.V., 1981. Triploidy in reciprocal hybrids obtained from crucian carp and carp. Genetika (Moscow), 17(6): 1136–1139.

Chevassus, B., 1979. Hybridization in salmonids: results and perspectives. Aquaculture, 17: 113–128.

Chevassus, B., 1983. Hybridization in fish. Aquaculture, 33: 245–262.

Chourrout, D., 1980. Thermal induction of diploid gynogenesis and triploidy in the eggs of rainbow trout (*Salmo gairdneri* Richardson). Reprod., Nutr., Dev., 20: 727–733.

Cimino, M.C., 1972. Egg production, polyploidization, and evolution in a diploid all-female fish of the genus *Poeciliopsis*. Evolution, 26: 294–306.

Clayton, J.W. and Tretiak, D.N., 1972. Amine-citrate buffers for pH control in starch-gel electrophoresis. J. Fish. Res. Board Can., 29: 1169–1172.

Cross, T.F. and O'Rourke, F.J., 1978. An electrophoretic study of the haemoglobins of some hybrid fishes. Proc. R. Irish Acad., Sect. B, 78: 171–178.

Gold, J.R., 1974. A fast and easy method for chromosome karyotyping in adult teleosts. Prog. Fish Cult., 36: 169–171.

Iwamatsu, T., Uwa, H., Inden, A. and Hirata, K., 1984. Experiments on interspecific hybridization between *Oryzia latipes* and *Oryzias celebensis*. Zool. Sci., 1: 653–663.

Johnson, K.R., 1984. Protein Variation in Salmonidae: Genetic Interpretations of Electrophoretic Banding Patterns, Linkage Associations Among Loci, and Evolutionary Relationships Among Species. Ph.D. thesis, The Pennsylvania State University, University Park, PA, 181 pp.

Johnson, K.R. and Wright, J.E., 1984. Linkage and pseudolinkage groups in salmonid fishes. In: S.J. O'Brien (Editor), Genetic Maps 1984: A Compilation of Linkage and Restriction Maps of Genetically Studied Organisms, Vol. 3. Cold Spring Harbor Laboratory, New York, NY, pp. 335–338.

Kirpichnikov, V.S., 1981. Genetic Bases of Fish Selection. Springer-Verlag, Berlin, Heidelberg, New York, NY, 410 pp.

Markert, C.L. and Faulhaber, I., 1965. Lactate dehydrogenase isozyme patterns of fish. J. Exp. Zool., 159: 319–332.

May, B., Wright, J.E. and Stoneking, M., 1979. Joint segregation of biochemical loci in Salmonidae: results from experiments with *Salvelinus* and review of the literature on other species. J. Fish. Res. Board Can., 36: 1114–1128.

May, B., Stoneking, M. and Wright, J.E., Jr., 1980. Joint segregation of biochemical loci in Salmonidae. II. Linkage associations from a hybridized *Salvelinus* genome (*S. namaycush* × *S. fontinalis*). Genetics, 95: 707–726.

Muller, W.P., 1977. Diplotene chromosomes of *Xenopus* hybrid oocytes. Chromosoma, 59: 273–282.

Nygren, A., Nilsson, B. and Jahnke, M., 1968. Cytological studies in Atlantic salmon (*Salmo salar*). Ann. Acad. Reg. Sci. Upsalien, 12: 21–52.

Nygren, A., Nilsson, B. and Jahnke, M., 1971. Cytological studies in *Salmo trutta* and *Salmo alpinus*. Hereditas, 67: 259–268.

Nygren, A., Nilsson, B. and Jahnke, M., 1972. Cytological studies in Atlantic salmon from Canada, in hubrids between Atlantic salmon from Canada and Sweden and in hybrids between Atlantic salmon and sea trout. Hereditas, 70: 295–306.

Nygren, A., Nyman, L., Svensson, K. and Jahnke, G., 1975. Cytological and biochemical studies in back-crosses between the hybrid Atlantic salmon × sea trout and its parental species. Hereditas, 81: 55–62.

Nyman, O.L., 1970. Electrophoretic analysis of hybrids between salmon (*Salmo salar* L.) and trout (*Salmo trutta* L.). Trans. Am. Fish. Soc., 99: 229–236.

Ohno, S., Muramoto, J., Klein, J. and Atkin, N.B., 1969. Diploid–tetraploid relationships in clupeoid and salmonoid fish. In: C.D. Darlington and K.R. Lewis (Editors), Chromosomes Today, Vol. II. Oliver and Boyd, Edinburgh, pp. 139–147.

Ojima, Y., Hayashi, M. and Ueno, K., 1975. Triploidy appeared in the backcross offspring from funa–carp crossings. Proc. Jpn. Acad., 51: 702–706.

Payne, R.H., Child, A.R. and Forrest, A., 1972. The existence of natural hybrids between the European trout and Atlantic salmon. J. Fish Biol., 4: 233–236.

Piggins, D.J., 1971. Salmon × sea trout hybrids (1969–1970). Salmon Res. Trust Ireland, Annu. Rep., 1970, 15: 41–58.

Rees, H., 1967. The chromosomes of *Salmo salar*. Chromosoma, 21: 472–474.

Refstie, T. and Gjedrem, T., 1975. Hybrids between Salmonidae species. Hatchability and growth rate in the freshwater period. Aquaclture, 6: 333–342.

Ridgway, G.J., Sherburne, S.W. and Lewis, R.D., 1970. Polymorphisms in the esterases of Atlantic herring. Trans. Am. Fish. Soc., 99: 147–151.

Roberts, F.L., 1970. Atlantic salmon (*Salmo salar*) chromosomes and speciation. Trans. Am. Fish. Soc., 99: 105–111.

Schultz, R.J., 1969. Hybridization, unisexuality and polyploidy in the teleost *Poeciliopsis* (Poeciliidae) and other vertebrates. Am. Nat., 103: 605–619.

Schultz, R.J., 1980. Role of polyploidy in the evolution of fishes. In: W.H. Lewis (Editor), Polyploidy: Biological Relevance. Plenum Press, New York, NY, pp. 313–340.

Scott, W.B. and Crossman, E.J., 1973. Freshwater Fishes of Canada. Bull. 184, Fish. Res. Board Can., Ottawa, Canada, 966 pp.

Solomon, D.J. and Child, A.R., 1978. Identification of juvenile natural hybrids between Atlantic salmon (*Salmo salar* L.) and trout (*Salmo trutta* L.). J. Fish Biol., 12: 499–501.

Stoneking, M., Wagner, D.J. and Hildebrand, A.C., 1981. Genetic evidence suggesting subspecific differences between northern and southern populations of brook trout (*Salvelinus fontinalis*). Copeia, 1981: 810–819.

Thorgaard, G.H., Jazwin, M.E. and Stier, A.R., 1981. Polyploidy induced by heat shock in rainbow trout. Trans. Am. Fish. Soc., 110: 546–550.

Thorgaard, G.H., Allendorf, F.W. and Knudsen, K.L., 1981. Gene-centromere mapping in rainbow trout: high interference over long map distances. Genetics, 103: 771–783.

Vuorinen, J. and Piironen, J., 1984. Electrophoretic identification of Atlantic salmon (*Salmo salar*), brown trout (*S. trutta*) and their hybrids. Can. J. Fish. Aquat. Sci., 41: 1834–1837.

Wright, D.A., Huang, C. and Chuoke, B.D., 1976. Meiotic origin of triploidy in the frog detected by genetic analysis of enzyme polymorphisms. Genetics, 84: 319–332.

Wright, J.E., Johnson, K., Hollister, A. and May, B., 1983a. Meiotic models to explain classical linkage, pseudolinkage, and chromosome pairing in tetraploid derivative salmonid genomes. In: G.S. Whitt, M.C. Ratazz and J. Scandalios (Editors), Isozymes: Current Topics in Biological and Medical Research, Vol. 10: Genetics and Evolution. Alan R. Liss, Inc., New York, NY, pp. 239–260.

Wright, J.E., Johnson, K.R. and May, B., 1983b. Determining gene-centromere distances for isozyme loci in triploid salmonids. Isozyme Bull., 16: 58.

ABSTRACTS OF POSTER PAPERS

A review of the production and quality control of triploid grass carp and progress in implementing 'sterile' triploids as management tools in the U.S.

S.K. ALLEN, Jr.[1] and ROBERT J. WATTENDORF[2]

[1] *School of Fisheries, University of Washington, Seattle, WA 98195 (U.S.A.)*
[2] *Florida Game and Fresh Water Fish Commission, Eustis, FL 32727 (U.S.A.)*

ABSTRACT

In 1983 triploid grass carp (*Ctenopharyngodon idella*) were produced in commercial quantities for the first time by J.M. Malone and Son, Lonake, Arkansas. Ploidy was analyzed by flow cytometry and karyology at the University of Washington; 84% triploids resulted from Malone's treatments, but this proportion could be increased to 98% by hand sorting. Criteria for hand sorting were based on similarities of some treated fish to highly inbred carp, suggesting that some diploids arise by gyno- or androgenesis.

Triploids are now routinely produced for use as a biological weed-control tool obviating previous concerns that (1) escaped diploid grass carp would become established where not wanted, and (2) the triploid hybrid between grass carp and bighead carp are not efficient weed eaters. Quality control of 100% triploidy (expected to be the adopted policy in most States) is the most important issue facing the implementation of these fish as management tools. A fast efficient screening technique has been developed using a Coulter Counter with Channelizer; as many as 1500 fish per day can be assayed. Many regions of the U.S. which have weed-control problems (such as the Southeast and California) are currently modifying their restrictive policies toward the use of grass carp to accommodate the triploid.

Chemically and pressure-induced triploidy in the Pacific oyster *Crassostrea gigas*

S.K. ALLEN, JR.[1], S.L. DOWNING[1], J. CHAITON[2] and J.H. BEATTIE[1]

[1] *School of Fisheries, University of Washington, Seattle, WA 98195 (U.S.A.)*
[2] *Coast Oyster Co., P.O. Box 327, Quilcene, WA 98376 (U.S.A.)*

ABSTRACT

Triploids *C. gigas*, if sterile, will likely mitigate some of the problems in survival and marketability of oysters harvested during the spawning season. Means of inducing triploidy were investigated by pursuing two independent lines of study: inhibiting meiosis with cytochalasin B or hydrostatic pressure.

Cytochalasin B (1 mg/ml) was administered to eggs at three intervals following fertilization: 0–15 min (I), 15–30 min (II) or 30–45 min (III). This experiment was repeated at four temperatures: 18, 20, 25, and 28°C. All trials were performed simultaneously on a single spawn of subdivided eggs (12 treatments); an untreated control was run at each temperature (four treatments) for a total of 16 treatments. Percent triploidy at each temperature for treatment intervals I, II, and III (listed respectively — sample size in parentheses) were: 18°C — 10% (30), 12% (25), 60% (25); 20°C — 4% (25), 40% (40), 92% (25); 25°C — 27% (30), 88% (25), 100% (25); 28°C — 38% (29), 20% (30), 23% (30). These trends show that the window of vulnerability of newly fertilized oyster eggs to cytochalasin treatments appears sooner at higher temperatures, probably reflecting increased spread of meiotic events.

Pressure treatments were also administered to newly fertilized eggs. Maximum proportions of triploids (60%) were obtained by pressures of 7200 psi applied 10 min after fertilization and lasting for 10 min at 29°C. Ploidy was analyzed in some pressure groups by flow cytometric analysis of 250 μm free-swimming larvae.

Effects of temperature, size and rations on the growth of strains of Arctic charr in intensive aquaculture

R.F. BAKER and G.B. AYLES

Fisheries and Oceans Canada, Winnipeg, Man. (Canada)

ABSTRACT

For several reasons, Arctic charr has been perceived as a highly suitable species for use in intensive aquaculture. This study examined the growth and bioenergetics of different sizes and strains (from Norway, Eastern North America and High Arctic) of Arctic charr under a wide range of temperatures and ration levels. Comparisons were made with a domestic rainbow trout strain and the feasibility of the use of Arctic charr as an intensive culture species was assessed. At low temperatures (7°C and 10°C) charr performed as well as rainbow trout; at 14°C rainbow trout performed significantly better; 19°C was well above the optimum for Arctic charr. There were also significant differences in performance betweeen the three strains of charr. It was concluded that Arctic charr is a suitable species for cool-water aquaculture.

Performance of two red tilapia hybrids in polyculture with channel catfish and freshwater prawn

L.L. BEHRENDS, J.B. KINGSLEY and A.H. PRICE

Agricultural Research, TVA, F4 NFDC, Muscle Shoals, AL 35660 (U.S.A.)

ABSTRACT

Replicated polyculture experiments involving tilapias, channel catfish, and freshwater prawns were conducted in earthen ponds during the summers of 1983 and 1984. One objective of these experiments was to evaluate the growth performance of two hybrid strains of red tilapia. In both studies, red hybrids were stocked as young-of-the-year fingerlings (YOY) as opposed to overwintered fingerlings. Fingerling *Tilapia aurea*, also YOY, were co-stocked with the red hybrids for comparative purposes. In preliminary experiments conducted during 1980, a commercial strain of red tilapia (Florida strain) derived from the cross (*T. hornorum* × *T. mossambica*, red mutant), grew only 50% as fast as *T. aurea*. The poor performance was attributed to (1) inbreeding depression, and (2) the genetic constitution of the parent lines. Both *T. hornorum* and *T. mossambica* are slow-growing species which mature reproductively at an early age (3-4 months from hatching).

During 1981-1983, an improved red hybrid was developed by crossing two F-1 hybrids (*T. aurea* × red tilapia, Florida strain) × (red tilapia, Florida strain × *T. nilotica*). The improved red-strain grew to an average weight of 302 g after 130 days of culture. Under identical conditions, *T. aurea* fingerlings grew to 252 g.

In a second study, a cold-tolerant population of red tilapia was developed by crossing the Florida red tilapia to *T. aurea*, followed by two generations of backcrossing to *T. aurea*. Backcross populations exhibit cold-tolerance qualities of *T. aurea* and results indicate that their growth performance is similar to *T. aurea*. Breeding techniques used to develop true-breeding populations of red tilapia are discussed with respect to breeding compatibility and possible commercial applications.

Characterization of genetic variability within and between seven distinct population-groups of rainbow trout (*Salmo gairdneri* Richardson)

WILLIAM J. BERG

Dept. Animal Science, University of California, Davis, CA 95616 (U.S.A.)

ABSTRACT

Electrophoretic analysis of well over 100 populations of native rainbow trout has revealed the presence of seven genetically distinct groups of populations. Eighty-three electrophoretic alleles, electromorphs, were found to be segregating at 38 genetic loci. Based on the frequency distribution of these alleles, diagnostic genetic profiles were developed for each population-group. Unweighted gene diversity analysis indicated that approximately 90% of the total genetic heterozygosity was attributable to the within population component with most of the remaining heterozygosity assigned to the between population-group component. Three indirect methods of analysis estimate high levels of gene flow among populations of the Coastal rainbow trout group. It is arguable that both naturally occurring straying and human intervention in the distribution of anadromous rainbow trout have resulted in the relative genetic homogeneity, in spite of large geographical distances, seen in Coastal rainbow trout. A model describing the genetic differentiation among population-groups suggests that recent geographic isolation and subsequent genetic drift has produced the genetic variability observed in rainbow trout.

Polyploidy induced by heat shock in channel catfish

C.A. BIDWELL[1], C.L. CHRISMAN[2] and G.S. LIBEY[3]

[1]Dept. Animal Science, University of California, Davis, CA 95616 (U.S.A.)
[2]Dept. Animal Sciences, Purdue University, West Lafayette, IN 47907 (U.S.A.)
[3]Forestry and Natural Resources, Purdue University, West Lafayette, IN 47907 (U.S.A.)

ABSTRACT

Female channel catfish were induced to ovulate in fiberglass raceways by administering a total dose of 11 mg/kg of carp pituitary extract in two injections. Eggs were handstripped and fertilized with minced testis from a donor male. First cleavage division was visible and occurred around 90 min post-fertilization. Eggs were heat shocked in 56-l aquaria at 80, 85, or 90 min post-fertilization at temperatures of 40, 41, 42, or 43°C for a duration of 1, 2, 3, or 4 min.

Eggs subjected to heat shock produced tetraploids, triploids, diploids and mosaics. Ploidy level was determined by chromosome preparations of five embryos per treatment combination sampled prior to hatching. Eggs stressed at 42 and 43°C had nearly complete mortality in shocks longer than 1 min. Forty-one degrees and 3 min duration induced tetraploidy in 62% of the embryos sampled. In shocks of 3 and 4 min at 40 and 41°C, post-fertilization time was not significant. Hatchability was low in treatments that induced tetraploidy but no abnormalities were found among sampled embryos. Live fry from these groups are being raised for further study.

Genetic variability in the European oyster, *Ostrea edulis*: geographic variation between local French stocks

FRANCOISE BLANC[1], PAUL PICHOT[2] and JEAN ATTARD[3]

[1]University Paul Valery, Montpellier (France)
[2]IFREMER, 34200 Sète (France)
[3]Pêches et Océans, 901 Cap Diamont, Que. G1K 7Y7 (Canada)

ABSTRACT

The European Oyster, *Ostrea edulis*, is native to the Atlantic and Mediterranean coasts of France. Valuable harvesting began about 1850 with efficient spat collection on natural beds and has been carried out until now with erratic performances due to epizootic diseases. IFREMER is developing a research program for evaluation of genetic variability among some natural beds currently used for spat collection. Genetic variability is reported here for five local stocks from the Atlantic coast (Cancale, Odet, Belle-ile, Auiberon), including one deep-water population called "Pied de cheval" (Saint Quay-Portrieux) and two from the Mediterranean coast (Leucate and Thau lagoons). Sample size for each population was about 75 individuals. Allozymic variation was analyzed for 11 enzymatic systems coded by 19 presumptive structural gene loci. Adenylate-kinase (AK), aspartate amino-transferase (AAT-1, AAT-2), isocitrate dehydrogenase (IDH-1,

IDH-2), glucosephosphate isomerase (GPI), malate dehydrogenase (MDH-1, MDH-2), malic enzyme (ME), phosphoglucomutase (PGM) and 6-phosphogluconate dehydrogenase (6-PGDH) were analyzed by horizontal starch-gel electrophoresis. Superoxide dismutase (SOD-1, SOD-2) and esterase (EST-1, EST-2) were analyzed using polyacrylamide gel electrophoresis while isoelectric focusing was used for leucine amino-peptidase (LAP-1, LAP-2, LAP-3, LAP-4).

Seven loci were monomorphic: MDH-1, ME, IDH-2, SOD-1, SOD-2, AAT-2, and LAP-2. Five loci, GPI, IDH-1, 6-PGDH, AK and LAP-1, were weakly polymorphic with mean expected heterozygosity per locus below 0.07. The remaining loci (PGM, MDH-2, EST-1, EST-2, AAT-1, LAP-3, and LAP-4) were highly polymorphic, with a mean expected heterozygosity per locus between 0.11 and 0.63. Mean proportion of polymorphic loci per population (P) estimated directly from actual data, was between 47 and 58% with a mean value of 56% for all populations. Due to numerous rare alleles, P with a 95% criterium was 26–42% with a mean value of 37%. Similarly, the mean number of alleles per locus (A) ranged from 1.68 to 1.84, and with a 95% confidence interval from 1.32 to 1.53. Mean expected heterozygosity ranged from 0.117 to 0.143 with mean observed heterozygosity ranging from 0.117 to 0.147. Heterozygote deficiencies from Hardy—Weinberg equilibrium were found for LAP-3 and LAP-4 in the Leucate and Thau samples (G-test; $P<0.01$). These estimates are similar to those reported by B. Magennis et al. (1982) for three Irish natural beds analyzed at nine loci. However, they appear higher than those of one sample imported from the Netherlands and bred in Maine (U.S.A.) based on an examination of 29 structural loci by N.E. Buroker.

Nei's genetic distances indicate a close relationship between this set of French local stocks and allows the inclusion of the deep-water "Pied de cheval" population in *Ostrea edulis* s.s. Only one (Leucate lagoon) of the two Mediterranean stocks appears slightly apart from the other six, whereas our data suggest that the second Mediterranean sample (Thau lagoon) may be derived from a mixed population imported from Atlantic beds.

The effect of inbreeding on reproduction, growth and survival of channel catfish

K. BONDARI[1] and R.A. DUNHAM[2]

[1]*Coastal Plain Experiment Station, University of Georgia, Tifton, GA 31793 (U.S.A.)*
[2]*Fisheries and Allied Aquacultures, Auburn University, Auburn, AL 36849 (U.S.A.)*

ABSTRACT

Fitness-related traits, such as litter size or lactation in mammals and fertility and egg production in poultry, are reported to decline with intense inbreeding. The inbreeding coefficient can, however, increase more rapidly, even with random mating, in fish species than in farm animals. This study was conducted to investigate the effects of two generations of full-sib matings on growth, reproduction, and survival of the Tifton strain of channel catfish (*Ictalurus punctatus*).

First-generation inbred channel catfish were produced from brother—sister matings in 1979 (I1 line). A randomly mated control line (C1) was propagated from the same base population. The inbred and control lines were comprised of five full-sib families each. Second-generation inbred (I2) and control (C2) lines were produced in 1981 by mating each male catfish from the I1 or C1 line to two females in sequence, one from the I1 and one from the C1 line. The design also produced two reciprocal lines, I1 male × C1 female and I1 female × C1 male to be compared with contemporary I2 and C2 lines. Repeat matings (10 spawns/line or cross) were accomplished. The coefficient of inbreeding for the inbred line increased from 0.25 in the first generation to 0.375 in the second generatrion. The inbreeding coefficient was zero for all other lines. Fish from both generations were performance tested in Tifton, GA and Auburn, AL.

The 1979 results indicated that one generation of full-sib mating reduced average egg weight by 17% but did not significantly influence spawn weight, hatchability score, or the number of days required for eggs to hatch. Two generations of full-sib mating reduced 4-week body weight significantly but did not influence survival rates to various ages.

Respawn of the brood fish in 1982 indicated that one generation of inbreeding did not depress spawning rate, eggs/kg female body weight or egg hatchability. Two generations of inbreeding did not depress vulnerability to capture by seine, growth rate during winter, tolerance of low oxygen or survival in the pond environment. However, two generations of inbreeding resulted in a 19% depression of 15-month body weight for channel catfish grown in ponds. Growth rate of reciprocals, I1 male × C1 female and I1 female × C1 male, was not different from the control, indicating that outcrossing restored the depressed growth performance.

One generation of full-sib mating reduced performance for some reproductive traits in channel catfish and two generations of full-sib mating reduced body weight in channel catfish. Inbreeding depression could be even more pronounced in other populations of channel catfish since the breeding history of Tifton strain indicates a high degree of heterozygosity at the start of the experiment.

Cryopreservation of oyster sperm

S. BOUGRIER[1] and L.D. RABENOMANANA[2]

[1]*IFREMER, B.P. 26, 56470 La Trinité Sur Mer (France)*
[2]*Marine Fisheries Direction, B.O. 291, Mahajanga (Madagascar)*

ABSTRACT

Gamete preservation allows us to realize matings normally impossible because of incompatibilities in species biology, and to study their genetic consequences. The aim of this work is to perfect a preservation technique for the oyster, *Crassostrea gigas*, by adaptation of the method used for the sea-fish, *Dicentrarchus labrax*.

The sperm, obtained by genital gland pressure, was diluted with a cryoprotectant diluent solution, then equilibrated and frozen. Freezing was initiated 5 cm above liquid nitrogen for 3 min. Liquid nitrogen was used for storage. Thawing was done in 1 min and fertilization rate was measured on 4-h embryos.

Two cryoprotectants have been tested: glycerol and DMSO. Glycerol was rejected because it caused agglutination of the spermatozoa after thawing. After testing two diluents, it was noticed that DCSB4 ($NaCl$, $MgSO_4$, $CaCl_2$ and DMSO in a Tris-HCl buffer, pH 8.2) was better than water. Fertilization rates were 61% and 27% for DCSB4 and the control, respectively. Three equilibration times have been tested: 0, 5 and 10 min. Instantaneous freezing produced a significantly higher fertilization rate ($P \leq 0.001$) than delayed freezing. After testing three DMSO concentrations (5, 10 and 15%) the optimum was found to be 10%. Of six dilution rates tested (from 1/5 to 1/20) the best results were obtained with 1/12.5 and 1/15, yielding results within 5% of each other.

Thus, it is recommended that cryopreservation of sperm of the oyster, *C. gigas*, be carried out as follows: obtain sperm by genital gland pressure, dilute 1 volume of sperm in 15 volumes of DCSB4 containing 10% DMSO, and use instantaneous equilibration. Using this technique, percentage fertilization compared to the control was 92% with sperm preserved for 3 days. The validity of these data will have to be checked using longer preservation periods. In addition, it will be necessary to study the success of embryogenesis, larvae and juveniles obtained with frozen sperm.

Crossing influence on shape and structure of oyster shells

S. BOUGRIER[1], H. GRIZEL[1] and J.P. DELTREIL[2]

[1]IFREMER, B.P. 26, 56470 La Trinité Sur Mer (France)
[2]IFREMER, 63 Boulevard Deganne, 33120 Arcachon (France)

ABSTRACT

In Brittany (West of France) the outside shape of the oyster, *Crassostrea gigas*, is generally wavy (Type W), especially in open bays such as the Quiberon bay. On the other hand, in the husbandries situated next to the harbor areas, if the wavy aspect is maintained, the antifouling-paint action disturbs the growth of *C. gigas* by a hollow accumulation which leads to an increase in thickness caused by tributyl tin (TBT). By studying the growth of different lots of oysters produced by a commercial hatchery, we observed, in November 1982, a low percentage of individuals exhibiting a shell of the smooth original type (Type S). The two types of oysters were crossed in a factorial plan to check the possible genetic nature of the smooth character. The four resulting groups have been bred at two sites where the environment parameters induce different growing types. In the St. Philibert river (Kerinis), the oysters are considered to grow under good conditions and their shape is suitable for the market. On the other hand, in the Crac'h river (Les Presses), the development of *C. gigas* is badly disturbed by activities of a nearby pleasure sailing harbor.

Almost all progeny from crossing female W × male W and female W × male S were W types while the progeny from female S × male W and female S × male S were S types. These results observed at both sites reveal important maternal effects. Regardless of male type, the individuals showed the mother shell aspect. The thickness coefficient of Imai and Sakai, a good parameter for the study of shell malformations, was employed. At Kerinis, the coefficient value was nearly constant in time, ranging from 20 to 25, and did not differ from one cross to another. At Les Presses, from July, two types of evolution according to the mother type were observed. For S, growth was comparable to that observed at Kerinis, and individuals did not present the thickening phenomenon. On the contrary, the coefficient value was higher for the W type ($C \sim 34$ for female W × male S, $C \sim 37$ for female W × male W). The presence of hollows inside the shell showed that these individuals were sensitive to TBT action.

Extensive karylogical investigations permitting the choice between different methods of gynogenesis and polyploidy induction in rainbow trout

D. CHOURROUT

INRA, Laboratoire de Physiologie des Poissons, Jouy en Josas (France)

ABSTRACT

Many researchers recently undertook the production and testing of gynogenetic and triploid strains for genetic improvement. However, accurate karyology is required to check the cleanliness of the different methods to be used. Remnants of sperm chromatin fragments have already been detected in gynogenetic embryos produced by gamma rays, and incomplete retention of the second polar body may occur at a high rate with pressure shocks that are too weak. In this study, karyotypes of 600 embryos resulting from various experiments have been carefully examined in order to compare two methods of sperm inactivation (dimethylsulphate and ultraviolet rays), to study the transition from haploid to diploid and from diploid to triploid by heat shocks, and to use a new heterologous sperm (grayling *Thymallus thymallus*).

Dimethylsulphate (15×10^{-3} M, $10°C$, dilution $5\times$) resulted in a Hertwig effect with all haploid embryos from 80 to 120 min of treatment. However, sperm chromatin fragments were encountered in the cells of embryos in most of this dose range; their freqency was appreciably decreased only at doses just lower than those destroying sperm motility.

Increasing durations (7–20 min) of $26°$ heat shocks applied 25 min after normal insemination provided results different from similar pressure experiments. No increase in the rate of abnormal embryos was detected all along the transition from 0 to 100% triploidy. Karyology showed a direct jump from diploidy to triploidy (without aneuploidy) as well as from haploidy to diploidy after use of irradiated sperm.

Insemination of rainbow trout eggs by grayling males resulted in embryos dying before the eyed stage. When a heat shock was added, triploid hybrids ($3n \sim 110$) were obtained but died between the eyed and hatching stage. These two observations resulted in the use of UV irradiated grayling sperm for producing rainbow trout gynogenetics. Other common salmonid species usually provide viable triploid hybrids with female rainbow trout.

Coregonid fish life history strategies

K.R. DABROWSKI

Tokyo University of Fisheries, Konan 4-5-7, Tokyo 108 (Japan)

ABSTRACT

Data on *Coregonus* sp. from three European locations, collected in the course of studies on fish growth and body chemical composition, provided the material for the quantitative evaluation of fish growth, metabolism and reproduction.

In summer, when most growth of fish occurs, respiratory energy contributes 40–66% of the energy consumed by the female. Energy partitioning into reproductive tissue attained its maximum before spawning in November–December, 20–27% of energy consumed. Such expression of reproductive effort in coregonid fish is not accurate since fish lose substantial amounts of their body reserves in populations situated in the north.

An attempt is made to compare the reproductive effort of coregonid fish domesticated in Japan to natural populations.

Genetic differentiation of isozymes in domesticated color pattern varieties and a wild population of the guppy, *Poecilia reticulata*

A.A. FERNANDO and V.P.E. PHANG

Department of Zoology, National University of Singapore, Kent Ridge 0511 (Singapore)

ABSTRACT

The guppy, *Poecilia reticulata*, is the predominant species of freshwater ornamental fish cultured in Singapore. It shows marked sexual dimorphism, males being smaller with polymorphic color patterns, while females are larger with drab coloration. Wild populations were introduced into Singapore before 1940 and have been under domestication for more than 30 years or about 100 generations. At present about 30 color pattern varieties are bred on farms practicing monoculture. Guppy farmers continuously select for vivid body and tail color patterns in males, and large body and fin size in both sexes.

The scope of this study was to determine genetic variability within, and differentiation between, two domesticated varieties, Red Snakeskin (RSs) and blond Tuxedo (bTu), and a wild-caught population (WT). Thirteen electrophoretically detectable enzyme loci were examined in skeletal muscle, liver, whole eye, ovary, testis, gills, heart, brain, spleen, and intestine. Nine loci, Ldh-A, Ldh-B, s-Mdh-B, G6pd-A, G6pd-B, 6-Pgd-A, Sdh-A, and Xdh-A, were monomorphic in all three populations. The 6PGD, SDH and XDH enzymes were expressed as a single invariant band in all fish examined. For G6PD the three stocks showed two bands which represented the G6pd-A and G6pd-B monomers. However, in 0.5% of RSs fish, a third band very close to the A monomer was detected and this was probably a conformational isozyme. There was variation in expression of LDH isozymes in the eye, with homotetramers (A_4, A_2B_2, B_4, B_2B_2, and C_4) evident in WT, while heterotetramers were also present in RSs and bTu. The three stocks were polymorphic for s-Mdh-A, with s-Mdh-A^1 subunit migrating faster. The frequency of this allele was 0.200 in WT, 0.375 in RSs, and 0.500 in bTu. A cathodally migrating isozyme of s-Mdh-C locus was observed in the liver of RSs and bTu. ADH was detected as a single band (Adh-A) in the eye of the three stocks, and the liver of WT and bTu, whereas in the liver of RSs, a variant allele (Adh-A^1) with greater anodal mobility than Adh-A was observed at a frequency of 0.125. In addition, a cathodally migrating band presumably coded by another locus, Adh-B, was manifested in the liver of 2.5% RSs fish.

The two color pattern varieties appeared to have differentiated from the wild population. The least genetic variation occurred in WT, in which only s-Mdh-A locus was polymorphic and the cathodally migrating bands coded by s-Mdh-C and Adh-B were not detected. RSs fish showed the greatest intra-strain genetic variation with four polymorphic loci (s-Mdh-A, s-Mdh-C, Adh-A, and Adh-B) while bTu was polymorphic for S-Mdh-A and s-Mdh-C. It was suggested that this higher level of heterozygosity in the two varieties was due to occasional outcrossing between strains and perhaps also to artificial selection favoring heterozygotes. In addition, guppy farmers use large numbers of broodstock to reduce inbreeding. Wild guppies occur in small pockets in streams and canals, often cut off from the next population. This could have led to inbreeding and hence erosion of genetic variability. No correlation was found between color pattern phenotypes of the domesticated varieties and isozyme variation for these seven gene—enzyme systems.

Alternative breeding methods in bivalve culture

PATRICK GAFFNEY

Dept. of Ecology and Evolution, S.U.N.Y., Stony Brook, NY 11733 (U.S.A.)

ABSTRACT

The use of modern breeding schemes in bivalve culture has been very limited to date. Most commercial operators use wild-caught or hatchery-reared animals for broodstock, without intentional selection; as such, they are in a position to benefit greatly from the application of various genetic techniques commonly used in agriculture today. The question remains however: which methods are best suited to bivalve culture?

Conventional long-term mass selection has proved to be of limited use thus far with commercial bivalves, largely due to the lack of continuous programs and adequate facilities. The long generation time of most species also militates against the use of conventional selection schemes.

Cytogenetic manipulations have been explored extensively in fish, and may be productively applied to bivalves as well. These methods include the production of various polyploids and gynogenetic diploids. Of the latter, suppression of the first mitotic division of haploid eggs to generate homozygous diploid clones may be the most promising method for commercial application. Its virtues include: (1) the complete elimination of all lethal and sublethal genes in a single stroke; (2) production of large numbers of distinct diploid homozygotes, which may then be cloned to yield isogenic lines; (3) such inbred lines may be desirable in themselves or for their combining abilities; (4) the production of high-performance hybrid seed from hatchery maintained clones provides an economic incentive for the development and application of such breeding schemes, an element heretofore lacking.

Current work in this area using the clam *Mulinia lateralis* as a model system is discussed.

Estimates of phenotypic and genetic parameters for carcass quality traits in rainbow trout

BJARNE GJERDE

Institute of Aquaculture Research, Agricultural University of Norway, Ås-NLH (Norway)

ABSTRACT

The data were derived at slaughter from three year-classes of rainbow trout at the Research Station for Salmonids. The analysis was based on records from 659 fish, 34 sires and 110 dams. The traits studied were: water, fat and protein percentage, gonad weight, meat color (score), visceral fat (score), belly thickness (score), and three different traits that meassure the shape of the carcass, i.e., the ratio between the height, width and circumference at the cranial base point of the dorsal fin to the body length of the fish.

Significant estimates of heritabilities were obtained for all traits except protein percentage. The phenotypic and genetic correlations between water and fat percentages were estimated at -0.84 and -1.03, respectively. The phenotypic correlation between fat percentage and body weight was positive ($r_p = 0.19$) while the genetic correlation was negative ($r_G = -0.51$). The genetic correlations between fat percentage and dressing percentage, gonad weight, meat color and visceral fat were estimated at 0.53, -0.49, -0.98, and -0.76, respectively. There was a very high genetic correlation between dressing percentage and belly thickness ($r_G = 0.94$). Both the phenotypic and genetic correlations between meat color and gonad weight were low ($r_p = 0.04$, $r_G = 0.22$). The phenotypic and genetic correlations between fat percentage and condition factor were both low ($r_p = 0.15$, $r_G = -0.10$). The phenotypic correlation between condition factor and the traits that measure the shape of the carcass varied from 0.69 to 0.81 and the genetic correlation from 0.89 to 1.10. The results are discussed in relation to possibilities for improving carcass quality traits in rainbow trout through selective breeding.

Genetic variation in reproductive traits in Atlantic salmon and rainbow trout

TRYGVE GJEDREM, ENDRE HAUS and VIGGO HALSETH

Institute of Aquaculture Research, Agricultural University of Norway, 1432 Ås-NLH (Norway)

ABSTRACT

The data were derived from six year-classes of Atlantic salmon and rainbow trout at the Research Station for Salmonids. The traits studied were: egg diameter, egg volume, egg number, body weight and body length. Genetic analysis was based on records from 854 and 717 fish, 91 and 51 sires and 218 and 124 dams of Atlantic salmon and rainbow trout, respectively.

Significant estimates of heritabilities based on sire components were obtained for all traits except egg diameter in rainbow trout and egg volume in Atlantic salmon. The dam components compared with the sire components were particularly high for egg diameter and body weight in rainbow trout and egg volume in Atlantic salmon.

There were very high genetic and phenotypic correlations between egg volume and egg number in both species. In Atlantic salmon there were quite high negative genetic correlations between egg size and egg volume ($r_G = 0.56$) and between egg size and egg number ($r_G = 0.71$). In Atlantic salmon, egg volume and egg number were highly correlated genetically with body weight and length, while these correlations were not significant in rainbow trout.

The results are discussed in relation to estimates given in the literature. It is concluded that the reproductive traits should not be included in the breeding goal; however, observations should be made in order to discern possible correlated responses.

Selection of fingerling rainbow trout for high and low tolerance to high temperature

PETER E. IHSSEN

Ministry of Natural Resources, Maple, Ont. (Canada)

ABSTRACT

An outbred strain of rainbow trout was selected for two generations for high and low tolerance to high temperature. The selection procedure was independent culling of individuals. Critical thermal maximum (thermal narcosis) was used as a measure of temperature tolerance. Almost all selected fish survived this selection procedure and, hence, could be reared to parent subsequent generations. Acclimation and test temperatures were chosen to limit the duration of each selection experiment to about 8 h (acclimation temperature $\sim 10°C$, test temperature $\sim 25.5°C$). Controls were maintained and tested under precisely the same environmental conditions throughout these experiments. After one generation of selectioen, a highly significant response was observed for the high tolerance group but not the low tolerance line. The realized heritabilities were 0.48 and 0.03 for the high and low lines, respectively. A second generation of selection resulted in no further increase in tolerance of the high line relative to the controls, whereas the tolerance of the low line decreased significantly with a realized heritability of 0.65. Maternal half-sibs were used to estimate the heritability for temperature tolerance in the controls and the two selected lines. The heritability estimates were 0.84 ± 0.40, 0.13 ± 0.08, and 0.25 ± 0.11 for the controls and the high and low tolerance lines, respectively, indicating a significant reduction in the genetic variability of the selected lines relative to the controls.

Life history considerations in broodstock management

A.R.D. KAPUSCINSKI

Fisheries and Wildlife, University of Minnesota, St. Paul, MN 55108 (U.S.A.)

ABSTRACT

The influence and relative importance of life history factors, including fecundity, age structure and immigration, on the probability distribution of fitness of broodstocks is discussed. The response of stock fitness to inadvertent or intentional changes in these variables is examined via the sensitivity analysis of equations developed for the following management goal: maintenance of a stock's probability distribution of fitness to allow it to perpetuate in a dynamic environment. This goal applied equally to extensive aquaculture and the maintenance of naturally reproducing stocks for gene banks.

Recommendations for broodstock management are given based on the sensitivity analysis results and our present understanding of life history patterns. Future research priorities are also suggested.

Strains of five trout species used in the management of U.S. fisheries

HAROLD L. KINCAID[1] and CHARLES R. BERRY[2]

[1]*U.S. Fish and Wildlife Service, RD 4, Box 63, Wellsboro, PA 16901 (U.S.A.)*
[2]*Utah Cooperative Fisheries, Utah State University, Logan, UT 84322 (U.S.A.)*

ABSTRACT

Surveys of trout broodstocks maintained by federal and state fisheries agencies during 1980, 1982, and 1984 identified broodstocks currently used in fishery management programs. Information obtained on each broodstock included broodstock origin, hatchery performance, field performance, disease resistance, management utilization, and breeding practices employed to maintain each broodstock. The number of trout broodstocks maintained in state and federal programs during 1984 were: rainbow trout (75 and 18), brook trout (36 and 4), brown trout (35 and 6), cutthroat trout (33 and 5), and lake trout (16 and 7). Broodstock numbers have declined since 1980 in the rainbow trout (-10.5%), brook trout (-14.6%), brown trout (-7.0%) and cutthroat trout (-7.3%) but have increased in the lake trout ($+64.3\%$). This trend toward fewer broodstocks is partially attributed to an increased awareness by fisheries personnel of strain performance differences which has led to the discontinuance of poorer broodstocks. The impact of this increased "efficiency" on long-term reductions in species genetic diversity is discussed.

General breeding practices followed to perpetuate broodstocks were similar in all five species and could be partitioned into three general types: random mating within year classes of a broodstock (33.8%), random mating between different year classes of a broodstock (37.0%), and mating according to a specialized selection system (29.2%). Traits included in selection programs for individual broodstocks were: body conformation (15), growth rate (10), egg quality (10), color (7), disease resistance (6), spawning time (4), and longevity (2). Implications of observed trends in management usage and breeding methods are discussed.

Distribution of genetic variation within and among hatchery strains of lake trout (*Salvelinus namaycush*)

J. ELLEN MARSDEN, CHARLES C. KRUEGER and BERNIE MAY

Dept. of Natural Resources, Cornell University, Ithaca, NY 14853 (U.S.A.)

ABSTRACT

Lake trout are currently the focus of intensive stocking programs in the Great Lakes and in other smaller inland lakes in North America. These programs propagate lake trout by collecting gametes from either wild populations or hatchery broodstocks. The application of electrophoretic techniques to the study of the processes of genetic change within and among populations has been difficult due to the lack of reported isozyme variability. This study was conducted to (1) examine the electrophoretic expression of enzymes previously undescribed in the literature, (2) identify polymorphic loci potentially useful in future population studies, and (3) describe the distribution of genetic variation within and among hatchery strains.

An additional 25 enzymes not previously examined were resolved and data can now be scored from an estimated 100 loci in lake trout. Among these loci 26 were observed to be polymorphic with 18 being useful for population studies among the lake trout strains examined within this study. Polymorphisms in Fum-1 and 2 and Pgm-5 and 6, previously undescribed in salmonids, were resolved. The lake trout strains examined were (omission of the work "lake" indicates a broodstock): Lake George, Raquette Lake, Clearwater Lake, Lake Manitou, Killala Lake, Superior, Seneca, Lake Ontario "strain" (1983 and 1984 year classes), and Gull Island Shoal. Clustering of strains based on Nei's genetic distances matched predictions based on strain lineages. Progress is underway to determine the relative contribution of each of the stocked strains to naturally produced fry in Lake Ontario.

Inheritance of body and tail coloration in two domesticated varieties of the guppy, *Poecilia reticulata*

V.P.E. PHANG, A.A. FERNANDO and O.K. CHOW

Department of Zoology, National University of Singapore, Kent Ridge 0511 (Singapore)

ABSTRACT

Several color pattern varieties of *Poecilia reticulata*, a popular ornamental fish, are cultured in Singapore. The inheritance of body and tail color of the blue tail and red tail varieties of the guppy was studied by reciprocal outcrossing with the wild type straion. Body color was determined by scale melanophore, xanthophore and erythrophore counts. It was demonstrated that body coloration was determined by an autosomal gene, with the wild type dominant over the blue tail and red tail color varieties. The tail colors of the blue tail and red tail varieties were determined by X-linked, sex-limited genes. Tail coloration was found to be affected by modifier genes.

Stock identification in lake trout using genetic markers

RUTH B. PHILLIPS[1] and PETER E. IHSSEN[2]

[1]*Department of Biological Sciences, University of Wisconsin, Milwaukee, WI (U.S.A.)*
[2]*Ontario Ministry of Natural Resources, Maple, Ont. (Canada)*

ABSTRACT

Data on the frequency of isozyme alleles and chromosome banding polymorphisms were obtained from six different lake trout populations. Chromosome banding polymorphisms scored include quinacrine (Q) band size and intensity variants and nucleolar organizer region (NOR) polymorphisms as determined by chromomycin A3 (CMA3) staining. CMA3 appeared to stain both active and inactive NORs in salmonid fish. CMA3-NOR polymorphisms scored included average number of CMA3-NORs/cell, average total CMA3 band length/cell, chromosomal location of CMA3-NORs, size of the CMA3-NORs and presence of adjacent Q bands. The size of chromosome bands was determined using a microcomputer-assisted image analysis system for chromosome measurement. The length of the short arm of chromosome 2 was measured in each cell and used as a standard for all other measurements.

The six stocks could be divided into three groups: a Lake Superior group, a group originally from Lake Michigan and an eastern stock from Seneca Lake, New York. These three groups could be distinguished on the basis of differences in frequencies of isozyme alleles, frequencies of negatively staining Q band chromosome variants and in the total number of CMA3-NORs/cell. Each stock had a unique combination of CMA3-NOR banding variants.

Genetic resources for tilapia culture

R.S.V. PULLIN[1], J.M. MACARANAS[2] and N. TANIGUCHI[3]

[1]ICLARM, P.O. Box 1501, Makati, Metro Manila (Philippines)
[2]University of the Philippines, Diliman, Quezon City (Philippines)
[3]Dept. of Cultural Fisheries, Kochi University, Kochi 783 (Japan)

ABSTRACT

Tilapia culture worldwide uses a small number of species and hybrids. The most important are *Oreochromis aureus*, *O. mossambicus*, *O. niloticus*, *O. spilurus*, and various interspecific F_1 hybrid crosses which yield all or nearly all male progeny. There is lesser but significant interest in the culture of *Tilapia rendalli*, *T. zillii*, *O. andersonii* and *O. macrochir*. The red tilapias, a heterogenous group of hybrids, are also attracting attention.

Despite worldwide interest in these tilapias, there are few reliable sources from which culturists can obtain founder stocks. Culture collections derived directly from founder stocks collected in Africa and checked by electrophoresis are maintained at only a few locations, for example, in Israel and the United Kingdom. Philippine and other Asian tilapia populations are currently being studied to assess their status, by electrophoresis, and their culture performance.

There is disturbing evidence of interbreeding within captive populations. For example, 11 cultured populations in the Philippines, assumed to be *O. niloticus*, are introgressed hybrids with *O. mossambicus*. Using mean allelic frequency at six electrophoretic marker loci to estimate introgression, the range (values for least and most introgressed populations) was 0.067 ± 0.019 SE to 0.301 ± 0.054 SE. The mean genetic distance (Nei) between the 11 populations and Philippine *O. mossambicus* is 0.2483 ± 0.0058. Samples from single Taiwanese and Thai populations of *O. niloticus*, kept for research, show negligible introgression: the corresponding mean allelic frequency was 0.012 ± 0.009 SE for both and their mean genetic distance from Philippine *O. mossambicus* was 0.3631 ± 0.0058.

Ongoing studies suggest that stable, well-characterized populations are rare throughout the Indo-Pacific. This emphasizes the need for the establishment of more culture collections to provide reliable material for research and production programs and the need for conservation of the wild tilapia resources of Africa and their habitats.

Genetic differences in stress response in Atlantic salmon and rainbow trout

T. REFSTIE

Institute of Aquaculture Research, Agricultural University of Norway, 1432 Ås-NLH (Norway)

ABSTRACT

Three year-classes of farmed Atlantic salmon (*Salmo salar*) and rainbow trout (*Salmo gairdneri*) were investigated for genetic differences in stress response. Each year-class contains 120–180 full- and half-sib groups from each species. The fish were exposed to a standardized stress condition and stress response was measured as cortisol and glucose level in blood. Significant differences in stress response were found between species, between sires, and between dams. Heritabilities are calculated for this trait together with correlations between stress response and disease resistance.

Growth rate variation between populations of juvenile coho salmon (*Oncorhynchus kisutch*)

B.E. RIDDELL and D.A. SORENSEN

Pacific Biological Station, Nanaimo, B.C. V9R 5K6 (Canada)

ABSTRACT

Coho salmon is an anadromous, semelparous species in the family Salmonidae. In the northeast Pacific Ocean coho are highly adaptable in their natural habitat and are extensively cultured in hatcheries. Over 90% of coho from central British Columbia southward mature at 3 years of age but significant variability in size at maturity is exhibited. This variation may be attributed to genetic variation for growth traits, timing of spawning migrations, and/or ocean distribution. We have been studying juvenile growth rate to evaluate genetic variation within and between coho populations. These studies have application in determining hatchery practices and in broodstock development for mariculture.

Over 3 years we have compared juvenile growth rates of eight to 10 families from each of eight coho populations (three hatchery and five natural). The 1984 study involved two hatchery and two natural populations plus four hybrid crosses, and each of the eight test groups consisted of eight half-sib families. Heritability could be estimated from the "pure" strain groups in the latter study but estimates had large standard errors.

Little between-population variability was observed. Variability between families within populations consistently and substantially exceeded between-population variation. Extreme variation in specific growth was observed in one Big qualicum River family which grow at five to six times the rate of other families and attained an average smolt size of 320 g.

Heritability of juvenile growth rate was high and supported by intermediacy of hybrids to parent stocks. Family selection should be used to initiate coho broodstocks but selection for increased juvenile growth rate will probably result in early maturity at smaller body size, as observed in the rapidly growing Big Qualicum family. Selection for juvenile growth alone is therefore unlikely to give desirable results in mariculture.

Survival in diploid and triploid Pacific salmon hybrids

JAMES SEEB[1], GARY THORGAARD[2], WILLIAM K. HERSHBERGER[1] and FRED M. UTTER[3]

[1]*School of Fisheries, University of Washington, Seattle, WA 98195 (U.S.A.)*
[2]*Dept. of Zoology, Washington State University, Pullman, WA 99164 (U.S.A.)*
[3]*National Marine Fisheries Service, Seattle, WA 98112 (U.S.A.)*

ABSTRACT

Historical emphasis has been placed on the use of hybrid organisms in agriculture and aquaculture. Useful characteristics of hybrids can include both the combination of desirable genotypes from both parental species into one organism and their frequent sterility which can provide advantages to intensive and extensive culture. Research has shown that triploid interspecific hybrids can have higher survival rates than corresponding diploid hybrids. This paper examines the viability of various diploid and triploid hybrid Pacific salmon.

All possible hybrid and control crosses between chinook (*Oncorhynchus tschawytscha*), coho (*O. kisutch*) and chum (*O. keta*) salmon were made. Approximately half of the eggs from each mating were heat shocked to induce triploidy. Survival rates were monitored until 30 days postfeeding, hybrid genotype was determined by the simultaneous expression of diagnostic maternal and paternal electrophoretic alleles, and ploidy was confirmed using flow cytometry.

Two crosses provided interesting results. First, coho salmon female × chinook salmon male diploid and triploid lots survived to conspecific control levels, contrary to results reported by other workers using other stocks of the same species. Second, and of particular interest, was that chum salmon × chinook salmon diploid hybrids did not survive while the triploid hybrids survived to control levels.

Adult chinook salmon have a high value per pound to commercial fishermen but require intensive rearing before smoltification. Chum salmon, on the other hand, have a relatively low value per pound but require little or no fresh-water rearing. Our studies showed that chinook—chum hybrids have the osmoregulatory capacity of chum salmon controls. Continuing studies will determine the aquacultural potential of these hybrids.

Induced diploid gynogenesis by mitotic interference in rainbow trout

D. THOMPSON and C.E. PURDOM

Ministry of Agriculture, Fisheries and Food, Directorate of Fisheries Research, Fisheries Laboratory, Lowestoft, Suffolk NR33 0HT (Great Britain)

ABSTRACT

Recent research has shown that the inbreeding potential of induced diploid gynogenesis is limited when diploidy is restored by interference with the second meiotic division of the egg. Electrophoretic analyses have demonstrated high levels of recombination at several enzyme loci indicating that loci distal to the centromers will retain heterozygosity throughout many generations of such gynogenesis. However, the restoration of diploidy by interference with the first mitotic division of developing haploid embryos can produce completely homozygous individuals. Such individuals were produced in rainbow trout by fertilizing eggs with genetically inert spermatozoa followed by a heat treatment, 10 min at 28°C, administered between 4.5 and 5.5 h after fertilization with the eggs incubated at a temperature of 9°C. The use of genetically controlled enzyme markers, with consistently high recombination rates, can identify these homozygous individuals and distinguish between these and heterozygous diploid gynogenomes produced by meiotic interference. Any genetical contribution from the inert spermatozoa may also be identified. Ideally chromosome manipulation experiments should be made between individual parents using females heterozygous for at least two loci where very high recombination rates are observed in meiotic gynogenomes and spermatozoa from a male homozygous for a third allele. Thus distinctions could be made between meiotic and mitotic gynogenomes, and the presence of any male contribution could be detected.

A study of strain selection of *Tilapia nilotica*

S. URAIWAN and V. PHANITCHAI

National Inland Fisheries Institute, Kasetsart University Bangkhen, Bangkok 10900 (Thailand)

ABSTRACT

The *Tilapia nilotica* Chitrada and Israeli strains were introduced into Thailand in 1965 and 1983, respectively. Two experiments were conducted to compare growth and survival of the two *Tilapia* strains. First, one experiment observed the differences in growth and survival of these two strains and their hybrids, and then, offspring of the two purebred strains from the first part were used to compare growth and survival in the second experiment. The first experiment started in May 1983 and ended in March 1984; the second experiment started in June 1984 and ended in April 1985. The results of the first experiment showed that the *Tilapia* Chitrada strain had the highest growth followed by the *Tilapia* Israeli strain and their hybrids. At termination, mean weights of the pure *Tilapia* Chitrada, the pure *Tilapia* Israeli, the crossbreed Chitrada × Israeli and the crossbreed Israeli × Chitrada were 152, 120, 112 and 83, respectively. The results of the second experiment are also discussed.

Electrophoretic variation in six rainbow trout strains

JUKKA VUORINEN and LIISA KUUSIPALO

University of Joensuu, Box 111, SF-80101, Joensuu (Finland)

ABSTRACT

Genetic variation among six domestic rainbow trout strains was studied by electrophoresis of 13 enzymes representing 30 presumptive loci. The stocks were of varying origin and were cultivated in Finland for at least six generations. The mean heterozygosity ranged from 6.6% to 9.3%; these values being of the same magnitude as reported for other domestic and also for wild populations.

Genetic distances between stocks ranged from 0.002 to 0.027, indicating that these stocks were genetically quite homogeneous. The differences were most evident for stocks cultivated in France or California. This might indicate an adaptation to the cooler climate of Finland.

Electrophoretic identification of largemouth bass using fin tissue — management of production broodstocks

J. HOLT WILLIAMSON[1], GARY J. CARMICHAEL[1], MAUREEN E. SCHMIDT[2] and DON C. MORIZOT[2]

[1]*National Fish Hatchery and Technology Center, San Marcos, TX (U.S.A.)*
[2]*University of Texas Science Park, Smithville, TX (U.S.A.)*

ABSTRACT

Protein electrophoresis is a contemporary biochemical technique that enables aquaculturists to more accurately assign individual largemouth bass, *Micropterus salmoides*, to subspecific or intergrade taxa. Electrophoresis becomes an even more effective tool, particularly where valuable broodfish or endangered species are involved, if the necessary data can be obtained without killing or seriously injuring the fish. Heretofore, obtaining tissues necessary for appropriate electrophoretic analyses required killing the fish or injuring it through invasive biopsy. Such procedures at best reduced the reliability of the fish as broodstock. In the present study, electrophoretic analysis of four stocks of largemouth bass was accomplished using a series of low-risk tissue samples: red blood cells, plasma, pelvic fin, caudal fin, and skin-epithelium (scale scrapings). One stock represented the northern subspecies, *M. s. salmoides*, and another stock represented the Florida subspecies, *M. s. floridanus*. The two remaining stocks were domestic, intergrade stocks. We assayed products of 64 loci and found 40 could be resolved from caudal fin tissue. Products of an additional three loci could be resolved from red blood cells. No additional information was gained from analysis of plasma (three loci), pelvic fin (39 loci), or skin-epithelium (38 loci) with the exception of glucosidase (E.C. 3.2.1.20) which was resolved using plasma or pelvic fin tissue.

Our work, along with that of others, has resulted in the electrophoretic analysis of 70 structural loci in largemouth bass. Of these 70 loci, 23 (33%) were found to be polymorphic. Loci from at least 16 (70%) of the 23 polymorphic loci can be resolved from fin tissue. Of the 64 loci which we examined, the following list consists of the 44 enzymes or proteins which can be resolved by vertical starch-gel electrophoresis from low-risk tissue samples. Each enzyme is represented by its "Commission of Biological Nomenclature" number. Those which have been found to be polymorphic and can be resolved in fin tissue are in parentheses: ACP, 3.1.3.2; ACON1, 4.2.1.3; (ADA, 3.5.4.4); AK, 2.7.4.3; (CKA, 2.7.3.2); (CKB, 2.7.3.2); (CKC, 2.7.3.2); ENO, 4.2.1.11; FUM, 4.2.1.2; G6PD, 1.1.1.49; (GPI1, 5.3.1.9); (GPI2, 5.3.1.9); GLU, 3.2.1.20; GUS, 3.2.1.31; GOTA, 2.6.1.1; (GOTM, 2.6.1.1); GS, 6.3.1.2; GAPD1, 1.2.1.12; GLO, 4.4.1.5; GUK2, 2.7.4.8; Hemoglobin – 1; Hemoglobin – 2; ITP, 3.6.1.19; (IDHB, 1.1.1.42); (LDHA, 1.1.1.27); LDHB, 1.1.1.27; MAN, 3.2.1.24; MDH2, 1.1.1.37; (MDH3, 1.1.1.37); ME, 1.1.1.40; (MPI, 5.3.1.8); NP2, 2.4.2.1; PEP1, 3.4.11; PEP2, 3.4.11; (PEP3, 3.4.11); (PGM, 2.7.5.1); (6PGD, 1.1.1.44); PGK 2.7.2.3; PK1, 2.7.1.40; PK2, 2.7.1.40; SOD1, 1.15.1.1; (SOD2, 1.15.1.1); TPI1, 5.3.1.1; (TPI2, 5.3.1.1).

Taking low-risk tissues from largemouth bass to use in electrophoretic analysis does not appear to adversely affect growth, reproduction, survival, or behavior of the fish. In addition, the cost in terms of equipment, skill, and time required to take the tissues — particular fin — is minimal, especially when compared to earlier biopsy techniques. The use of low-risk tissues in electrophoretic analysis is a useful technique in genetics and breeding programs, particularly where genetic identification of valuable, rare or endangered individuals is concerned.

On the introduction of spat-rearing and experimental culture of bay scallop, *Argopecten irradians* (Lamarck)

FUSUI ZHANG, HE YICHAO, LIU XIANGSHENG and MA JIANGHU

Institute of Oceanology, Academica Sinica, Shandong Province (China)

ABSTRACT

For years we have planned the introduction of bay scallops from the east coast of the U.S.A. into China for experimental culture to try to shorten the economic turnover from 2 years needed for culturing the local scallops, *Chlamys farreri*, to 1 year by introducing the bay scallop. Parent bay scallops from the U.S.A. were carried to Qingdao, China on 20 December 1982, stocked in indoor tanks at conditioning temperature, and fed with a mixture of *Phaeodictylum tricornutum*, *Pyramimonas* sp., *Platymonas* sp. and *Chlorella* sp. They spawned on 26 January 1983. The hatched larvae were reared at 18-21°C, and fed with *Isochrysis galbana*, *Pyramimonas* sp. and *Chlorella* sp. In a week's growth, the spats averaged 827 μm and on 9 May attained 6.9 mm. In the middle of May, the seed scallops were transferred to Luoyuan Bay, Fujian Province, Jiaozhou Bay, and off Cape Taiping near Qingdao, Shandong Province, for experimental culture, where they were stored in plastic netcages suspended on a single line from a raft.

Bay scallops cultured in Luoyuan Bay grew to an average shell height of 41 mm and 16.9 g average weight by 15 August, with a monthly mean increase of 10.4 mm. In Jiaozhou Bay and Cape Taiping, they grew to an average shell height of 50 mm (marketable size) and 26 average weight in late September and attained 59 mm average shell height (range 39-75 mm) and 46 g average weight in late December. The ovary and testis in the gonad could be distinguished by color in August. Eggs and sperm were collected in the laboratory in early September, from which the second generation of seed scallops were reared successfully and grown and bred normally. They will be harvested within 1 year, counting from egg fertilization. We consider it a suitable species for mariculture in the Yellow Sea and East China Sea. We have enlarged the culture area to 1/3 ha and grown about 500 000 bay scallops for social economic tests with seed scallops reared last April. They may be ready for harvesting at the end of 1984.

AUTHOR INDEX

Allen, S.K., Jr. (Seattle, WA, U.S.A.) and Wattendorf, R.J. (Eustis, FL, U.S.A.)
   A review of the production and quality control of triploid grass carp and progress in
   implementing 'sterile' triploids as management tools in the U.S. (Abstract) 359
Allen, S.K., Jr., Downing, S.L. (Seattle, WA, U.S.A.), Chaiton, J. (Quilcene, WA, U.S.A.)
   and Beattie, J.H. (Seattle, WA, U.S.A.)
   Chemically and pressure-induced triploidy in the Pacific oyster *Crassostrea gigas*
   (Abstract) 359
Allendorf, F.W., see Scheerer, P.D. et al.
Attard, J., see Blanc, F. et al.
Ayles, G.B., see Baker, R.F. and Ayles, G.B.
Bailey, J.K. and Loudenslager, E.J. (St. Andrews, N.B., Canada)
   Genetic and environmental components of variation for growth of juvenile Atlantic
   salmon (*Salmo salar*) 125
Baker, R.F. and Ayles, G.B. (Winnipeg, Man., Canada)
   Effects of temperature, size and rations on the growth of strains of Arctic charr in
   intensive aquaculture (Abstract) 360
Beattie, J.H., see Allen, S.K., Jr. et al.
Beaumont, A.R. (Menai Bridge, Great Britain)
   Genetic aspects of hatchery rearing of the scallop, *Pecten maximus* (L.) 99
Behrends, L.L., Kingsley, J.B. and Price, A.H. (Muscle Shoals, AL, U.S.A.)
   Performance of two red tilapia hybrids in polyculture with channel catfish and
   freshwater prawn (Abstract) 360
Berg, W.J. (Davis, CA, U.S.A.)
   Characterization of genetic variability within and between seven distinct population-
   groups of rainbow trout (*Salmo gairdneri* Richardson) (Abstract) 361
Berry, C.R., see Kincaid, H.L. and Berry, C.R.
Bidwell, C.A. (Davis, CA, U.S.A.), Chrisman, C.L. and Libey, G.S. (West Lafayette, IN,
   U.S.A.)
   Polyploidy induced by heat shock in channel catfish (Abstract) 362
Blanc, F. (Montpellier, France), Pichot, P. (Sète, France) and Attard, J. (Cap Diamont,
   Que., Canada)
   Genetic variability in the European oyster, *Ostrea edulis*: geographic variation between
   local French stocks (Abstract) 362
Bondari, K. (Tifton, GA, U.S.A.)
   Response of channel catfish to multi-factor and divergent selection of economic traits 163
Bondari, K. (Tifton, GA, U.S.A.) and Dunham, R.A. (Auburn, AL, U.S.A.)
   The effect of inbreeding on reproduction, growth and survival of channel catfish
   (Abstract) 363
Bougrier, S. (La Trinité Sur Mer, France) and Rabenomanana, L.D. (Mahajanga,
   Madagascar)
   Cryopreservation of oyster sperm (Abstract) 364
Bougrier, S., Grizel, H. (La Trinité Sur Mer, France) and Deltreil, J.P. (Arcachon, France)
   Crossing influence on shape and structure of oyster shells (Abstract) 365

Busack, C.A., see Gall, G.A.E. and Busack, C.A.
Busch, R.A., see Parsons, J.E. et al.
Bye, V.J. and Lincoln, R.F. (Lowestoft, Great Britain)
    Commercial methods for the control of sexual maturation in rainbow trout (*Salmo gairdneri* R.) 299
Carmichael, G.J., see Williamson, J.H. et al.
Chaiton, J., see Allen, S.K., Jr. et al.
Chatry, M., see Foltz, D.W. and Chatry, M.
Chourrout, D. (Jouy en Josas, France)
    Extensive karylogical investigations permitting the choice between different methods of gynogenesis and polyploidy induction in rainbow trout (Abstract) 366
Chow, O.K., see Phang, V.P.E. et al.
Chrisman, C.L., see Bidwell, C.A. et al.
Cobolli Sbordoni, M., see Sbordoni, V. et al.
Dabrowski, K.R. (Tokyo, Japan)
    Coregonid fish life history strategies 366
Deltreil, J.P., see Bougrier, S. et al.
De Matthaeis, E., see Sbordoni, V. et al.
Dickhoff, W.W., see Johnson, O.W. et al.
Dickie, L.M., see Mallet, A.L. et al.
Downing, S.L., see Allen, S.K., Jr. et al.
Doyle, R.W. and Talbot, A.J. (Halifax, N.S., Canada)
    Effective population size and selection in variable aquaculture stocks 27
Doyle, R.W., see also Uraiwan, S. and Doyle, R.W.; Villegas, C.T. and Doyle, R.W.
Dunham, R.A., Smitherman, R.O., Goodman, R.K. and Kemp, P. (Auburn, AL, U.S.A.)
    Comparison of strains, crossbreeds and hybrids of channel catfish for vulnerability to angling 193
Dunham, R.A., see also Bondari, K. and Dunham, R.A.
Fernando, A.A. and Phang, V.P.E. (Kent Ridge, Singapore)
    Genetic differentiation of isozymes in domesticated color pattern varieties and a wild population of the guppy, *Poecilia reticulata* (Abstract) 367
Fernando, A.A., see also Phang, V.P.E. et al.
Foltz, D.W. (Baton Rouge, LA, U.S.A.) and Chatry, M. (Grand Isle, LA, U.S.A.)
    Genetic heterozygosity and growth rate in Louisiana oysters (*Crassostrea virginica*) 261
Fredeen, H. (Lacombe, Alta., Canada)
    Monitoring genetic change 1
Freeman, K.R., see Mallet, A.L. et al.
Fricke, H., see Hörstgen-Schwark, G. et al.
Gaffney, P. (Stony Brook, NY, U.S.A.)
    Alternative breeding methods in bivalve culture (Abstract) 368
Gall, G.A.E. (Davis, CA, U.S.A.) and Busack, C.A. (University, MS, U.S.A.)
    Prologue and acknowledgments ix
Gibor, A., see Polne-Fuller, M. and Gibor, A.
Gjedrem, T. (Ås, Norway)
    Breeding plan for sea ranching 77
Gjedrem, T., Haus, E. and Halseth, V. (Ås, Norway)
    Genetic variation in reproductive traits in Atlantic salmon and rainbow trout (Abstract) 369

Gjerde, B. (Ås, Norway)
  Growth and reproduction in fish and shellfish 37
  Estimates of phenotypic and genetic parameters for carcass quality traits in rainbow
    trout (Abstract) 368
Goodman, R.K., see Dunham, R.A. et al.
Grizel, H., see Bougrier, S. et al.
Halevy, A., see Hulata, G. et al.
Halseth, V., see Gjedrem, T. et al.
Harger, B.W.W., see Lewis, R.J. et al.
Haus, E., see Gjedrem, T. et al.
He Yichao, see Zhang Fusui et al.
Hershberger, W.K., see Iwamoto, R.N. et al.; Seeb, J. et al.
Hörstgen-Schwark, G., Fricke, H. and Langholz, H.-J. (Göttingen, Federal Republic of
  Germany)
  The effect of strain crossing on the production performance in rainbow trout 141
Hulata, G., Wohlfarth, G.W. and Halevy, A. (Hof Hacarmel, Israel)
  Mass selection for growth rate in the Nile tilapia (*Oreochromis niloticus*) 177
Ihssen, P.E. (Maple, Ont., Canada)
  Selection of fingerling rainbow trout for high and low tolerance to high temperature
    (Abstract) 370
  see also Phillips, R.B. and Ihssen, P.E.
Iwamoto, R.N., Myers, J.M. and Hershberger, W.K. (Seattle, WA, U.S.A.)
  Genotype-environment interactions for growth of rainbow trout, *Salmo gairdneri* 153
Johnson, K.R. and Wright, J.E. (University Park, PA, U.S.A.)
  Female brown trout × Atlantic salmon hybrids produce gynogens and triploids when
    backcrossed to male Atlantic salmon 345
Johnson, O.W., Dickhoff, W.W. and Utter, F.M. (Seattle, WA, U.S.A.)
  Comparative growth and development of diploid and triploid coho salmon,
    Oncorhynchus kisutch 329
Jørstad, K.E. (Bergen, Norway)
  Genetic studies connected with artificial propagation of cod (*Gadus morhua* L.) 227
Kapuscinski, A.R.D. (St. Paul, MN, U.S.A.)
  Life history considerations in broodstock management (Abstract) 370
  see also Lannan, J.E. and Kapuscinski, A.R.D.
Kemp, P., see Dunham, R.A. et al.
Kijima, A., see Taniguchi, N. et al.
Kincaid, H.L. (Wellsboro, PA, U.S.A.) and Berry, C.R. (Logan, UT, U.S.A.)
  Strains of five trout species used in the management of U.S. fisheries (Abstract) 371
Kingsley, J.B., see Behrends, L.L. et al.
Knudsen, K.L., see Scheerer, P.D. et al.
Koljonen, M.-L. (Helsinki, Finland)
  The enzyme gene variation of ten Finnish rainbow trout strains and the relation between
    growth rate and mean heterozygosity 253
Krueger, C.C., see Marsden, J.E. et al.
Kuusipalo, L., see Vuorinen, J. and Kuusipalo, L.
La Rosa, G., see Sbordoni, V. et al.
Langholz, H.-J., see Hörstgen-Schwark, G. et al.
Lannan, J.E. (Newport, OR, U.S.A.) and Kapuscinski, A.R.D. (St. Paul, MN, U.S.A.)
  Application of a genetic fitness model to extensive aquaculture 81

Lewis, R.J. (Goleta, CA, U.S.A.), Neushul, M. (Santa Barbara, CA, U.S.A.) and Harger, B.W.W. (Goleta, CA, U.S.A.)
  Interspecific hybridization of the species of *Macrocystis* in California 203
Libey, G.S., *see* Bidwell, C.A. et al.
Lincoln, R.F., *see* Bye, V.J. and Lincoln, R.F.
Liu Xiangsheng, *see* Zhang Fusui et al.
Loudenslager, E.J., *see* Bailey, J.K. and Loudenslager, E.J.
Ma Jianghu, *see* Zhang Fusui et al.
Macaranas, J.M., *see* Pullin, R.S.V. et al.
Mallet, A.L., Freeman, K.R. and Dickie, L.M. (Dartmouth, N.S., Canada)
  The genetics of production characters in the blue mussel *Mytilus edulis*. I. A preliminary analysis 133
Marsden, J.E., Krueger, C.C. and May, B. (Ithaca, NY, U.S.A.)
  Distribution of genetic variation within and among hatchery strains of lake trout (*Salvelinus namaycush*) (Abstract) 371
Mattoccia, M., *see* Sbordoni, V. et al.
May, B., *see* Marsden, J.E. et al.
Morizot, D.C., *see* Williamson, J.H. et al.
Myers, J.M. (Seattle, WA, U.S.A.)
  Tetraploid induction in *Oreochromis* spp 281
  see also Iwamoto, R.N. et al.
Neushul, M., *see* Lewis, R.J. et al.
Newkirk, G.F. (Halifax, N.S., Canada)
  Controlled mating of the European oyster, *Ostrea edulis* 111
Panelay, P.J., *see* Quillet, E. and Panelay, P.J.
Parsons, J.E., Busch, R.A. (Buhl, ID, U.S.A.), Thorgaard, G.H. and Scheerer, P.D. (Pullman, WA, U.S.A.)
  Increased resistance of triploid rainbow trout × coho salmon hybrids to infectious hematopoietic necrosis virus 337
Phang, V.P.E., Fernando, A.A. and Chow, O.K. (Kent Ridge, Singapore)
  Inheritance of body and tail coloration in two domesticated varieties of the guppy, *Poecilia reticulata* (Abstract) 372
Phang, V.P.E., *see also* Fernando, A.A. and Phang, V.P.E.
Phanitchai, V., *see* Uraiwan, S. and Phanitchai, V.
Phillips, R.B. (Milwaukee, WI, U.S.A.) and Ihssen, P.E. (Maple, Ont., Canada)
  Stock identification in lake trout using genetic markers (Abstract) 372
Pichot, P., *see* Blanc, F. et al.
Polne-Fuller, M. and Gibor, A. (Santa Barbara, CA, U.S.A.)
  Calluses, cells, and protoplasts in studies towards genetic improvement of seaweeds 117
Price, A.H., *see* Behrends, L.L. et al.
Pullin, R.S.V. (Manila, The Philippines), Macaranas, J.M. (Quezon City, The Philippines) and Taniguchi, N. (Kochi, Japan)
  Genetic resources for tilapia culture (Abstract) 373
Purdom, C.E., *see* Thompson, D. and Purdom, C.E.
Quillet, E. and Panelay, P.J. (Brest, France)
  Triploidy induction by thermal shocks in the Japanese oyster, *Crassostrea gigas* 271
Rabenomanana, L.D., *see* Bougrier, S. and Rabenomanana, L.D.
Refstie, T. (Ås, Norway)
  Genetic differences in stress response in Atlantic salmon and rainbow trout (Abstract) ... 374
Riddell, B.E. and Sorensen, D.A. (Nanaimo, B.C., Canada)
  Growth rate variation between populations of juvenile coho salmon (*Oncorhynchus kisutch*) (Abstract) 374

Sbordoni, V., De Matthaeis, E., Cobolli Sbordoni, M., La Rosa, G. and Mattoccia, M. (Rome, Italy)
    Bottleneck effects and the depression of genetic variability in hatchery stocks of *Penaeus japonicus* (Crustacea, Decapoda) 239

Scheerer, P.D., Thorgaard, G.H. (Pullman, WA, U.S.A.), Allendorf, F.W. and Knudsen, K.L. (Missoula, MT, U.S.A.)
    Androgenetic rainbow trout produced from inbred and outbred sperm sources show similar survival 289

Scheerer, P.D., *see also* Parsons, J.E. et al.

Schmidt, M.E., *see* Williamson, J.H. et al.

Seeb, J. (Seattle, WA, U.S.A.), Thorgaard, G. (Pullman, WA, U.S.A.), Hershberger, W.K. and Utter, F.M. (Seattle, WA, U.S.A.)
    Survival in diploid and triploid Pacific salmon hybrids (Abstract) 375

Shelton, W.L. (Norman, OK, U.S.A.)
    Broodstock development for monosex production of grass carp 311

Shultz, F.T. (Sonoma, CA, U.S.A.)
    Developing a commercial breeding program 65

Siitonen, L. (Jokioinen, Finland)
    Factors affecting growth in rainbow trout (*Salmo gairdneri*) stocks 185

Smitherman, R.O., *see* Dunham, R.A. et al.

Smoker, W.W. (Corvallis, OR, U.S.A.)
    Variability of embryo development rate, fry growth, and disease susceptibility in hatchery stocks of chum salmon 219

Sorensen, D.A., *see* Riddell, B.E. and Sorensen, D.A.

Takegami, K., *see* Taniguchi, N. et al.

Talbot, A.J., *see* Doyle, R.W. and Talbot, A.J.

Tallman, R.F. (Nanaimo, B.C., Canada)
    Genetic differentiation among seasonally distinct spawning populations of chum salmon, *Oncorhynchus keta* 211

Tamura, T., *see* Taniguchi, N. et al.

Taniguchi, N., Kijima, A., Tamura, T., Takegami, K. and Yamasaki, I. (Nankoku, Japan)
    Color, growth and maturation in ploidy-manipulated fancy carp 321

Taniguchi, N., *see also* Pullin, R.S.V. et al.

Thompson, D. and Purdom, C.E. (Lowestoft, Great Britain)
    Induced diploid gynogenesis by mitotic interference in rainbow trout (Abstract) 376

Thorgaard, G.H. (Pullman, WA, U.S.A.)
    Ploidy manipulation and performance 57
    see also Parsons, J.E. et al.; Scheerer, P.D. et al.; Seeb, J. et al.

Uraiwan, S. (Bangkok, Thailand) and Doyle, R.W. (Halifax, N.S., Canada)
    Replicate variance and the choice of selection procedures for tilapia (*Oreochromis niloticus*) stock improvement in Thailand 93

Uraiwan, S. and Phanitchai, V. (Bangkok, Thailand)
    A study of strain selection of *Tilapia nilotica* (Abstract) 376

Utter, F.M., *see* Johnson, O.W. et al.; Seeb, J. et al.

Villegas, C.T. (Tigbauan, The Philippines) and Doyle, R.W. (Halifax, N.S., Canada)
    Duration of feeding and indirect selection for growth of tilapia 89

Vuorinen, J. and Kuusipalo, L. (Joensuu, Finland)
    Electrophoretic variation in six rainbow trout strains (Abstract) 377

Wada, K.T. (Nansei, Japan)
    Genetic selection for shell traits in the Japanese pearl oyster, *Pinctada fucata martensii* 171

Wattendorf, R.J., *see* Allen, S.K., Jr. and Wattendorf, R.J.
Williamson, J.H., Carmichael, G.J. (San Marcos, TX, U.S.A.), Schmidt, M.E. and Morizot, D.C. (Smithville, TX, U.S.A.)
    Electrophoretic identification of largemouth bass using fin tissue – management of production broodstocks (Abstract) 377
Wohlfarth, G.W., *see* Hulata, G. et al.
Wright, J.E., *see* Johnson, K.R. and Wright, J.E.
Yamasaki, I., *see* Taniguchi, N. et al.
Zhang Fusui, He Yichao, Liu Xiangsheng and Ma Jianghu (Shandong Province, China)
    On the introduction of spat-rearing and experimental culture of bay scallop, *Argopecten irradians* (Lamarck) (Abstract) 378